Scientific Computation

Editorial Board

J.-J. Chattot, Davis, CA, USA
P. Colella, Berkeley, CA, USA
Weinan E, Princeton, NJ, USA
R. Glowinski, Houston, TX, USA
M. Holt, Berkeley, CA, USA
Y. Hussaini, Tallahassee, FL, USA
P. Joly, Le Chesnay, France
H. B. Keller, Pasadena, CA, USA
D. I. Meiron, Pasadena, CA, USA
O. Pironneau, Paris, France
A. Quarteroni, Lausanne, Switzerland
J. Rappaz, Lausanne, Switzerland
R. Rosner, Chicago, IL, USA
J. H. Seinfeld, Pasadena, CA, USA
A. Szepessy, Stockholm, Sweden
M. F. Wheeler, Austin, TX, USA

Springer
Berlin
Heidelberg
New York
Hong Kong
London
Milan
Paris
Tokyo

Physics and Astronomy ONLINE LIBRARY

http://www.springer.de/phys/

Vladimir D. Liseikin

A Computational Differential Geometry Approach to Grid Generation

With 37 Figures

 Springer

Professor Vladimir D. Liseikin
Institute of Computational Technologies
Siberian Branch of the Russian Academy of Sciences
Pr. Lavrentyeva 6
630090 Novosibirsk 90, Russia

Library of Congress Cataloging-in-Publication Data applied for
A catalog record for this book is available from the Library of Congress.

Bibliographic information published by Die Deutsche Bibliothek.
Die Deutsche Bibliothek lists this publication in the Deutsche Nationalbibliografie; detailed bibliographic data is available in the Internet at http://dnb.ddb.de.

ISSN 1434-8322
ISBN 3-540-14008-5 Springer-Verlag Berlin Heidelberg New York

This work is subject to copyright. All rights are reserved, whether the whole or part of the material is concerned, specifically the rights of translation, reprinting, reuse of illustrations, recitation, broadcasting, reproduction on microfilm or in any other way, and storage in data banks. Duplication of this publication or parts thereof is permitted only under the provisions of the German Copyright Law of September 9, 1965, in its current version, and permission for use must always be obtained from Springer-Verlag. Violations are liable for prosecution under the German Copyright Law.

Springer-Verlag Berlin Heidelberg New York
a member of BertelsmannSpringer Science+Business Media GmbH

http://www.springer.de

© Springer-Verlag Berlin Heidelberg 2004
Printed in Germany

The use of general descriptive names, registered names, trademarks, etc. in this publication does not imply, even in the absence of a specific statement, that such names are exempt from the relevant protective laws and regulations and therefore free for general use.

Typesetting: Camera-ready copy from the author
Cover design: *design & production*, Heidelberg
Printed on acid-free paper   55/3141/tr - 5 4 3 2 1 0

Preface

Grid technology whose achievements have significant impact on the efficiency of numerical codes still remains a rapidly advancing field of computational and applied mathematics. New achievements are being added by the creation of more sophisticated techniques, modification of the available methods, and implementation of more subtle tools as well as the results of the theories of differential equations, calculas of variations, and Riemannian geometry being applied to the formulation of grid models and analysis of grid properties.

The development of comprehensive differential and variational grid generation techniques reviewed in the monographs of J.F. Thompson, Z.U.A. Warsi, C.W. Mastin, P. Knupp, S. Steinberg, V.D. Liseikin has been largely based on a popular concept in accordance with which a grid model realizing the required grid properties should be formulated through a linear combination of basic and control grid operators with weights. A typical basic grid operator is the operator responsible for the well-posedness of the grid model and construction of unfolding grids, e.g. the Laplace equations (generalized Laplace equations for surfaces) or the functional of grid smoothness which produces fixed nonfolding grids while grid clustering is controlled by source terms in differential grid formulations or by an adaptation functional in variational models. However, such a formulation does not obey the fundamental invariance laws with respect to parameterizations of physical geometries. It frequently results in cumbersome governing grid equations whose choice of weight and control functions provide conflicting grid requirements. In particular, the choice for providing well-posedness, grid nondegeneracy, and adaptation is not readily verified but is largely based on nebulous theoretical assumptions borrowed from one-dimensional models.

The current book revises this popular concept and pursues a more updated and somewhat revolutionary one based on the general fact that an arbitrary one-to-one, smooth multidimensional coordinate transformation deriving a numerical grid in a domain or on a surface is realized by a solution of a system of the generalized Laplace equations in a suitable monitor metric specified in the physical geometry. The system can be interpreted as the multidimensional equidistribution principle in which the monitor metric tensor is an extension of a scalar-valued weight function. With this interpretation for a mathematical model for generating grids in domains or on surfaces only the generalized Laplace equations need be chosen, without any need for complementary control operators which worsen the model. The required grid properties are realized through the specification of suitable metric tensors.

Thus the single generalized Laplace mathematical model provides a real foundation for the solution of the problem of the development of a comprehensive grid generator. Consequently the efforts of research should be directed towards implementing this model into grid technology by developing approaches for the construction of metrics in physical geometries and establishing the necessary relations between them and the required grid properties for the purpose of setting up an adequate control of the grid quality by the choice of the suitable metric.

One natural approach for formulating metric tensors and corresponding tensor-valued weight functions is based on the notion of a monitor surface over the physical domain or surface which undergoes a gridding process. The monitor surface is defined as the graph of some (in general vector-valued) monitor function that takes into account the behavior of the physical solution. This monitor surface having an inherent metric tensor that can be considered as the very tensor-valued weight function is suitable for generating adaptive grids with the use of a smoothness functional (which is the functional of energy) whose Euler–Lagrange equations are, in fact, equivalent to the generalized Laplace equations in the metric of the monitor surface. The resulting grid derived by this metric tends to cluster its nodes in the zones of the large gradient of the monitor function. The approach for formulating the adaptive metric is readily extended to define more general monitor metrics in domains or on surfaces, thus turning them into Riemannian manifolds whose implementation in grid technology allows one to generate real grids satisfying the most broad mesh quality requirements.

In order to control the required grid properties by the monitor metrics one needs a knowledge of geometric characteristics of the monitor geometries and their relations to the resulting grid behavior. This knowledge can be attained with the aid of the theory of multidimensional differential geometry of Riemannian manifolds adjusted to the features of grid technology. The theory of multidimensional differential geometry really is one of the promising branches of the pure mathematical field of science capable of pushing grid technology to a more advanced level in its development. Indeed, many notions and characteristics of common surfaces such as metric tensors, their invariants, first and second groundforms, curvatures and torsions of lines, the mean and Gauss curvatures, and Christoffel symbols have already been used by many authors as natural elements in defining grid quality measures and formulating appropriate variational and differential grid techniques in a unified manner regardless of the geometry of the physical domains and surfaces. A theory of more general geometric objects such as regular multidimensional surfaces and Riemannian manifolds implemented for generating grids with necessary properties is expected to become a highly beneficial tool for boosting grid technology. The known relations and techniques of differential geometry also present an efficient means for transforming and modernizing the physical and grid equations into a suitable form. It is presumable that the science of differential geometry will play in numerical grid technology the same role to that played by the science of matrices in the theory of difference approximations of boundary value problems.

Therefore there is a need for a monograph which is essentially aimed at providing a deep and balanced insight into the fields of grid science, multidimensional geometry adjusted to grid technology, and the up-to-date achievements of the applications of geometric tools to the creation of advanced grid techniques. With this background the reader will be able to formulate and develop well-posed grid models and algorithms and analyze grid properties with geometry related tools, thus taking part in the solution of the very challenging problem of the development of advanced comprehensive grid generators.

This monograph gives an account of the geometrization of popular comprehensive grid methods and presents an important extension to the methods, related to the application of the technique of Riemannian manifolds to the formulation of grid equations by developing some procedures for the construction of monitor metric tensors. Contrary to classical geometric studies which analyze geometric features and characteristics of specified Riemannian manifolds, the problem of finding appropriate monitor metrics producing grid systems with the required properties is somewhat an inverse problem of the creation of Riemannian manifolds with desirable geometric characteristics. In accordance with the concept of the inverse problem the monograph discusses rather thoroughly some new techniques aimed at the construction of special monitor metrics in physical geometries. The techniques are designed by generalizing the projection approach in which the monitor metric in an n-dimensional physical geometry is borrowed from a natural metric of the n-dimensional surface derived by a height monitor function over the geometry. By this technology the required metric is defined through the original metric of the physical geometry and several monitor and weight functions.

The book establishes and reviews some of the relations of the Riemannian geometry for the purpose of obtaining new equations with implemented metric characteristics aimed at facilitating the control of the generation of grids with the required properties. Taking advantage of the relations established, the equations are converted into a compact form convenient for numerical treatment via the available algorithms.

The technique of multidimensional differential geometry is also applied to study the qualitative effect of a general class of monitor metrics on the resulting mesh. For this purpose a new characteristic of grid clustering is formulated. Certain relations between this measure and some geometric characteristics of grid hypersurfaces and the monitor functions forming the monitor metrics are established. The well-known results for grids generated by Laplace equations about node-clustering near concave boundary segments of domains and node-rarefaction near convex boundary segments are extended, using the relations, to arbitrary boundary segments and to more general elliptic equations formulated for generating grids. On the basis of the formulas established monitor functions are readily estimated in the popular Poisson, diffusion, and generalized Laplace elliptic models of grid equations to provide grid clustering or, if it is reasonable, grid rarefaction near arbitrary segments of physical geometries.

Some relations of the mean curvature of the monitor surfaces to the generalized Laplace equations for grid generation are exhibited. The book also includes a chapter devoted to the implementation of the comprehensive grid equations into a numerical code.

Though it is evident to the author that the approach advocated in the book has a brilliant prospect for boosting grid technology, there is a need for some critical mass in the knowledge of its tools, opportunities it provides, and spectacular results open to it. There is also a need to increase the availability of the information about the approach, here described completely and compactly, in order that the process of its development becomes attractive and indispensable for other researchers. Perhaps this book may appear to be that sufficient contribution by which the necessary critical mass is attained, thus arousing an interest of mathematicians in attacking grid generation problems with promising and advancing geometric techniques.

Since grid technology has widespread applications to nearly all field problems, this monograph will be useful for a broad range of readers, including teachers, students, researchers, and practitioners in applied mathematics, mechanics, and physics interested in the numerical analysis of multidimensional field problems with complicated physical geometries and complex solutions.

The book is divided into two parts. Part I of the monograph including Chaps. 1–4 introduces the reader to multidimentional differential geometry for the purpose of the implementation of its techniques into advanced grid generation technology. Chapter 1 gives a general introduction to the subject of numerical grids. It discusses structured grids that are commonly obtained by mapping a standard grid into the physical region with a transformation from a reference computational domain. The most popular structured grids are coordinate grids. The cells of such grids are curvilinear hexahedrons, and the identification of neighboring points is done by incrementing coordinate indices. The geometric implementation in grid technology pursued in the book assumes the development of robust techniques for producing suitable metrics by monitor functions in both physical domains and surfaces thus converting them into Riemannian manifolds. The metrics should guarantee generation of grids with the necessary properties through popular mathematical models. The chapters establish relations between the monitor functions and geometric characteristics of the Riemannian manifolds produced and the coordinate lines and surfaces generated by a corresponding mathematical model for the purpose of realization of grid control through a suitable specification of the monitor functions.

Part II of the book is devoted to the implementation of geometric tools into the development of grid techniques and codes. It contains Chaps. 5–9. Chapter 5 deals with fundamental elliptic grid models formulated through the operator of Beltrami, in particular, popular Poisson, diffusion, and generalized Laplace equations. Two-dimensional generalized Laplace equations in a natural metric of a physical surface were originally proposed by Warsi for generating fixed grids on the surface. The ordinary Laplace equations in the Euclidean metric were applied to generate fixed grids in domains by Crowley and Winslow. The grid approaches using the Poisson equations were originally considered by Godunov and Prokopov and further generalized, developed, and implemented for practical applications by Thompson, Thames, Mastin, and others. One justification of the Beltramian operators is demonstrated by the proof of the statement that an arbitrary nondegenerate smooth transformation of a physical domain or surface is realized as a solution of the Dirichlet boundary value problem for a system of comprehensive grid equations in some appropriate metric. The chapter also discusses some variational and harmonic interpretations of the generalized Laplace equations, in particular, a variational approach for generating harmonic maps through the minimization of energy functionals, which was suggested by Dvinsky.

Chapter 6 establishes some relations between the geometric characteristics of original physical geometries and monitor manifolds over them. These relations provide efficient means to control grid properties by monitor functions.

With the help of the geometric relations, established in Chap. 6, the grid equations introduced in Chap. 5 are transformed in Chap. 7 to equations in invariant forms with respect to independent logical variables. A special metric on two-dimensional surfaces is designed that results in more simple transformed equations even in comparison with the equations that have been used for generating fixed grids. Besides this it is found how the mean curvature of the monitor surface is included in the comprehensive multidimensional grid equations.

Analysis and control of grid properties with geometric tools is carried out in Chap. 8. For this purpose a general formula for the change of relative grid spacing is established. The well-known results for grids generated by Laplace equations about node clustering near concave boundary segments of domains and node rarefaction near convex ones are extended, using this formula, to arbitrary surfaces and boundary segments and to more general elliptic equations formulated for generating grids. On the basis of the formula some algorithms are developed for specifying monitor metrics on domains and surfaces to provide grid clustering or, if it is reasonable, grid rarefaction near arbitrary boundary segments of the physical geometries.

The final chapter gives a description of a numerical code for generating grids with the numerical solution in the logical domain of the elliptic equations obtained by changing mutually dependent and independent variables in the generalized Laplace equations.

The book ends with a list of references.

The author is very conscious of helpful suggestions in geometry, algebra, and numerical techniques made by his colleagues, Professors Borisov, Churkin, Kuzminov, Lebedev, Sharafutdinov, and Shvedov. He also thanks Galya Liseikin and Alesha Liseikin who assisted him with preparing figures and text of the manuscript in LATEX code.

Novosibirsk
June 2003 *Vladimir D. Liseikin*

Contents

Part I. Geometric Background to Grid Technology

1. **Introductory Notions** 5
   1.1 Representation of Physical Geometries 5
   1.2 General Concepts Related to Grids 8
       1.2.1 Grid Cells 8
       1.2.2 Requirements Imposed on Cells and Grids 10
   1.3 Grid Generation Models 15
       1.3.1 Mapping Approach 16
       1.3.2 Requirements Imposed on Mathematical Models 20
       1.3.3 Algebraic Methods 22
       1.3.4 Differential Methods 23
       1.3.5 Variational Methods 27
   1.4 Comprehensive Codes 30

2. **General Coordinate Systems in Domains** 33
   2.1 Jacobi Matrix 33
   2.2 Coordinate Lines, Tangential Vectors, and Grid Cells ... 34
   2.3 Coordinate Surfaces and Normal Vectors 36
   2.4 Representation of Vectors Through the Base Vectors 38
   2.5 Metric Tensors 40
       2.5.1 Covariant Metric Tensor 40
       2.5.2 Line Element 41
       2.5.3 Contravariant Metric Tensor 42
       2.5.4 Relations Between Covariant
             and Contravariant Elements 43
   2.6 Cross Product 44
       2.6.1 Geometric Meaning 44
       2.6.2 Relation to Volumes 45
       2.6.3 Relation to Base Vectors 46
   2.7 Relations Concerning Second Derivatives 47
       2.7.1 Christoffel Symbols of Domains 47
       2.7.2 Differentiation of the Jacobian 49
       2.7.3 Basic Identity 50

3. Geometry of Curves 53
3.1 Curves in Multidimensional Space 53
3.1.1 Definition 53
3.1.2 Basic Curve Vectors 54
3.2 Curves in Three-Dimensional Space 55
3.2.1 Basic Vectors 55
3.2.2 Curvature 56
3.2.3 Torsion 57

4. Multidimensional Geometry 59
4.1 Tangent and Normal Vectors and Tangent Plane 59
4.2 First Groundform 61
4.2.1 Covariant Metric Tensor 61
4.2.2 Contravariant Metric Tensor 62
4.3 Generalization to Riemannian Manifolds 65
4.3.1 Definition of the Manifolds 65
4.3.2 Example of a Riemannian Manifold 67
4.3.3 Christoffel Symbols of Manifolds 68
4.4 Tensors 72
4.4.1 Definition 72
4.4.2 Examples of Tensors 73
4.4.3 Tensor Operations 76
4.5 Basic Invariants 77
4.5.1 Beltrami's Differential Parameters 77
4.5.2 Measure of Relative Spacing 78
4.5.3 Measure of Relative Clustering 80
4.5.4 Mean Curvature 80
4.6 Geometry of Hypersurfaces 81
4.6.1 Normal Vector to a Hypersurface 81
4.6.2 Second Fundamental Form 85
4.6.3 Surface Curvatures 85
4.6.4 Formulas of the Mean Curvature 86
4.7 Relations to the Principal Curvatures
of Two-Dimensional Surfaces 101
4.7.1 Second Fundamental Form 101
4.7.2 Principal Curvatures 101

Part II. Application to Advanced Grid Technology

5. Comprehensive Grid Models 111
5.1 Formulation of a Differential Grid Generator 112
5.1.1 Beltramian Operator 112
5.1.2 Boundary Value Problem for Grid Equations 113

	5.1.3	Interpretation as a Multidimensional Equidistribution Principle 116
	5.1.4	Examples of Familiar Grid Equations 116
	5.1.5	Realization of Specified Grids 118
	5.1.6	Formulation of Monitor Metrics 122
5.2	Variational Formulation 125	
	5.2.1	Functional of Grid Smoothness...................... 125
	5.2.2	Geometric Interpretation 127
	5.2.3	Relation to Harmonic Functions 130
	5.2.4	Application to Adaptive Grid Generation............ 132
5.3	Conclusion 134	

6. Relations to Monitor Manifolds........................... 137

6.1 Computation of Geometric Characteristics 138
    6.1.1 Recursive Representation of the Monitor Metric 138
    6.1.2 Jacobian of the Covariant Metric Tensor 139
    6.1.3 Contravariant Metric Tensor 140
    6.1.4 Beltrami's Mixed and First Differential Parameters ... 142
    6.1.5 Christoffel Symbols 142
    6.1.6 Mixed Derivatives 143
    6.1.7 Beltrami's Second Differential Parameter 144
    6.1.8 Geometric Characteristics for a Spherical Monitor Metric 144
    6.1.9 Computation of Geometric Characteristics of a Monitor Manifold 145

6.2 Geometric Characteristics of Monitor Surfaces 147
    6.2.1 Computation of the Elements of the Contravariant Metric Tensor 148
    6.2.2 Beltrami's Mixed and First Differential Parameters ... 153
    6.2.3 Christoffel Symbols 154
    6.2.4 Mixed Derivatives 155
    6.2.5 Beltrami's Second Differential Parameter 156

6.3 Particular Two-Dimensional Case......................... 158
    6.3.1 Preliminary Results................................. 159
    6.3.2 Contravariant Metric Tensor 162
    6.3.3 Beltrami's Mixed and First Differential Parameters ... 164

7. Grid Equations with Respect to Intermediate Transformations 169

7.1 Relations to Comprehensive Equations 169
7.2 Resolved Grid Equations 171
    7.2.1 Basic Elliptic Operator............................. 171
    7.2.2 General Grid Equations 171
    7.2.3 Equations for a Spherical Monitor Metric............ 172

    7.2.4 Equations for a Spherical Monitor Metric
           Over a Surface 173
    7.2.5 Domain Grid Equations
           with Respect to a Monitor Surface 174
    7.2.6 Domain Grid Equations
           with Respect to a Monitor Metric 176
    7.2.7 Surface Grid Equations
           with Respect to a Monitor Surface 179
    7.2.8 Surface Grid Equations
           with Respect to a Monitor Metric 182
  7.3 Role of the Mean Curvature in the Grid Equations 183
  7.4 Practical Grid Equations 185
    7.4.1 Equations for Generating Grids on Curves 186
    7.4.2 Equations for Generating Grids
           on Two-Dimensional Surfaces 188
    7.4.3 Equations for Generating Grids in Domains 191

8. **Control of Grid Clustering** 195
  8.1 Fundamental Formula for Grid Clustering 196
    8.1.1 Relative Spacing Between Coordinate Surfaces 197
    8.1.2 Rate of Change of the Relative Spacing 197
    8.1.3 Relations to Geometry Characteristics 199
    8.1.4 Basic Relation to Grid Coordinates 205
    8.1.5 Remarks 205
    8.1.6 Grid Behavior near Boundary Segments
           of a Monitor Surface 208
  8.2 Application of Theorem to Popular Elliptic Models 210
    8.2.1 Control of Grids near Boundaries of Domains 210
    8.2.2 Diffusion Equations 227
    8.2.3 Control of Grid Spacing near Boundaries
           of Physical Surfaces 229

9. **Numerical Implementation of Grid Generator** 241
  9.1 One-Dimensional Equation 241
    9.1.1 Numerical Algorithm 241
  9.2 Two-Dimensional Equations 243
    9.2.1 Algorithms for Generating Grids
           in Two-Dimensional Domains 244
    9.2.2 Algorithm for Generating Grids
           on Two-Dimensional Surfaces 250
  9.3 Three-Dimensional Equations 253

References 255

Index 263

Part I

Geometric Background to Grid Technology

The science of differential geometry is a powerful and helpful tool capable for significantly boosting the development of grid technology. In particular, many notions and characteristics of two-dimensional surfaces bounding physical domains such as metric tensors, their invariants, the first and second groundforms, curvatures and torsions of lines, mean and Gauss curvatures of surfaces, and Christoffel symbols have already been reviewed in the monographs of Warsi (1981), Thompson, Warsi, and Mastin (1985), Knupp and Steinberg (1993), and Liseikin (1999) as natural elements in defining grid quality measures and formulating appropriate variational and differential grid techniques in a unified manner regardless of the geometry of domains and surfaces.

The papers of Dvinski (1991) and Liseikin (1991) gave rise to the implementation of more general geometric objects such as multidimensional regular surfaces and Riemannian manifolds into two- and three-dimensional adaptive grid generation techniques. The known relations and techniques of multidimensional differential geometry also present a useful tool for transforming and modernizing the physical and grid equations into a suitable form. It appears that the science of differential geometry will play in grid technology the same role to that played by matrices in the theory of difference approximations of boundary value problems.

This part of the book, containing Chaps. 1–4, gives an introduction to the geometric and tensor theories necessary in advanced grid technology. Chapter 1 acquaints the reader with the most general aspects of grid generation. It expounds the most popular approaches and mapping techniques that have been developed for generating numerical grids in domains and on surfaces. The elementary theories of curves, multidimensional regular surfaces, and Riemannian manifolds are reviewed in Chaps 2–4. These geometries play a crucial role in formulating comprehensive grid equations and in controlling grid properties in domains and on their boundaries. The metric tensors and basic geometric characteristics, in particular the mean curvature of n-dimensional surfaces, domains, and Riemannian manifolds necessary for grid technology, are discussed in detail.

1. Introductory Notions

This chapter gives an introduction to the subject of numerical grid technology. It deliniates the notion of grid cells, requirements imposed on grid properties, and the most popular methods of the mapping approach.

1.1 Representation of Physical Geometries

The goal of advanced grid technology is concluded with developing robust algorithms and codes for generating grids in arbitrary spatial domains. A spatial domain

$$X^3 \subset R^3, \quad X^3 = \{x \in R^3, x = (x^1, x^2, x^3)\},$$

is defined by a specification of a number of two-dimensional boundary patches that bound it (Fig. 1.1). Let S^{x2} be some such patch that is a surface in R^3. There are two ways for describing points of the boundary surface S^{x2}: 1) implicit and 2) explicit. By the implicit way the points of the patch S^{x2} are found from some equation $F(x^1, x^2, x^3) = 0$. In the explicit approach the patch is specified by a nondegenerate parametrization

$$\boldsymbol{x}(\boldsymbol{s}) : S^2 \to X^3, \quad \boldsymbol{s} = (s^1, s^2), \quad \boldsymbol{x}(\boldsymbol{s}) = [x^1(\boldsymbol{s}), x^2(\boldsymbol{s}), x^3(\boldsymbol{s})], \quad (1.1)$$

which is a one-to-one map between a two-dimensional domain $S^2 \subset R^2$, referred to as a parametric domain, and S^{x2}.

In the same way there is specified a boundary curve S^{x1} of a patch: implicitly by two equations $F_1(x^1, x^2, x^3) = 0$ and $F_2(x^1, x^2, x^3) = 0$ or explicitly through a nondegenerate parametrization

$$\boldsymbol{x}(s) : [a, b] \to X^3, \quad \boldsymbol{x}(s) = [x^1(s), x^2(s), x^3(s)], \quad (1.2)$$

between the parametric interval $[a, b]$ and S^{x1} (Fig. 1.1).

Analogous schemes (implicit and explicit) are applied to describing boundaries of two-dimensional domains.

Note the domain X^3 itself is readily representated in the form (1.1) and (1.2), as

$$\boldsymbol{x}(\boldsymbol{s}) : S^3 \to X^3, \quad \boldsymbol{s} = (s^1, s^2, s^3), \quad (1.3)$$

6   1. Introductory Notions

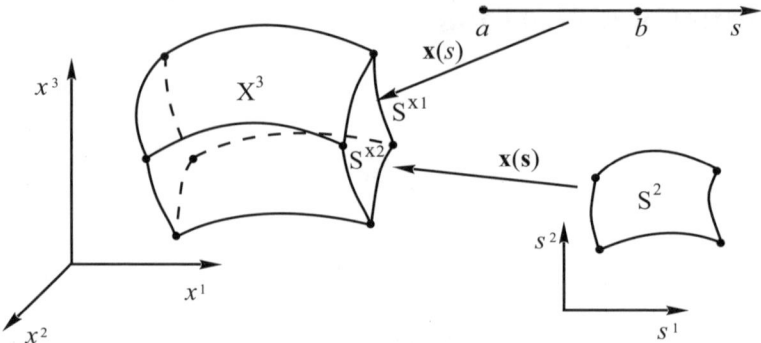

Fig. 1.1. Scheme for a representation of a domain as a curvilinear hexahedron

where S^3 is X^3 while $\boldsymbol{x}(\boldsymbol{s})$ is the identical transformation $\boldsymbol{x}(\boldsymbol{s}) \equiv \boldsymbol{s}$. However the domain may also have a general form of its representation (1.3) with the parametric domain S^3 different from X^3. The representations (1.1–1.3) give the full set of the explicit specifications of the domain X^3 and its boundary.

Of these two ways for describing physical geometries the modern grid technology relies only on the explicit specifications in the forms (1.1–1.3).

Thus the first inevitable step in the process of grid generation in a domain $X^3 \subset R^3$ is concluded with preparing a working place for the application of grid techniques, i.e. one must present an explicit representation of X^3, the boundary of X^3 by choosing a set of surface patches covering this boundary and the parametrizations of these patches and their boundary curves.

If the domain X^3 is diffeomorphic to a three-dimensional cube then one of the typical approaches to the specification of X^3 concludes with considering X^3 in the form of a curvelinear hexahedron with nonlinear faces, i.e. as a deformed cube (Fig. 1.1). Thus there are chosen 6 patches on the boundary of X^3 which represent 6 faces of the curvelinear hexahedron. The intersecion of two contiguous patches forms an edge of the hexahedron. As a result there are 12 edges. The intersections of contiguous edges form 8 vertices. The domain X^3, its faces, and edges should by specified by corresponding parametrizations (1.1–1.3).

In another consideration a domain X^3 may be viewed as a curvilinear tetrahedron, i.e. as a deformed three-dimensional simplex as in Fig. 1.2 (left). Consequently the boundary patches are triangular surfaces while their parametric domains may naturally be triangles. In a similar way the domain X^3 may be interpreted as one more solid, for example, as a curvilinear prism (Fig. 1.2) (right). Analogously a two-dimensional domain X^2 or surface S^{x2} diffeomorph to a square may be viewed as a curvilinear quadrilateral or triangle.

If a physical domain is not diffeomorph to an n-dimesional cube then, for its representation, it is either divided into several subdomains called blocks,

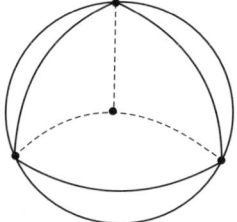 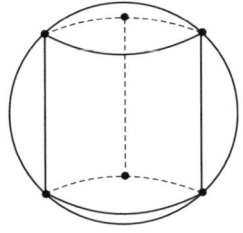

Fig. 1.2. Scheme for a representation of a globe as both a curvilinear tetrahedron and prism

each of which is diffeomorph to the cube, or imaginable faces are introduced. Figure 1.3 demonstrates such representations of a two-dimensional ring.

Relying on the explicit representation of an arbitrary physical geometry a physical domain, surface, or curve can be considered in a unified manner as a collection of geometric objects referred to as regular surfaces locally parametrized as

$$\boldsymbol{x}(\boldsymbol{s}) : S^n \to \mathbb{R}^{n+k}, \ \boldsymbol{x} = (x^1, \ldots, x^{n+k}), \ \boldsymbol{s} = (s^1, \ldots, s^n), \ n \geq 1, \quad (1.4)$$

where S^n is an n-dimensional parametric domain (an interval if $n = 1$), while $\boldsymbol{x}(\boldsymbol{s})$ is a smooth vector-valued function of rank n at all points $\boldsymbol{s} \in S^n$. We shall designate by S^{xn} the regular surface parametrized by (1.4). Note, when $k = 0$ then S^{xn} is a domain $Y^n \subset \mathbb{R}^n$.

With such consideration the process of grid generation can be carried out uniformly both for the boundary of a physical geometry and for its interior part.

We further assume throughout this book that we deal with an arbitrary geometry S^{xn} represented explicitly by the parametrization (1.4).

Using the specification (1.4) of the physical geometry S^{xn} allows one to generate grid points uniformly first on the parametric domain and then transfoming them through the parametrization of S^{xn}. This scheme of grid generation transpires the natural requirement for grid techniques that the grids generated in the physical geometry for different parametrizations should be the same, i.e. the grid algorithms should be invariant of parametrizations of S^{xn}.

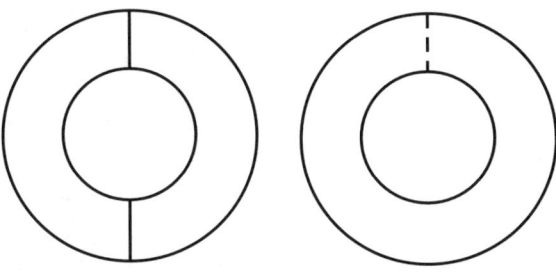

Fig. 1.3. Representations of a two-dimensional ring as both two and one curvilinear quadrilaterals

1.2 General Concepts Related to Grids

By a grid in a physical geometry S^{xn} there is understood an algorithmically described collection of elemental standard n-dimensional volumes covering or approximating the necessary area of the geometry. The standard volumes are referred to as grid cells. The cells have boundaries that are divided into a few segments which are $(n-1)$-dimensional cells. Therefore the cells can be formulated successively from one dimension to higher dimensions.

The boundary points of the one-dimensional cells are the cell vertices. These vertices are called the grid nodes.

This section discusses some general concepts related to grid cells and grids.

1.2.1 Grid Cells

In general the grid cells are small curvilinear volumes having simple standard shapes (Fig. 1.4 below). These curvilinear volumes are obtained by deforming reference cells. Such reference cells common in practical applications are demonstrated in Fig. 1.4 up.

Reference Cells. In one dimension the reference cell is a closed straight segment, whose boundary is composed of two points referred to as the cell vertices.

A two-dimensional reference cell is a two-dimensional equilateral polygon whose boundary segments are one-dimensional reference cells referred to as the edges of the cell. Commonly, the two-dimensional reference cells are considered in the form of equilateral triangles or squares. The boundary of a triangular cell is composed of three edges, while the boundary of a square is represented by four edges. These edges are the one-dimensional reference cells.

By a reference three-dimensional cell there is meant a three-dimensional equilateral polyhedron whose boundary is partitioned into a finite number of two-dimensional cells called its faces. In practical applications, three-dimensional reference cells typically have the shape of either equilateral tetrahedrons or cubes. The boundary of a tetrahedral cell is composed of four triangular cells, while a cube is bounded by six squares. Some applications also use for three-dimensional reference cells the cells, in the form of prisms having triangular top and bottom faces. Such a prism has two triangular and three square faces, nine edges, and six vertices. Note bees also prefer cells in the form of prisms however with hexagonal top and bottom faces.

Standard Cells. The edges and the faces of the reference cells are linear. The standard grid cells being deformed reference cells have, as a rule, nonequilateral edges and besides this they may be curvilinear. Thus, in general, the standard cells have the form of curves and curvilinear triangles, quadrelaters, tetrahedrons, hexahedrons, and prisms as shown in Fig. 1.4 below.

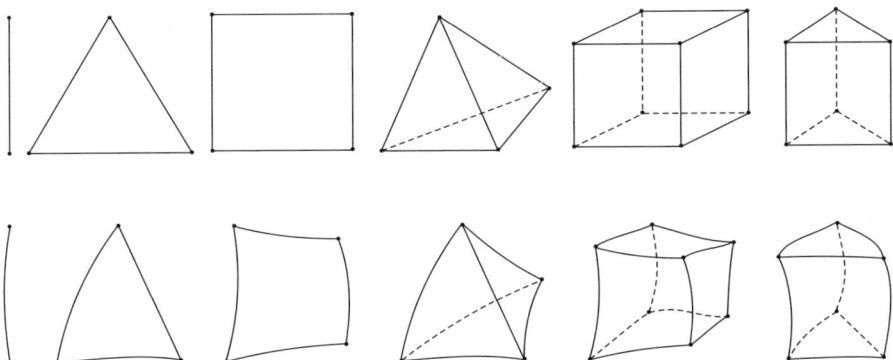

Fig. 1.4. Typical reference (up) and corresponding standard (below) grid cells

The selection of the shapes shown in Fig. 1.4 to represent the reference and standard cells is justified, first, by their geometrical simplicity and, second, because the existing approaches for the numerical simulation of physical problems are largely based on approximations of partial differential equations using these elemental volumes. The choice of cell shape depends on the geometry and physics of the specific problem and on the method of solution. In particular, tetrahedrons (triangles in two dimensions) are well suited for finite-element methods, while hexahedrons are commonly used for finite-difference techniques.

The major advantage of hexahedral cells (quadrilaterals in two dimensions) is that their faces (or edges) may be aligned with the coordinate surfaces (or curves). In contrast, no coordinates can be aligned with tetrahedral meshes. However, strictly hexahedral meshes may be ineffective near boundaries with sharp corners.

Prismatic cells are generally placed near boundary surfaces which have previously been triangulated. The surface triangular cells serve as faces of prisms, which are grown out from these triangles. Prismatic cells are efficient for treating boundary layers, since they can be constructed with a high aspect ratio in order to resolve the layers, but without small angles, as would be the case for tetrahedral cells.

Triangular cells are the simplest two-dimensional volumes and can be produced from quadrilateral cells by constructing interior edges. Analogously, tetrahedral cells are the simplest three-dimensional volumes and can be derived from hexahedrons and prisms by constructing interior faces. The strength of triangular and tetrahedral cells is in their applicability to virtually any type of domain configuration. The drawback is that the integration of the physical equations becomes a few times more expensive with these cells in comparison with quadrilateral or hexahedral cells.

The vertices of the cells define grid points which approximate a physical geometry. Alternatively, the grid points in the physical geometry may have been generated previously by some other process. In this case the construction of the grid cells requires special techniques.

1.2.2 Requirements Imposed on Cells and Grids

The grid should discretize a physical geometry by displacing the standard cells in such a manner that the computation of the physical quantities on the geometry is carried out as efficiently as desired. The accuracy, which is one of the components of the efficiency of the computation, is influenced by a number of grid factors, such as grid size, grid topology, cell shape and size, and consistency of the grid with the geometry and with the solution. A general consideration of these grid factors is reviewed in this subsection.

Grid Size and Cell Size. The grid size is indicated by the number of grid points, while the cell size implies the maximum value of the lengths of the cell edges. Grid generation requires techniques which possess the intrinsic ability to increase the number of grid nodes. At the same time the edge lengths of the resulting cells should be reduced in such a manner that they approach zero as the number of nodes tends to infinity.

Small cells are necessary to obtain more accurate solutions and to resolve physical quantities on small scales, such as transition layers and turbulence. Also, the opportunity to increase the number of grid points and to reduce the size of the cells enables one to study the convergence rate of a numerical code and to improve the accuracy of the solution by multigrid approaches.

Cell and Grid Deformation. The cell deformation characteristics can be formulated as some measures of the departure of the cell from a reference one. Cells with low deformity are preferable from the point of view of simplicity and uniformity of the construction of the algebraic equations approximating the differential equations.

Typically, cell deformation is characterized through the aspect ratio, the angles between the cell edges, and the volume (area in two dimensions) of the cell.

The major requirement for the grid cells is that they must not be folded or degenerate at any points or lines, as demonstrated in Fig. 1.5 (2,3,4). Unfolded cells are obtained from the reference cells by a one-to-one deformation. Commonly, the value of any grid generation method is judged by its ability to yield unfolded grids in regions with complex geometry. A more tough condition imposed on grid techniques also requires the generation of linear and convex cells only.

The grid deformity is also characterized by the rate of the change of the geometrical features of contiguous cells. Grids whose neighboring cells do not change abruptly are referred to as smooth grids.

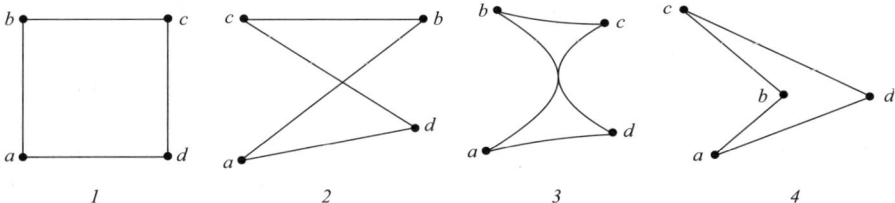

Fig. 1.5. Normal (*1*), singular (*2,3*), and badly deformed (*4*) quadrilateral cells

Grid Consistency. By a consistent grid or a consistent discretization there is meant a collection of n-dimensional strongly convex cells satisfying the following condition: if two different cells intersect, then the region of the intersection is a common face for both cells. This definition does not admit the fragments of discretizations depicted in Fig. 1.6 (2,3,4).

If the union of the cells of the consistent discretization constitutes a simply connected n-dimensional geometry S^{xn}, i.e. a geometry which is homeomorphic to an n-dimensional cube, then, in accordance with the Euler theorem,

$$\sum_{i=0}^{n-1}(-1)^i N_i = 1 + (-1)^{n-1},$$

where N_i, $i > 0$, is the number of i-dimensional boundary faces of the domain discretization, while N_0 is the number of boundary vertices. In particular, N_1 is the number of boundary edges. This relation can be used to verify the consistency of a generated grid.

Grid Organization. There also is a requirement on grids to have some organization of their nodes, faces, and cells, which is aimed at facilitating the procedures for formulating and solving the algebraic equations substituted for the differential equations. This organization should identify neighboring points and cells. The grid organization is especially important for that class of numerical methods whose procedures for obtaining the algebraic equations consist of substituting differences for derivatives. To a lesser degree, this

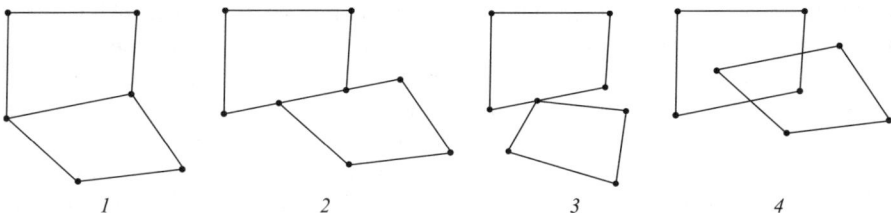

Fig. 1.6. Admitted (*1*) and nonadmitted (*2, 3, 4*) intersections of neighboring quadrilateral cells

organization is needed for finite-volume methods because of their inherent compatibility with irregular meshes.

Consistency with Geometry. The accuracy of the numerical solution of a partial differential equation and of the interpolation of a discrete function is considerably influenced by the degree of compatibility of the mesh with the shape of the physical geometry. First of all, the grid nodes must adequately approximate the original geometry, that is, the distance between any point of the physical geometry and the nearest grid node must not be too large. Moreover, this distance must approach zero when the grid size tends to infinity. This requirement of adequate geometry approximation by the grid nodes is indispensable for the accurate computation and interpolation of the solution over the whole physical geometry.

The second requirement for consistency of the grid with the physical geometry is concerned with the approximation of its boundary by the faces (edges in two dimensions) of the boundary grid cells, i.e. there is to be a sufficient number of nodes which can be considered as boundary ones, so that a set of edges (in two dimensions) and cell faces (in three dimensions) formed by these nodes models efficiently the boundary. In this case, the boundary conditions may be applied more easily and accurately. If these points lie on the boundary of the geometry, then the grid is referred to as a boundary-fitting or boundary-conforming grid.

Consistency with Solution. It is evident that distribution of the grid points and the form of the grid cells should be dependent on the features of the physical solution. In particular, it is better to generate the cells in the shape of hexahedrons or prisms in boundary layers. Often, the grid points are aligned with some preferred directions, e.g. streamlines. Furthermore, a nonuniform variation of the solution requires clustering of the grid point in regions of high gradients, so that these areas of the physical geometry have finer resolution. Local grid clustering is needed because the uniform refinement of the entire domain may be very costly for multidimensional computations. It is especially true for problems whose solutions have localized regions of very rapid variation (layers). Without grid clustering in the layers, some important features of the solution can be missed, and the accuracy of the solution can be degraded. Problems with boundary and interior layers occur in many areas of application, for example in fluid dynamics, combustion, solidification, solid mechanics and wave propagation. Typically the locations where the high resolution is needed are not known beforehand but are found in the process of computation. Consequently, a suitable mesh, tracking the necessary features of the physical quantities, as the solution evolves, is required.

A local grid refinement is accomplished in two ways: (a) by moving a fixed number of grid nodes, with clustering of them in zones where this is necessary, and coarsening outside of these zones, and (b) by inserting new points in the zones of the physical geometry where they are needed. Local grid refinement in zones of large variation of the solution commonly results in the following improvements:

(1) the solution at the grid points is obtained more accurately;
(2) the solution is interpolated over the whole region more precisely;
(3) oscillations of the solution are eliminated;
(4) larger time steps can be taken in the process of computing solutions of time-dependent problems.

Independence of Parametrizations of Geometries. If a grid algorithm uses parametrizations of a physical geometry S^{xn} in the process of grid generation then, inevitably, this algorithm should be independent of the choice of a parametrization. To clarify this we consider one popular equidistribution approach for generating grids on curves (Fig. 1.7). Let a curve S^{x1} in R^n be specified by two parametrizations

$$\boldsymbol{x}_1(s) : [0,1] \to R^n \, , \quad \boldsymbol{x}_1 = (x_1^1, \ldots, x_1^n) \, , \tag{1.5}$$

and

$$\boldsymbol{x}_2(t) : [0,1] \to R^n \, , \quad \boldsymbol{x}_2(t) = \boldsymbol{x}_1[s(t)] \, , \tag{1.6}$$

where

$$s(t) : [0,1] \to [0,1]$$

is a smooth one-to-one function connecting these parametrizations. The popular universal approach, based on the parametrization (1.5), for generating grid nodes on S^{x1}, uses a solution of the following two-point boundary value problem

$$\frac{\mathrm{d}}{\mathrm{d}\xi}\left[\frac{\mathrm{d}s}{\mathrm{d}\xi} w_1(s)\right] = 0 \, , \quad 0 < \xi < 1 \, , \\ s(0) = 0 \, , \quad s(1) = 1 \, , \tag{1.7}$$

where $w(s) > 0$ is some function called a weight function. If $s(\xi)$ is a solution of this problem then the grid nodes $\boldsymbol{x}_i$, $i = 0, 1, \ldots, N$, on the curve S^{x1}, obtained by the method, are defined as follows:

$$\boldsymbol{x}_i = \boldsymbol{x}_1[s(ih)] \, , \quad i = 0, \ldots, N \, , \quad h = 1/N \, .$$

Let now the parametrization $\boldsymbol{x}_2(t)$ specified by (1.6) be used in the problem (1.7) with some weight function $w_2(t)$ for the generation of the same grid nodes on S^{x1}. In this case the function $t_1(\xi)$ for which

$$\boldsymbol{x}_2(t_1(ih)) = \boldsymbol{x}_1[s(t_1(ih))] = \boldsymbol{x}_i = \boldsymbol{x}_1(s(ih)) \, , \quad i = 0, 1, \ldots, n \, ,$$

must coincide with $t[s(\xi)]$, where $s(\xi)$ is the solution of (1.7), while $t(s)$ is the inverse of $s(t)$. Therefore the function $t_1(\xi)$ is a solution of the boundary value problem

14   1. Introductory Notions

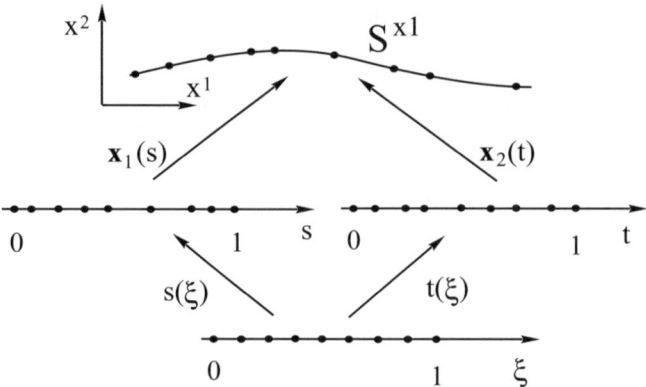

Fig. 1.7. Illustration for independence of grid generation of the choice of parametrizations

$$\frac{d}{d\xi}\left[\frac{dt_1}{d\xi}\frac{ds}{dt}w_1[s(t_1)]\right] = 0, \quad 0 < \xi < 1,$$

$$t_1(0) = 0, \quad t_1(1) = 1.$$

Since the weight functions $w_1(s)$ and $w_2(t)$ for defining the same grid nodes on S^{x1} by the model (1.7) through the parametrizations (1.5) and (1.6), respectively, are not independent but they should be connected by the following relation

$$w_2(t) = w_1[s(t)]\frac{ds}{dt}. \tag{1.8}$$

If this relation is not satisfied, the grid nodes obtained with the help of the solution of (1.7) may vary for different parametrizations of S^{x1}.

It appears that if we take for the weight functions related to the parametrization (1.5) and (1.6) the corresponding functions $w_1(s) = \sqrt{g^s}$ and $w_2(t) = \sqrt{g^t}$ where

$$g^s = \frac{d\mathbf{x}_1}{ds} \cdot \frac{d\mathbf{x}_1}{ds} \quad \text{and} \quad g^t = \frac{d\mathbf{x}_2}{dt} \cdot \frac{d\mathbf{x}_2}{dt}$$

is the covariant metric tensor of S^{x1} in the coordinate s and t, respectively, then the equation (1.8) holds since there is valid the obvious equation

$$g^t = g^s \left(\frac{ds}{dt}\right)^2.$$

The consideration given for the curvilinear curve also is actual for surfaces. As well as in the one-dimensional case, the application of the elements of the metric tensors of n-dimensional surfaces allows one to formulate grid equations which produce the same grid nodes for different surface parametrizations.

Compatibility with Numerical Methods. The locations of the zones of local refinement are also dependent on the numerical approximation to the physical equations. In particular, the areas of high solution error require more refined grid cells. However, the error is estimated through the derivatives of the solution and the size of the grid cells. Thus, ultimately, the grid point locations are to be defined in accordance with the derivatives of the solution.

In general, numerical methods for solving partial differential equations can be divided into two classes: methods based on direct approximations of the derivatives in the differential equation and methods that approximate the solution of the continuum differential equation by linear combinations of trial functions. Finite-difference methods belong to the first class. This difference in methods has a direct impact on the construction of the numerical grid. For the finite-difference methods it is desirable to locate the grid points along directions of constant coordinates in the physical region in order to provide a natural approximation of the derivatives: on the other hand, the methods in the second class that approximate the solution with trial functions do not impose such a restriction on the grid, since the approximate derivatives are obtained after substitution of the approximate solution.

1.3 Grid Generation Models

There are two fundamental classes of grid popular in the numerical solution of boundary value problems in multidimensional regions: structured and unstructured. These classes differ in the way in which the mesh points are locally organized. In the most general sense, this means that if the local organization of the grid points and the form of the grid cells do not depend on their position but are defined by a general rule, the mesh is considered as structured. When the connection of the neighboring grid nodes varies from point to point, the mesh is called unstructured. As a result, in the structured case the connectivity of the grid is implicitly taken into account, while the connectivity of unstructured grids must be explicitly described by an appropriate data structure procedure.

Detailed descriptions of the most popular structured methods and their theoretical and logical justifications and numerical implementations were given in the monographs by Thompson, Warsi, and Mastin (1985), Knupp and Steinberg (1993), and Liseikin (1999). Particular issues concerned with the generation of one-dimensional moving grids, the stretching technique for the numerical solution of singularly perturbed equations, nonstationary grid techniques, and equidistribution methods for wave propogation problems were considered in the books by Alalykin et al. (1970), Liseikin (2001), Zegeling (1993), and Khakimzyanov et al. (2001), respectively.

A considerable number of general structured grid generation methods were reviewed in surveys by Thompson, Warsi, and Mastin (1982), Thompson (1984a, 1996), Eiseman (1985), Liseikin (1991b), and Thompson and Weatherill (1993).

Adaptive structured grid methods were first surveyed by Anderson (1983) and Thompson (1984b, 1985). Then a series of surveys on general adaptive methods was presented by Eiseman (1987), Hawken, Gottlieb, and Hansen (1991), Liseikin (1996b), and Baker (1997). Adaptive techniques for moving grids were described by Hedstrom, Rodrigue (1982) and Zegeling (1993).

Methods for unstructured grids were reviewed by Thacker (1980), Ho-Le (1988), Shephard et al. (1988a), Baker (1995, 1997), Field (1995), Carey (1997), George and Borouchaki (1998), and Krugljakova et al. (1998). An exhaustive survey of both structured and unstructured techniques has been given by Thompson and Weatherill (1993).

The two fundamental classes of mesh give rise to three additional subdivisions of grid types: block-structured, overset, and hybrid. These kinds of mesh possess to some extent the features of both structured and unstructured grids, thus occupying an intermediate position between the purely structured and unstructured grids.

The multiblock strategy for generating grids around complicated shapes was originally proposed by Lee et al. (1980); however, the idea of using different coordinates in different subregions of the domain can be traced back to Thoman and Szewezyk (1969). Some of the first applications of block-structured grids to the numerical solution of three-dimensional fluid-flow problems in realistic configurations were demonstrated by Rizk and Ben-Shmuel (1985), Sorenson (1986), Atta, Birchelbaw, and Hall (1987), and Belk and Whitfield (1987).

The overset grid approach was introduced by Atta and Vadyak (1982), Berger and Oliger (1983), Benek, Steger, and Dougherty (1983), Miki and Takagi (1984), and Benek, Buning, and Steger (1985). The concept of blocks with a continuous alignment of grid lines across adjacent block boundaries was described by Weatherill and Forsey (1984) and Thompson (1987). Thomas (1982) and Eriksson (1983) applied the concept of continuous line slope, while a discontinuity in slope was discussed by Rubbert and Lee (1982).

A shape recognition technique based on an analysis of a physical domain and an interactive construction of a computational domain with a similar geometry was proposed by Takahashi and Shimizu (1991) and extended by Chiba et al. (1998). The embedding technique was considered by Albone and Joyce (1990) and Albone (1992).

1.3.1 Mapping Approach

The process of grid generation on the physical geometry S^{xn} locally represented by (1.4) is generally turned to finding an intermediate transformation

$$s(\boldsymbol{\xi}) : \Xi^n \to S^n \ , \quad \boldsymbol{\xi} = (\xi^1, \ldots, \xi^n) \tag{1.9}$$

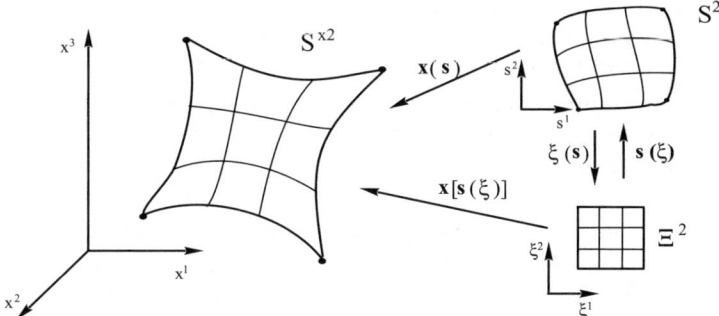

Fig. 1.8. Illustration for quadrilateral grid generation by the mapping approach

from a suitable computational (logical) domain Ξ^n to the parametric domain S^n. Consequently the mesh points on S^{xn} are generated as images through

$$x[s(\boldsymbol{\xi})] : \Xi^n \to \mathbb{R}^{n+k}$$

of the nodes of a reference grid in Ξ^n which can be either structured or unstructured (see Figs. 1.8–1.10).

Realization of Grid Requirements. The notion of using an intermediate transformation to generate a mesh is very helpful. The idea is to choose a computational domain Ξ^n with a simpler geometry than that of the parametric domain S^n and then to find a transformation $s(\boldsymbol{\xi})$ between these domains which eliminates the need for a complicated mesh when approximating the physical quantities. That is, if the computational area and the transformation are well chosen, the transformed physical boundary value problem can be accurately represented by a small number of mesh points. Emphasis is placed on a small number of points, because any transformed problem (provided only that the transformation is nonsingular) may be accurately approximated with a sufficiently fine, simple mesh. In practice, there will be a trade-off between the difficulty of finding the transformation and the number of grid nodes required to find the solution to a given accuracy.

The idea of using mappings to generate grids is extremely appropriate for finding the conditions that the grid must satisfy for obtaining accurate solutions of partial differential equations in the physical geometry S^{xn}, because these conditions can be readily defined in terms of the transformations. For example, the grid requirements described in Sect. 1.3.2 are readily formulated through the transformation concept.

Since a solution which is a linear function is computed accurately at the grid points and is approximated accurately over the whole region, an attractive possible method for generating grids is to find a transformation $s(\boldsymbol{\xi})$ such that the solution is linear in Ξ^n. Though in practice this requirement for the transformation is not attained even theoretically (except in the case of strongly monotonic univariate functions), it is useful in the sense of an ideal that the developers of grid generation techniques should bear in mind. One modification of this requirement which can be practically realized consists of the requirement of a local linearity of the solution in Ξ^n.

The requirements imposed on the grid and the cell size are realized by the construction of a uniform grid in Ξ^n and a smooth function $s(\boldsymbol{\xi})$. The grid cells are not folded if $s(\boldsymbol{\xi})$ is a one-to-one mapping. Consistency with the geometry is satisfied with a transformation $\boldsymbol{x}(\boldsymbol{\xi})$ that maps the boundary of Ξ^n onto the boundary of S^n. Grid concentration in zones of large variation of a function $\boldsymbol{u}(\boldsymbol{x})$ is accomplished with a mapping $\boldsymbol{x}[s(\boldsymbol{\xi})]$ which provides variations of the function $\boldsymbol{u}[\boldsymbol{x}(\boldsymbol{\xi})]$ in the domain Ξ^n that are not large.

Coordinate Grids. Among grids, coordinate grids in which the nodes and cell faces are defined by the intersection of lines and surfaces of a coordinate system in S^{xn} are very popular in finite-difference methods. The range of values of this system defines a computation region Ξ^n in which the cells of the uniform grid are rectangular n-dimensional parallelepipeds, and the coordinate values define the function $\boldsymbol{x}[s(\boldsymbol{\xi})] : \Xi^n \to S^{xn}$.

In the case S^{xn} is a domain $X^n \subset R^n$ the simplest of such grids are the Cartesian grids obtained by the intersection of the Cartesian coordinates in X^n. The cells of these grids are rectangular parallelepipeds (rectangles in two dimensions). The use of Cartesian coordinates avoids the need to transform the physical equations. However, the nodes of the Cartesian grid do not coincide with the curvilinear boundary, which leads to difficulties in implementing the boundary conditions with second-order accuracy.

Boundary-Conforming Grids. An important subdivision of grids is the boundary-fitted or boundary-conforming grids. These grids are obtained from one-to-one transformations $s(\boldsymbol{\xi})$ which map the boundary of the domain Ξ^n onto the boundary of S^n.

The most popular of these, for finite-difference methods, have become the coordinate boundary-fitted grids whose points are formed by intersection of the coordinate lines, while the boundary of S^n is composed of a finite number of coordinate surfaces (lines in two dimensions) $\xi^i = \xi_0^i$. Consequently, in this case the computation region Ξ^n is a rectangular domain, the boundaries of which are determined by $(n-1)$-dimensional coordinate planes in R^n, and the uniform grid in Ξ^n is the Cartesian grid. Thus the physical region is represented as a deformation of a rectangular domain and the generated grid as a deformed lattice. These grids give a good approximation to the boundary of the region and are therefore suitable for the numerical solution of problems with boundary singularities, such as those with boundary layers in which the solution depends very much on the accuracy of the approximation of the boundary conditions.

The requirements imposed on boundary-conforming grids are naturally satisfied with the coordinate transformations $s(\boldsymbol{\xi})$.

The algorithm for the organization of the nodes of boundary-fitted coordinate grids consists of the trivial identification of neighboring points by incrementing the coordinate indices, while the cells are curvilinear hexahedrons. This kind of grid is very suitable for algorithms with parallelization.

Its design makes it easy to increase or change the number of nodes as required for multigrid methods or in order to estimate the convergence rate and error, and to improve the accuracy of numerical methods for solving boundary value problems.

With boundary-conforming grids there is no necessity to interpolate the boundary conditions of the problem, and the boundary values of the region can be considered as input data to the algorithm, so automatic codes for grid generation can be designed for a wide class of regions and problems.

In the case of unsteady problems the most direct way to set up a moving grid is to do it via a coordinate transformation. These grids do not require a complicated data structure, since they are obtained from uniform grids in simple fixed domains such as rectangular ones, where the grid data structure remains intact.

Shape of Computational Domains. The idea of the mapping approach is to transform a complex physical geometry S^{xn} to a simpler domain Ξ^n with the help of the parametrization $\boldsymbol{x}[\boldsymbol{s}(\boldsymbol{\xi})]$. The region Ξ^n in (1.9), which is called the computational or logical domain, can be either rectangular (Fig. 1.8) or of a different shape matching qualitatively the shape of the physical geometry; in particular, it can be triangular for $n = 2$ (Fig. 1.9) or tetrahedral for $n = 3$. Using such approach, a numerical solution of a partial differential equation in a physical region of arbitrary shape can be carried out in a standard computational domain, and codes can be developed that require only changes in the input.

Shape of a Reference Grid. The cells of the reference grid in the computational domain Ξ^n can be rectangular or of a different shape. Schematic

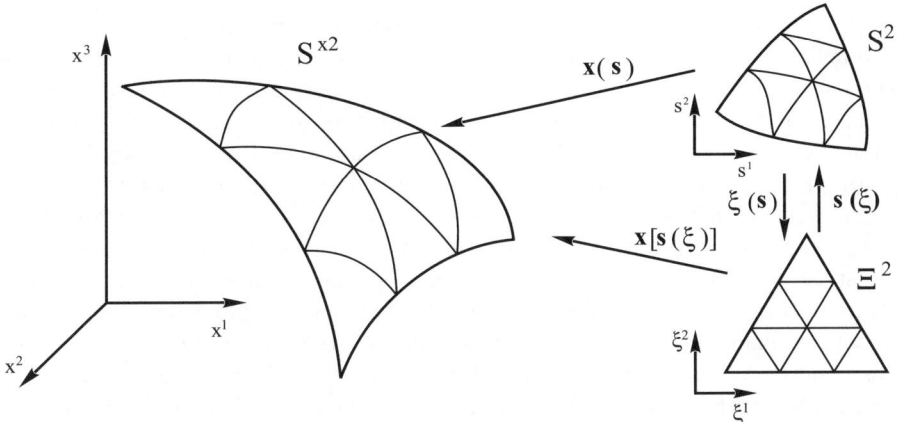

Fig. 1.9. Illustration for triangular grid generation by the mapping approach

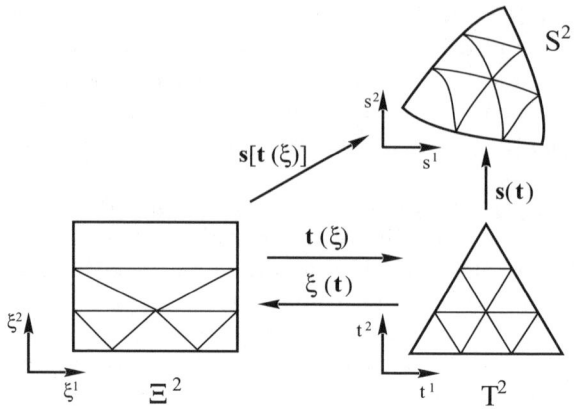

Fig. 1.10. Illustration for generating triangular cells by the mapping approach with a nonstructured reference grid

illustration of grid cells is presented in Fig. 1.1. Note that regions in the form of curvilinear triangles, such as that shown in Fig. 1.9, may be more suitable for gridding by triangular cells than by quadrilateral ones.

The triangular grid such as in Fig. 1.9 can also be obtained by mapping a rectangle Ξ^2 onto the triangular parametric domain S^2 provided the reference grid in Ξ^2 is not a structured one but it is obtained from a uniform grid in the triangle T^2 which undergoes a deformation $\boldsymbol{\xi}(t)$ (Fig. 1.10). This deformation is the inverse of the contraction $\boldsymbol{t}(\boldsymbol{\xi})$ of the rectangle Ξ^2 along the horisontal lines to transform it to the triangle T^2.

Analogous scheme with a nonstructured reference grid in the logical domain Ξ^3 having the shape of a three-dimensional rectangular parallelepiped can be applied to the generation of tetrahedral grids.

1.3.2 Requirements Imposed on Mathematical Models

The chief practical difficulty facing methods for gridding general physical geometries is that of formulating satisfactory techniques which can realize the user's requirements. Grid generation technology should develop methods that can help in handling problems with multiple variables, each varying over many orders of magnitude. These methods also should be capable of generating grids whose node displacement is independent of parametrizations of a physical geometry. The methods should incorporate specific control tools, with simple and clear relationships between these control tools and characteristics of the grid such as mesh spacing, skewness, smoothness, and aspect ratio, in order to provide a reliable way to influence the efficiency of the computation. And finally, the methods should be computationally efficient and easy to code. Thus with the mapping approach a mathematical model for generating grids through intermediate transformations between the computational and parametric domains, which is claimed to be a foundation of a robust comprehensive grid generator, should satisfy the following fundamental properties:

(1) well-posedness of the mathematical problem formulated by this model for the grid generator;
(2) independence of the grid construction of a parametrization of the geometry;
(3) malleability to a numerical implementation into an automatic code;
(4) existence of a straightforward means for efficient control of the grid quality;
(5) ability to obtain in a unified manner the domain and surface grids required in practice.

A number of techniques for generating intermediate transformation have been developed. Every method has its strengths and its weaknesses. Therefore, there is also the question of how to choose the most efficient method for the solution of any specific problem, taking into account the geometrical complexity, the computing cost for generating the grid, the grid structure, and other factors.

The goal of the development of these methods is to provide effective and acceptable grid generation systems.

The most efficient numerical grids are boundary-conforming grids. The generation of these grids can be performed by a number of approaches and techniques. Many of these methods are specifically oriented to the generation of grids for the finite-difference method.

A boundary-fitted coordinate grid in the physical geometry S^{xn} is commonly generated first on the chosen edges of S^{xn}, then on its faces, and finally in its interior. Thus at each step the intermediate transformation $s(\boldsymbol{\xi})$ in the mapping approach is known at the boundary of the corresponding logical domain and this boundary map is extended from the boundary to the domain interior. This process is analogous to the interpolation of a function from a boundary or to the solution of the Dirichlet boundary value problem. On this basis there have been developed three basic groups of methods for finding the intermediate transformation $s(\boldsymbol{\xi}) : \Xi^n \to S^n$ provided there exists a boundary transformation

$$\partial s(\boldsymbol{\xi}) : \partial \Xi^n \to \partial S^n \ .$$

These methods are

(1) algebraic methods, which use various forms of interpolation or special functions;
(2) differential methods, based mainly on the solution of elliptic, parabolic, and hyperbolic equations in a selected computational domain Ξ^n;
(3) variational methods, based on optimization of grid quality properties.

1.3.3 Algebraic Methods

In the algebraic approach the intermediate transformation $s(\boldsymbol{\xi})$ which extends the boundary mapping $\partial s(\boldsymbol{\xi}): \partial \Xi^n \to \partial S^n$ found on the previous step is commonly computed through the formulas of transfinite interpolation. There are two types of transfinite interpolations popular with grid generation: Lagrange and Hermite.

In particular the three-dimensional Lagrange intermediate transformation $s(\boldsymbol{\xi})$ between the interior of the unit cube $0 \leq \xi^i \leq 1$, $i = 1, 2, 3$, and the parametric domain S^3 is defined by the following recursive formula

$$\boldsymbol{F}_1(\boldsymbol{\xi}) = \alpha_0^1(\xi^1)\boldsymbol{s}(0, \xi^2, \xi^3) + \alpha_1^1(\xi^1)\boldsymbol{s}(1, \xi^2, \xi^3) \,,$$

$$\boldsymbol{F}_2(\boldsymbol{\xi}) = \boldsymbol{F}_1(\boldsymbol{\xi}) + \alpha_0^2(\xi^2)\boldsymbol{s}(\xi^1, 0, \xi^3) + \alpha_1^2(\xi^2)\boldsymbol{s}(\xi^1, 1, \xi^3) \,, \quad (1.10)$$

$$\boldsymbol{s}(\boldsymbol{\xi}) = \boldsymbol{F}_2(\boldsymbol{\xi}) + \alpha_0^3(\xi^3)\boldsymbol{s}(\xi^1, \xi^2, 0) + \alpha_1^3(\xi^3)\boldsymbol{s}(\xi^1, \xi^2, 1) \,,$$

where the univariate functions $\alpha_k^i(\xi)$, $i = 1, 2, 3$, $k = 0, 1$, referred to as blending functions, are subject to the relations of consistency

$$\alpha_0^i(0) = 1 \,, \quad \alpha_0^i(1) = 0 \,,$$

$$\alpha_1^i(0) = 0 \,, \quad \alpha_1^i(1) = 1 \,.$$

Analogous formulas are held for the Hermite interpolation that matches at the points of the boundary of Ξ^n the values of both the function $s(\boldsymbol{\xi})$ and its first derivatives in the directions normal to the boundary segments. A detailed review of the Lagrange and Hermit techniques for generating algebraic grids is presented in the monograph of Liseikin (1999).

Algebraic methods are simple; they enable the grid to be generated rapidly and the spacing and slope of the coordinate lines to be controlled by the tangential derivatives at the boundary points and blending coefficients in the transfinite interpolation formulas. However, in regions of complicated shape the coordinate surfaces obtained by algebraic methods can become degenerate or the cells can overlap or cross the boundary. Moreover, they basically preserve the features of the boundary surfaces, in particular, discontinuties. Besides this the algebraic methods of transfinite interpolation do not guarantee the independents of grid nodes displacement on parametrizations of a physical geometry.

Algebraic approaches are commonly used to generate grids in regions with smooth boundaries that are not highly deformed, or as an initial approximation in order to start the iterative process of an elliptic grid solver.

The construction of intermediate transformations through the formulas of transfinite interpolation was originally formulated by Gordon and Hall (1973) and Gordon and Thiel (1982). The Hermite interpolation was presented by Smith (1982).

The multisurface method which allows for the specification of the intermediate transformation $s(\boldsymbol{\xi})$ at the points of some interior sections of the physical domain was described by Eiseman (1980). The blending functions were implicitly derived from global and/or local interpolants which result from an expression for the tangential derivative spanning between the exterior boundary surfaces. A two-boundary technique was introduced by Smith (1981). It is based on the description of two opposite boundary surfaces, tangential derivatives on the boundary surfaces which are used to compute normal derivatives, and Hermite cubic blending functions.

The construction of some special blending functions aimed at grid clustering at the boundaries of physical geometries was performed by Eriksson (1982), Smith and Eriksson (1987), and Liseikin (1989). A detailed description of various forms of blending functions was presented in monographs by Thompson, Warsi, and Mastin (1985) and Liseikin (1999).

1.3.4 Differential Methods

For gridding geometries with arbitrary boundaries, differential methods based on the solution of elliptic and parabolic equations are commonly used. Such equations generate smooth grids, allow for full specification of grid nodes on the boundary of a physical geometry, does not propagate boundary singularities into its interior, have less danger of producing cell overlapping, and can be solved efficiently using many well-developed codes. The use of parabolic and elliptic systems enables orthogonal and clustering coordinate lines to be constructed, while, in many cases, the maximum principle, which is typical for these systems, ensures that the intermediate transformations are nondegenerate. Elliptic equations are also used to smooth algebraic or unstructured grids.

Elliptic Equations. The most popular elliptic equations with differential grid approaches are the generalized Poisson equations formulated with respect to the components $\xi^i(s)$ of the transformation

$$\boldsymbol{\xi}(s) : S^n \to \Xi^n \tag{1.11}$$

that is the inverse of the intermediate transformation (1.9). The equations for generating grids of the physical geometry S^{xn} include coefficients defined by the elements

$$g_{ij}^{xs} = \boldsymbol{x}_{s^i} \cdot \boldsymbol{x}_{s^j} , \quad i,j = 1, \ldots, n , \tag{1.12}$$

of the covariant metric tensor of S^{xn} in the parametric coordinates $s^1, \ldots, s^n$. A general form of these generalized Poisson equations is as follows:

$$\Delta_B[\xi^i] = P^i(s) , \quad i = 1, \ldots, n . \tag{1.13}$$

Here Δ_B is the operator of Beltrami defined at a function $f(s)$ by the formula

$$\Delta_B[f] = \frac{1}{\sqrt{g^{xs}}} \sum_{j=1}^{n} \frac{\partial}{\partial s^j} \left(\sqrt{g^{xs}} \sum_{k=1}^{n} g_{sx}^{jk} \frac{\partial f(s)}{\partial s^k} \right) , \tag{1.14}$$

where $g^{xs} = \det(g_{ij}^{xs})$, while g_{sx}^{jk} is the (jk)th element of the contravariant metric tensor of S^{xn} in the parametric coordinates $s^1, \ldots, s^n$. The elements g_{sx}^{ij}, $i,j = 1, \ldots, n$, comprise a matrix that is the inverse of the matrix formed by the elements g_{ij}^{xs}, $i,j = 1, \ldots, n$. The terms $P^i(s)$ in (1.13) referred to as source terms or control functions are introduced to control the grid behavior.

A particular form of (1.13) for generating grids in a domain $S^n \subset R^n$ is given by the Poisson equations

$$\sum_{j=1}^{n} \frac{\partial^2 \xi^i}{\partial s^j \partial s^j} = P^i(s), \quad i = 1, \ldots, n. \tag{1.15}$$

The intermediate transformation $s(\boldsymbol{\xi}) = [s^1(\boldsymbol{\xi}), \ldots, s^n(\boldsymbol{\xi})]$ for generating grids on a physical geometry S^{xn} is found from the solution of the Dirichlet boundary value problem for the transformed equations obtained from (1.13) by changing mutually dependent and independent variables. These equations are of the form

$$\sum_{i,j=1}^{n} g_{\xi x}^{ij} \frac{\partial^2 s^k}{\partial \xi^i \partial \xi^j} = \Delta_B[s^k] - \sum_{i=1}^{n} P^i \frac{\partial s^k}{\partial \xi^i}, \quad k = 1, \ldots, n, \tag{1.16}$$

where $g_{\xi x}^{ij}$ is the (ij)th element of the contravariant metric tensor of S^{xn} in the grid coordinates $\xi^1, \ldots, \xi^n$.

A two-dimensional Laplace system which implied the parametric coordinates to be solutions in the logical domain Ξ^2 was introduced by Godunov and Prokopov (1967), Barfield (1970), and Amsden and Hirt (1973). A general two-dimensional elliptic system for generating structured grids was considered by Chu (1971). A two-dimensional system (1.15) with $P^i(s) \equiv 0$ using the logical coordinates ξ^i as dependent variables was proposed by Crowley (1962) and Winslow (1967).

Godunov and Prokopov (1972) obtained a system of the Poisson type (1.15) assuming that its solution is a composition of conformal and stretching transformations. The general Poisson system (1.15) was justified by Thompson, Thames, and Mastin (1974) and Tompson, Warsi, and Mastin (1985) in their monograph.

The algorithm aimed at grid clustering at a boundary and forcing grid lines to intersect the boundary in a nearly normal fashion through the source terms of the Poisson system (1.15) was developed by Steger and Sorenson (1979), Visbal and Knight (1982), and White (1990). Thomas and Middlecoff (1980) described a procedure to control the local angle of intersection between transverse grid lines and the boundary through the specification of the control functions. Control of grid spacing and orthogonality was performed by Tamamidis and Assanis (1991) by introducing a distortion function (the ratio of the diagonal metric elements) into the system of Poisson equations. Warsi (1982) replaced the source terms P^i in (1.15) by $g^{ii}P^i$ (i fixed) to improve the numerical behavior of the grid generator. As a result the modified system acquired the property of satisfying the maximum principle.

The technique based on setting to zero the off-diagonal elements of the elliptic system (1.16) was proposed by Lin and Shaw (1991) to generate nearly orthogonal grids, while Soni, et al. (1993) used a specification of the control functions for this purpose.

The use of generalized Laplace equations to generate surface grids was proposed by Warsi (1982), in analogy with the widely utilized Laplace grid generator of Crowley (1962) and Winslow (1967). Warsi (1990) has also justified these equations by using some fundamental results of differential geometry.

A surface grid generation scheme that uses a quasi-two-dimensional elliptic system, obtained by projecting the inverted three-dimensional Laplace system, to generate grids on smooth surfaces analytically specified by the equation $z = f(x, y)$ was proposed by Thomas (1982). The method was extended and updated by Takagi et al. (1985) and Warsi (1986) for arbitrary curved surfaces using a parametric surface representation. An adaptive surface grid technique based on control functions in (1.13) and parametric specifications was also considered by Lee and Loellbach (1989).

Liseikin (1991a, 1992) used an elliptic system derived from a variational principle to produce n-dimensional harmonic coordinate transformations which generate both uniform and adaptive grids on surfaces. Harmonic mapping was also used by Arina and Casella (1991) to derive a surface elliptic system. The conformal mapping technique for generating surface grids was formulated by Khamayseh and Mastin (1996).

Hyperbolic Equations. The most known hyperbolic equations are the first order partial differential equations of the Cauchy–Riemann type. In practice, two-dimensional hyperbolic equations with respect to the intermediate transformation $s(\xi)$ have the following form

$$A s_{\xi^1} + B s_{\xi^2} = f , \qquad (1.17)$$

where A and B some matrices. These equations are simpler then nonlinear elliptic (1.16) and enable marching methods to be used and an orthogonal system of coordinates to be constructed, while grid adaptation can be performed using the coefficients of the equations. However, methods based on the solution of hyperbolic equations are not always mathematically correct and they are not applicable to regions in which the complete boundary surface is strictly defined. Therefore hyperbolic methods are mainly used for simple regions which have several lateral faces for which no special nodal distribution is required. The marching procedure for the solution of hyperbolic equations allows one to decompose only the boundary geometry in such a way that neighboring boundary grids overlap.

The first systematic analysis of the use of two-dimensional hyperbolic equations to generate grids was made by Starius (1977) and Steger and Chaussee (1980), although hyperbolic grid generation can be traced back to McNally (1972). This system was generalized by Cordova and Barth (1988).

They developed a two-dimensional hyperbolic system with an angle-control source term which allows one to constrain a grid with more than one boundary. A combination of grids using the hyperbolic technique of Steger and Chaussee (1980), which starts from each boundary segment was generated by Jeng and Shu (1995). The extension to three dimensions was performed by Steger and Rizk (1985), Chan and Steger (1992), and Tai, Chiang, and Su (1996), who introduced grid smoothing as well.

The hyperbolic approach based on grid orthogonality was extended to surfaces by Steger (1991). An analogous technique for generating overset surface grids was presented by Chan and Buning (1995).

Parabolic Equations. The parabolic grid approach lies between the elliptic and hyperbolic ones.

The two-dimensional parabolic grid generation equation where the marching direction is ξ^2 may be written in the following form:

$$\boldsymbol{s}_{\xi^2} = A_1 \boldsymbol{s}_{\xi^1 \xi^1} - B_1 \boldsymbol{s} + \boldsymbol{P} \,, \tag{1.18}$$

where A_1 and B_1 are matrix coefficients, and $\boldsymbol{P}$ is a source vector-valued function that contains the information about the outer boundary configuration. Analogously, the three-dimensional parabolic equations may be written as follows:

$$\boldsymbol{s}_{\xi^3} = A_i \boldsymbol{s}_{\xi^i \xi^i} - B_1 \boldsymbol{s} + \boldsymbol{P} \,, \quad i = 1, 2 \,. \tag{1.19}$$

The generation of grids based on a parabolic scheme approximating the inverted Poisson equations was first proposed for two-dimensional grids by Nakamura (1982). A variation of the method of Nakamura was developed by Noack (1985) to use in space-marching solutions to the Euler equations. Extensions of this parabolic technique to generate solution adaptive grids were performed by Edwards (1985) and Noack and Anderson (1990).

Hybrid Grid Generation Scheme. The combination of the hyperbolic and parabolic schemes into a single scheme is attractive because it can use the advantages of both schemes. These advantages are; first, it is a noniterative scheme; second, the orthogonality of the grid near the initial boundary is well controlled; and third, the outer boundary can be prescribed.

A hybrid grid generation scheme in two dimensions for the particular marching direction ξ^2 can be derived by combining hyperbolic and parabolic equations, in particular, as the sum of equations (1.17) multiplied by B^{-1} and (1.18) with weights α and $1 - \alpha$, respectively:

$$\alpha(B^{-1} A \boldsymbol{s}_{\xi^1} + \boldsymbol{s}_{\xi^2}) + (1 - \alpha)(\boldsymbol{s}_{\xi^2} - A_1 \boldsymbol{s}_{\xi^1 \xi^1} + B_1 \boldsymbol{s})$$
$$= \alpha B^{-1} \boldsymbol{f} + (1 - \alpha) \boldsymbol{P} \,. \tag{1.20}$$

The parameter α can be changed as desired to control the proportions of the two methods. If α approaches 1, the system (1.20) becomes the hyperbolic grid system, while if α approaches zero it becomes the parabolic grid system. In practical applications α is set to 1, when the grid generation starts from the initial boundary curve $\xi^2 = 0$, but it gradually decreases and approaches zero when the grid reaches the outer boundary.

An analogous combination can be used to generate three-dimensional grids through a hybrid of parabolic and hyperbolic equations.

A combination of hyperbolic and parabolic schemes that uses the advantages of the two but eliminates the drawbacks of each was proposed by Nakamura and Suzuki (1987).

1.3.5 Variational Methods

Variational methods are widely used to generate grids which are required to satisfy several critical properties, e. g., mesh concentration in areas needing high resolution of the physical solution, mesh alignment to some prescribed vector fields, mesh nondegeneracy, smoothness, uniformity, and near-orthogonality that cannot be realized simultaneously with algebraic or differential techniques. Variational methods take into account the conditions imposed on the grid by constructing special functionals defined on a set of smooth or discrete transformations. A compromise grid, with properties close to those required, is obtained with the optimum transformation for a combination of these functionals.

The major task of the variational approach to grid generation is to describe all basic measures of the desired grid features in an appropriate functional form and to formulate a combined functional that provides a well-posed minimization problem. These functionals can provide mathematical feedback in an automatic grid procedure.

Commonly, in the calculus of variations, any functional over some admissible set of functions $\boldsymbol{f} : D^n \to R^m$ is defined by the integral

$$I[\boldsymbol{f}] = \int_{D^n} G(\boldsymbol{f}) \mathrm{d}V \; , \tag{1.21}$$

where D^n is a bounded n-dimensional domain, and $G(\boldsymbol{f})$ is some operator specifying, for each vector-valued function $\boldsymbol{f} : D^n \to R^m$, a scalar function $G(\boldsymbol{f}) : D^n \to R$. The admissible set is composed of those functions $\boldsymbol{f}$ which satisfy a prescribed boundary condition

$$\boldsymbol{f}\,|_{\partial D^n} = \boldsymbol{\phi}$$

and for which the integral (1.21) is limited.

In the application of the calculus of variations to grid generation this set of admissible functions is a set of sufficiently smooth invertible coordinate transformations (1.11) between the parametric domain S^n and the computational domain Ξ^n or, vice versa, a set of sufficiently smooth invertible intermediate transformations (1.9) from the computational domain Ξ^n onto the parametric region S^n. The integral (1.21) is defined over the domain S^n or Ξ^n, respectively.

In grid generation applications the operator G is commonly chosen as a combination of weighted local grid characteristics which are to be optimized. The choice depends, of course, on what is expected from the grid. Some forms of the weight functions and both local and integral grid characteristics were formulated in a monograph of Liseikin (1999) through the transformations (1.9) or (1.11) and their first and second derivatives. Therefore, for the purpose of grid generation, it can be supposed that the most widely acceptable formula for the operator G in (1.21) is one which is derived from some expressions containing the first and second derivatives of the coordinate transformations. Thus it is generally assumed that the functional (1.21), depending on the transformation $\boldsymbol{\xi}(\boldsymbol{s})$, is of the form

$$I[\boldsymbol{\xi}] = \int_{S^n} G(\boldsymbol{s}, \boldsymbol{\xi}, \boldsymbol{\xi}_{s^i}, \boldsymbol{\xi}_{s^i s^j}) d\boldsymbol{s} ,$$

where G is a smooth function of its variables.

Analogously, the functional (1.21) formulated over a set of invertible intermediate transformations $\boldsymbol{s}(\boldsymbol{\xi})$ has the form

$$I[\boldsymbol{s}] = \int_{\Xi^n} G_1(\boldsymbol{\xi}, \boldsymbol{s}, \boldsymbol{s}_{\xi^i}, \boldsymbol{s}_{\xi^i \xi^j}) d\boldsymbol{\xi} .$$

In one popular approach the functional formulated with respect to the intermediate mapping $\boldsymbol{s}(\boldsymbol{\xi})$ has the following form

$$I[\boldsymbol{s}] = \int_{\Xi^n} \left(\sqrt{g^{m\xi}} \sum_{i,j=1}^{n} g_{\xi m}^{ij} g_{ij}^{s\xi} \right) d\boldsymbol{\xi} , \qquad (1.22)$$

where $g_{\xi m}^{ij}$, $i,j = 1, \ldots, n$, are the elements of the contravariant tensor in the logical coordinates $\xi^1, \ldots, \xi^n$ of a monitor metric $g_{ij}^{m\xi}$ imposed on Ξ^n, $g^{m\xi} = \det(g_{ij}^{m\xi})$, while $g_{ij}^{s\xi}$ is the covariant Eucledian metric tensor of S^n in the coordinates $\xi^1, \ldots, \xi^n$. This functional was proposed for $n=2$ by Godunov and Prokopov (1967) for generating conformal and quasi-conformal grids in S^2. In their consideration the elements $g_{ij}^{m\xi}$, $i,j = 1, 2$, of the monitor metric should be dependent on $\boldsymbol{\xi}$ and some, in general vector-valued parameter $\boldsymbol{r}$. Belinskii et al. (1975) and Godunov, Romenskii, and Chumakov (1990) discussed the same two-dimensional functional of the form (1.22) with the following monitor metric introduced in Ξ^2

$$g_{ij}^{m\xi} = \begin{pmatrix} e^{2p(\boldsymbol{\xi})} & e^{p(\boldsymbol{\xi})+q(\boldsymbol{\xi})} \cos[\alpha(\boldsymbol{\xi}) - \beta(\boldsymbol{\xi})] \\ e^{p(\boldsymbol{\xi})+q(\boldsymbol{\xi})} \cos[\alpha(\boldsymbol{\xi}) - \beta(\boldsymbol{\xi})] & e^{2q(\boldsymbol{\xi})} \end{pmatrix} ,$$

where the functions $p(\boldsymbol{\xi})$, $q(\boldsymbol{\xi})$, $\alpha(\boldsymbol{\xi})$, and $\beta(\boldsymbol{\xi})$ are subject to the restrictions

$$p(\boldsymbol{\xi}) - q(\boldsymbol{\xi}) = \ln \sqrt{g_{11}^{s\xi}/g_{22}^{s\xi}} ,$$

$$\alpha(\boldsymbol{\xi}) - \beta(\boldsymbol{\xi}) = \arccos\left(g_{12}^{s\xi}/\sqrt{g_{11}^{s\xi} g_{22}^{s\xi}}\right) .$$

The grid approach based on the minimization of the functional (1.22) for $n = 2$ was also used by Chumakov and Chumakov (1998) for generating quasi-isometric grids by introducing a monitor metric in Ξ^2 borrowed from the metric of a surface of a constant Gauss curvature.

Note the functional (1.22) is twice the energy functional of the function $s(\boldsymbol{\xi}): \Xi^n \to S^n$ where Ξ^n is endowed by the monitor metric $g_{ij}^{m\xi}$, while S^n has the Eucledian metric.

The twice energy functional of the function $\boldsymbol{\xi}(s): S^n \to \Xi^n$ between S^n with an imposed monitor metric g_{ij}^s and Ξ^n with the Eucledian metric for generating adaptive grids was considered by Dvinsky (1991) and Liseikin (1991). This functional has the following form

$$I[\boldsymbol{\xi}] = \int_{S^n} \left(\sqrt{g^s} \sum_{i,j,k=1}^{n} g_s^{ij} \frac{\partial \xi^k}{\partial s^i} \frac{\partial \xi^k}{\partial s^j} \right) ds , \tag{1.23}$$

where $g^s = \det(g_{ij}^s)$, g_s^{ij}, $i,j = 1,\ldots,n$, are the elements of the contravariant monitor metric tensor.

The functionals are used to control and realize various grid properties. This is carried out by combining these functionals with weights in the form

$$I = \sum_i \lambda_i I_i , \qquad i = 1, \cdots, k . \tag{1.24}$$

Here λ_i, $i = 1, \cdots, k$, are specified parameters which determine the individual contribution of each functional I_i to I. The ranges of the parameters λ_i controlling the relative contributions of the functionals can be defined readily when the functionals I_i are dimensionally homogeneous. However, if they are dimensionally inhomogeneous, then the selection of a suitable value for λ_i presents some difficulties. A common rule for selecting the parameters λ_i involves making each component $\lambda_i I_i$ in (1.24) of a similar scale by using a dimensional analysis.

The most common practice in forming the combination (1.24) uses both the functionals of adaptation to the physical solution and the functionals of grid regularization. The first reason for using such a strategy is connected with the fact that the process of adaptation can excessively distort the form of the grid cells. The distortion can be prevented by functionals which impede cell deformation. These functionals are ones which control grid skewness, smoothness, and conformality. The second reason for using the regularization functionals is connected with the natural requirement for the well-posedness of the grid generation process. This requirement is achieved by the utilization of convex functionals in variational grid generators. The convex functionals are represented by energy-type functionals (1.23) producing harmonic maps and by the functionals of conformality.

The various functionals provide broad opportunities to control and realize the required grid properties, though problems still remain; these require more detailed studies of all properties of the functionals. The knowledge of these properties will allow one to utilize the functionals as efficient tools to generate high-quality grids.

Liseikin and Yanenko (1977), Danaev, Liseikin, and Yanenko (1978), Ghia, Ghia, and Shin (1983), Brackbill and Saltzman (1982), Bell and Shubin (1983), Huang, Ren, and Russell (1994), and Huang (2001) have each used the variational principle for grid adaptation. The variational formulation of grid properties was described by Warsi and Thompson (1990).

The functional measuring the alignment of the two-dimensional grid with a specified vector field was formulated by Giannakopoulos and Engel (1988). The extension of this approach to three dimensions was discussed by Brackbill (1993). A variational method optimizing cell aspect ratios was presented and analyzed by Mastin (1992). A dimensionally homogeneous functional of two-dimensional grid skewness was proposed by Steinberg and Roache (1986).

The introduction of the volume-weighted functional was originally proposed in two dimensions by Yanenko, Danaev, and Liseikin (1977).

The approach of determining functionals which depend on invariants of orthogonal transformations of the metric tensor $g_{ij}^{s\xi}$, to ensure that the problems are well-posed and to obtain more compact formulas for the Euler–Lagrange equations, was proposed by Jacquotte (1987). In his paper, the grids were constructed through functionals obtained by modeling different elastic and plastic properties of a deformed body.

The possibility of using harmonic function theory to provide a general framework for developing multidimensional mesh generators was discussed by Dvinsky (1991). A detailed survey of the theory of harmonic mappings was published by Eells and Lenaire (1988). The interpretation of the functional of diffusion as a version of the energy functional was presented by Brackbill (1993).

1.4 Comprehensive Codes

A comprehensive grid generation code is an effective system for generating structured and unstructured grids, as well as hybrid and overset combinations, in arbitrary physical geometries. The development of such codes is a considerable problem in its own right. The present comprehensive grid generation codes developed for the solution of multidimensional problems have to incorporate combinations of block-structured, hybrid, and overset grid methods and are still rather cumbersome, rely on interactive tools, and take too many man-hours to generate a complicated grid. Efforts to increase the efficiency and productivity of these codes are mainly being conducted in two interconnected research areas.

The first, the "array area", is concerned with the automation of those routine processes of grid generation which require interactive tools and a great deal of human time and effort. Some of these are:

(1) the decomposition of a domain into a set of contiguous or ovelapping blocks consistent with the distinctive features of the domain geometry, the singularities of the physical medium and the sought-for solution, and the computer architecture;
(2) numbering the set of blocks, their faces, and their edges with a connectivity hierarchy and determining the order in which the grids are constructed in the blocks and their boundaries;
(3) choosing the grid topology and the requirements placed on the qualitative and quantitative characteristics of the internal and boundary grids and on their communication between the blocks;
(4) selecting appropriate methods to satisfy the requirements put on the grid in accordance with a particular geometry and solution;
(5) assessment and enhancement of grid quality.

The second, more traditional, "methods area" deals with developing new, more reliable, and more elaborate methods for generating, adapting, and smoothing grids in domains in a unified manner, irrespective of the geometry of the domain or surface and of the qualitative and quantitative characteristics the grids should possess, so that these methods, when incorporated in the comprehensive codes, should ease the bottlenecks of the array area, in particular, by enabling a considerable reduction of the number of blocks required. The contents of the current monograph is aimed at advocating one such method based on the use of the Beltramian operator with respect to monitor metrics and the theory of multidimensional differential geometry. This method is a natural extention of the approaches proposed by Crowley (1962), Winslow (1967), Godunov and Prokopov (1967), Warsi (1981), Dvinski (1991), and Liseikin (1991). Some recent results of the method have been presented in the papers of Liseikin (2001b, 2002a, 2002b, 2003).

The overall purpose of the development of the comprehensive grid generation codes is to create a system which enables one to generate grids in a "black box" mode without or with only a slight human interaction. Currently, however, the user has to take active role and be occupied in the grid generation process. The user has to make conclusions about qualitative properties of the grid and undertake corrective measures when necessary. The present codes include significant measures to increase the productivity of such human activity, namely, graphical interactive systems and user-friendly interfaces. Efforts to eliminate the "human component" of the codes are directed towards developing new techniques, in particular, new grid generation methods and automated block decomposition techniques.

The first comprehensive grid codes were described by Holcomb (1987), Thompson (1987a), Thomas, Bache, and Blumenthal (1990), Widhopf et al. (1990), and Steinbrenner, Chawner, and Fouts (1990). These codes have stimulated the development of updated ones, reviewed by Thompson (1996). This paper also describes the current domain decomposition techniques developed by Shaw and Weatherill (1992), Stewart (1992), Dannenhoffer (1995), Wulf and Akrag (1995), Schonfeld, Weinerfelt, and Jenssen (1995), and Kim and Eberhardt (1995). The first attempts to overcome the problem of domain decomposition were discussed by Andrews (1988), Georgala and Shaw (1989), Allwright (1989), and Vogel (1990).

2. General Coordinate Systems in Domains

We consider here some notions and relations connected with smooth invertible coordinate transformations of the physical region $X^n \subset R^n$ from the parametric domain $\Xi^n \subset R^n$:

$$\boldsymbol{x}(\boldsymbol{\xi}) : \Xi^n \to X^n \,, \quad \boldsymbol{\xi} = (\xi^1, \cdots, \xi^n) \,, \quad \mathbf{x} = (x^1, \cdots, x^n) \,.$$

If Ξ^n is a standard logical domain then, in accordance with Chap. 1, such coordinate transformations are used to generate grids in X^n. Here and later R^n are presents the Euclidean space with the Cartesian basis of an orthonormal system of vectors $\boldsymbol{e}_1, \cdots, \boldsymbol{e}_n$, i.e.

$$\boldsymbol{e}_i \cdot \boldsymbol{e}_j = \delta^i_j \,, \quad i, j = 1, \ldots, n \,,$$

where δ^i_j is the Kronecker symbol:

$$\delta^i_j = 0 \quad \text{if} \quad i \neq j \,, \quad \delta^i_j = 1 \quad \text{if} \quad i = j \,.$$

Thus the position of a point $\boldsymbol{x}$ in R^n is determined unequivocally by the expansion

$$\boldsymbol{x} = x^1 \boldsymbol{e}_1 + \cdots + x^n \boldsymbol{e}_n \,.$$

The values x^i, $i = 1, \cdots, n$, are called the Cartesian coordinates of the point $\boldsymbol{x}$. Each coordinate transformation $\boldsymbol{x}(\boldsymbol{\xi})$ defines, in the domain X^n, new coordinates $\xi^1, \cdots, \xi^n$ which are called the curvilinear coordinates.

2.1 Jacobi Matrix

The matrix

$$\jmath = \left(\frac{\partial x^i}{\partial \xi^j}\right) \,, \quad i, j = 1, \cdots, n \,,$$

is referred to as the Jacobi matrix, and its Jacobian is designated by J:

$$J = \det\left(\frac{\partial x^i}{\partial \xi^j}\right) \,, \quad i, j = 1, \cdots, n \,.$$

The inverse transformation to the coordinate mapping $\boldsymbol{x}(\boldsymbol{\xi})$ is denoted by

$$\boldsymbol{\xi}(\boldsymbol{x}) : X^n \to \Xi^n , \quad \boldsymbol{\xi}(\boldsymbol{x}) = (\xi^1(\boldsymbol{x}), \ldots, \xi^n(\boldsymbol{x})) .$$

This transformation can be considered analogously as a mapping introducing a curvilinear coordinate system $x^1, \cdots, x^n$ in the domain $\Xi^n \subset R^n$. It is obvious that the inverse to the matrix J is

$$J^{-1} = \left(\frac{\partial \xi^i}{\partial x^j}\right), \quad i, j = 1, \cdots, n ,$$

and consequently

$$\det\left(\frac{\partial \xi^i}{\partial x^j}\right) = \frac{1}{J}, \quad i, j = 1, \cdots, n .$$

In the case of two-dimensional space the elements of the matrices $(\partial x^i/\partial \xi^j)$ and $(\partial \xi^i/\partial x^j)$ are connected by

$$\frac{\partial \xi^i}{\partial x^j} = (-1)^{i+j} \frac{\partial x^{3-j}}{\partial \xi^{3-i}} \Big/ J, \quad i, j = 1, 2 ,$$

$$\frac{\partial x^i}{\partial \xi^j} = (-1)^{i+j} J \frac{\partial \xi^{3-j}}{\partial x^{3-i}}, \quad i, j = 1, 2 . \tag{2.1}$$

Similar relations between the elements of the corresponding three-dimesional matrices have the form

$$\frac{\partial \xi^i}{\partial x^j} = \frac{1}{J}\left(\frac{\partial x^{j+1}}{\partial \xi^{i+1}} \frac{\partial x^{j+2}}{\partial \xi^{i+2}} - \frac{\partial x^{j+1}}{\partial \xi^{i+2}} \frac{\partial x^{j+2}}{\partial \xi^{i+1}}\right), \quad i, j = 1, 2, 3 ,$$

$$\frac{\partial x^i}{\partial \xi^j} = J\left(\frac{\partial \xi^{j+1}}{\partial x^{i+1}} \frac{\partial \xi^{j+2}}{\partial x^{i+2}} - \frac{\partial \xi^{j+1}}{\partial x^{i+2}} \frac{\partial \xi^{j+2}}{\partial x^{i+1}}\right), \quad i, j = 1, 2, 3 , \tag{2.2}$$

where for each superscript or subscript index, say l, $l \pm 3$ is equivalent to l. With this condition the sequence of indices $(l, l+1, l+2)$ is a cyclic permutation of $(1, 2, 3)$ and vice versa; the indices of a cyclic sequence (i, j, k) satisfy the relation $j = i + 1$, $k = i + 2$.

2.2 Coordinate Lines, Tangential Vectors, and Grid Cells

The value of the function $\boldsymbol{x}(\boldsymbol{\xi}) = [x^1(\boldsymbol{\xi}), \ldots, x^n(\boldsymbol{\xi})]$ in the Cartesian basis $\boldsymbol{e}_1, \cdots, \boldsymbol{e}_n$, i.e.

$$\boldsymbol{x}(\boldsymbol{\xi}) = x^1(\boldsymbol{\xi})\boldsymbol{e}_1 + \ldots + x^n(\boldsymbol{\xi})\boldsymbol{e}_n , \quad \boldsymbol{\xi} = (\xi^1, \ldots, \xi^n) , \tag{2.3}$$

is a position vector for every $\boldsymbol{\xi} \in \Xi^n$. If one variable ξ^i varies and the others ξ^j, $j \neq i$, are kept constant then the function (2.3) depends upon a single parameter and therefore describes a curve. This curve is referred to as the ξ^i curvilinear coordinate line.

2.2 Coordinate Lines, Tangential Vectors, and Grid Cells

The vector-valued function $\boldsymbol{x}(\boldsymbol{\xi})$ generates the nodes, edges, faces, etc. of the cells of the coordinate grid in the domain X^n. Each edge of the cell corresponds to a coordinate line ξ^i for some i and is defined by the vector

$$\Delta_i \boldsymbol{x} = \boldsymbol{x}(\boldsymbol{\xi} + h_i \boldsymbol{e}_i) - \boldsymbol{x}(\boldsymbol{\xi}) \; ,$$

where h_i is the step size of the uniform grid in the ξ^i direction in the logical domain Ξ^n. We have

$$\Delta_i \boldsymbol{x} = h_i \boldsymbol{x}_{\xi^i} + \boldsymbol{t} \; ,$$

where

$$\boldsymbol{x}_{\xi^i} = \left(\frac{\partial x^1}{\partial \xi^i}, \cdots, \frac{\partial x^n}{\partial \xi^i} \right) \tag{2.4}$$

is the vector tangential to the coordinate curve ξ^i, and $\boldsymbol{t}$ is a residual vector whose length does not exceed the following quantity:

$$\frac{1}{2} \max |\boldsymbol{x}_{\xi^i \xi^i}| (h_i)^2 \; , \quad i = 1, \ldots, n \; .$$

So the cells in the domain X^n whose edges are formed by the vectors $h_i \boldsymbol{x}_{\xi^i}$, $i = 1, \cdots, n$, are approximately the same as those obtained by mapping the uniform coordinate cells in the computational domain Ξ^n with the transformation $\boldsymbol{x}(\boldsymbol{\xi})$. The tangential vectors $\boldsymbol{x}_{\xi^i}$, $i = 1, \cdots, n$, form a parallelepiped (parallelogram when n=2) referred to as a basic parallelepiped. Thus the uniformly contracted basic parallelepiped spanned by the tangential vectors $\boldsymbol{x}_{\xi^i}$, $i = 1, \cdots, n$, represents to a high order of accuracy with respect to h_i the cell of the coordinate grid at the corresponding point in X^n (see Fig. 2.1 for $n = 2$). In particular, for the length l_i of the ith grid edge we have

$$l_i = h_i |\boldsymbol{x}_{\xi^i}| + O(h_i^2) \; , \quad i = 1, \ldots, n \; .$$

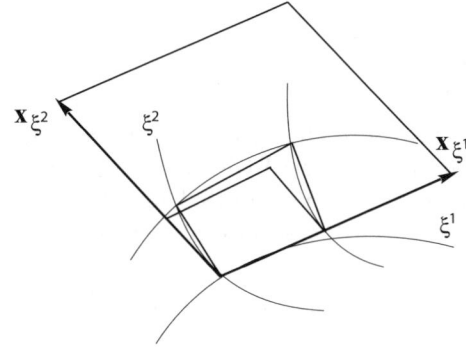

Fig. 2.1. Basic and contracted parallelograms and corresponding grid cell

36   2. General Coordinate Systems in Domains

The volume V_h (area in two dimensions) of the cell is expressed as follows:

$$V_h = \prod_{i=1}^n h_i V + O\left(\prod_{i=1}^n h_i \sum_{j=1}^n h_j\right),$$

where V is the volume of the n-dimensional basic parallelepiped determined by the tangential vectors $\boldsymbol{x}_{\xi^i}$, $i = 1, \cdots, n$.

The tangential vectors $\boldsymbol{x}_{\xi^i}$, $i = 1, \cdots, n$, are called the base covariant vectors since they comprise a vector basis. The sequence $\boldsymbol{x}_{\xi^1}, \cdots, \boldsymbol{x}_{\xi^n}$ of the tangential vectors has a right-handed orientation if the Jacobian of the transformation $\boldsymbol{x}(\boldsymbol{\xi})$ is positive. Otherwise, the base vectors $\boldsymbol{x}_{\xi^i}$ have a left-handed orientation.

The operation of the dot product on the tangential vectors produces elements of the covariant metric tensor. These elements generate the coefficients that appear in the transformed grid equations. Besides this, the metric elements play a primary role in studying and formulating various geometric characteristics of the grid cells in domains.

2.3 Coordinate Surfaces and Normal Vectors

If one curvilinear coordinate, say ξ^i, is fixed ($\xi^i = \xi_0^i$) then the position function (2.3) describes an $(n-1)$-dimensional surface which is called the coordinate hypersurface. Thus the coordinate hypersurface is defined by the equation $\xi^i = \xi_0^i$; i.e. along the surface all of the coordinates $\xi^1, \cdots, \xi^n$ except ξ^i are allowed to vary.

The inverse transformation

$$\boldsymbol{\xi}(\boldsymbol{x}) = [\xi^1(\boldsymbol{x}), \ldots, \xi^n(\boldsymbol{x})]$$

to (2.3) yields for each fixed i the base contravariant vector

$$\boldsymbol{\nabla}\xi^i = \left(\frac{\partial \xi^i}{\partial x^1}, \cdots, \frac{\partial \xi^i}{\partial x^n}\right), \qquad (2.5)$$

which is the gradient of $\xi^i(\boldsymbol{x})$ with respect to the Cartesian coordinates $x^1, \cdots, x^n$. The set of the vectors $\boldsymbol{\nabla}\xi^i$, $i = 1, \cdots, n$, is called the set of base contravariant vectors.

Similarly, as the tangential vectors relate to the coordinate curves, the contravariant vectors $\boldsymbol{\nabla}\xi^i$, $i = 1, \cdots, n$, are connected with their respective coordinate hypersurfaces (curves in two dimensions). Indeed for all of the tangent vectors $\boldsymbol{x}_{\xi^j}$ to the coordinate lines on the surface $\xi^i = \xi_0^i$ we have the obvious identity

$$\boldsymbol{x}_{\xi^j} \cdot \boldsymbol{\nabla}\xi^i = 0, \quad i \neq j,$$

and thus the vector $\boldsymbol{\nabla}\xi^i$ is a normal to the coordinate hypersurface $\xi^i = \xi_0^i$. Therefore the vectors $\boldsymbol{\nabla}\xi^i$, $i = 1, \cdots, n$, are also called the normal base vectors.

2.3 Coordinate Surfaces and Normal Vectors

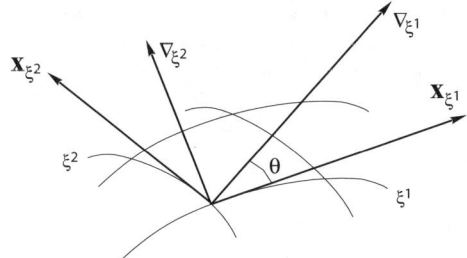

Fig. 2.2. Base tangential and normal vectors in two dimensions

Since

$$\boldsymbol{x}_{\xi^i} \cdot \boldsymbol{\nabla}\xi^i = 1$$

for each fixed $i = 1, \cdots, n$, the vectors $\boldsymbol{x}_{\xi^i}$ and $\boldsymbol{\nabla}\xi^i$ intersect each other at an angle θ which is less than $\pi/2$. Now, taking into account the orthogonality of the vector $\boldsymbol{\nabla}\xi^i$ to the hypersurface $\xi^i = \xi_0^i$, we find that these two vectors $\boldsymbol{x}_{\xi^i}$ and $\boldsymbol{\nabla}\xi^i$ are directed to the same side of the coordinate hypersurface (curve in two dimensions). An illustration of this fact in two dimensions is given in Fig. 2.2.

The length of any normal base vector $\boldsymbol{\nabla}\xi^i$ is linked to the distance d_i between the corresponding opposite boundary segments (joined by the vector $\boldsymbol{x}_{\xi^i}$) of the n-dimensional basic parallelepiped formed by the base tangential vectors, namely,

$$d_i = 1/|\boldsymbol{\nabla}\xi^i|, \quad |\boldsymbol{\nabla}\xi^i| = \sqrt{\boldsymbol{\nabla}\xi^i \cdot \boldsymbol{\nabla}\xi^i}, \quad i = 1, \ldots, n.$$

To prove this relation we recall that the vector $\boldsymbol{\nabla}\xi^i$ is a normal to all of the vectors $\boldsymbol{x}_{\xi^j}$, $j \neq i$, and therefore to the boundary segments of the parallelepiped formed by these $n-1$ vectors. Hence, the unit normal vector $\boldsymbol{n}_i$ to these segments is expressed by

$$\boldsymbol{n}_i = \boldsymbol{\nabla}\xi^i/|\boldsymbol{\nabla}\xi^i|.$$

Now, taking into account that

$$d_i = \boldsymbol{x}_{\xi^i} \cdot \boldsymbol{n}_i,$$

we readily obtain

$$d_i = \boldsymbol{x}_{\xi^i} \cdot \boldsymbol{\nabla}\xi^i/|\boldsymbol{\nabla}\xi^i| = 1/|\boldsymbol{\nabla}\xi^i|.$$

Let l_i denote the distance between a grid point on the coordinate hypersurface $\xi^i = c$ and the nearest point on the neighboring coordinate hypersurface $\xi^i = c + h$; then

$$l_i = hd_i + O(h^2) = h/|\boldsymbol{\nabla}\xi^i| + O(h^2).$$

This equation shows that the inverse length of the normal vector $\boldsymbol{\nabla}\xi^i$ multiplied by h represents with high accuracy the distance between the corresponding faces of the coordinate cells in the domain X^n.

Note that the volume of the parallelepiped spanned by the tangential vectors equals J, so we find that the volume of the n-dimensional parallelepiped defined by the normal vectors $\boldsymbol{\nabla}\xi^i$, $i = 1,\cdots,n$, is equal to $1/J$. Thus both the base normal vectors $\boldsymbol{\nabla}\xi^i$ and the base tangential vectors $\boldsymbol{x}_{\xi^i}$ have the same right-handed or left-handed orientation.

If the coordinate system $\xi^1,\cdots,\xi^n$ is orthogonal, i.e.

$$\boldsymbol{x}_{\xi^i} \cdot \boldsymbol{x}_{\xi^j} = p(\boldsymbol{\xi})\delta_j^i\,, \quad i,j=1,\cdots,n\,, \quad p(\boldsymbol{\xi}) \neq 0$$

then for each fixed $i = 1,\cdots,n$ the vector $\boldsymbol{\nabla}\xi^i$ is parallel to $\boldsymbol{x}_{\xi^i}$.

2.4 Representation of Vectors Through the Base Vectors

If there are n independent vectors $\boldsymbol{a}_1,\cdots,\boldsymbol{a}_n$ of the Euclidean space R^n then any vector $\boldsymbol{b}$ with components $b^1,\cdots,b^n$ in the Cartesian basis $\boldsymbol{e}_1,\cdots,\boldsymbol{e}_n$ is represented through the vectors $\boldsymbol{a}_i$, $i = 1,\cdots,n$, by

$$\boldsymbol{b} = a^{ij}(\boldsymbol{b}\cdot\boldsymbol{a}_j)\boldsymbol{a}_i, \quad i,j = 1,\cdots,n\,, \tag{2.6}$$

where a^{ij} are the elements of the matrix (a^{ij}) which is the inverse of the tensor (a_{ij}),

$$a_{ij} = \boldsymbol{a}_i \cdot \boldsymbol{a}_j\,, \quad i,j = 1,\cdots,n\,, \tag{2.7}$$

(Fig. 2.3).

It is assumed in (2.6) and later, unless otherwise noted, a popular geometric index convention that a summation is carried out over repeated indices in a product or single term, namely, a sign $\sum$ is understood whenever an index is repeated in the aforesaid cases.

The components of the vector $\boldsymbol{b}$ in the natural basis of the tangential vectors $\boldsymbol{x}_{\xi^i}$, $i = 1,\cdots,n$, are called contravariant. Let them be denoted by $\overline{b}^i$, $i = 1,\cdots,n$. Thus

$$\boldsymbol{b} = \overline{b}^1 \boldsymbol{x}_{\xi^1} + \cdots + \overline{b}^n \boldsymbol{x}_{\xi^n}\,.$$

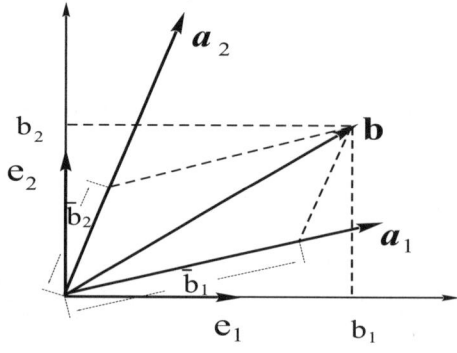

Fig. 2.3. Expansion of the vector $\boldsymbol{b}$ in the vectors $\boldsymbol{a}_1$ and $\boldsymbol{a}_2$

2.4 Representation of Vectors Through the Base Vectors

Assuming in (2.6) $\boldsymbol{a}_i = \boldsymbol{x}_{\xi^i}$, $i = 1, \cdots, n$, we obtain

$$\overline{b}^i = a^{mj}\left(b^k \frac{\partial x^k}{\partial \xi^j}\right) \frac{\partial x^i}{\partial \xi^m}, \quad i,j,k,m = 1, \cdots, n, \qquad (2.8)$$

where $b^1, \cdots, b^n$ are the components of the vector $\boldsymbol{b}$ in the Cartesian basis $\boldsymbol{e}_1, \cdots, \boldsymbol{e}_n$. Since (2.7)

$$a_{ij} = \boldsymbol{x}_{\xi^i} \cdot \boldsymbol{x}_{\xi^j} = \frac{\partial x^k}{\partial \xi^i} \frac{\partial x^k}{\partial \xi^j}, \quad i,j,k = 1, \cdots, n,$$

we have

$$a^{ij} = \frac{\partial \xi^i}{\partial x^k} \frac{\partial \xi^j}{\partial x^k}, \quad i,j,k = 1, \cdots, n.$$

Therefore, from (2.8),

$$\overline{b}^i = \frac{\partial \xi^i}{\partial x^k} \frac{\partial \xi^j}{\partial x^k} b^m \frac{\partial x^m}{\partial \xi^j} = b^j \frac{\partial \xi^i}{\partial x^j}, \quad i,j,k,m = 1, \cdots, n,$$

or, using the dot product notation

$$\overline{b}^i = \boldsymbol{b} \cdot \boldsymbol{\nabla} \xi^i, \quad i = 1, \cdots, n. \qquad (2.9)$$

Thus, in this case (2.6) has the form

$$\boldsymbol{b} = (\boldsymbol{b} \cdot \boldsymbol{\nabla} \xi^i) \boldsymbol{x}_{\xi^i}, \quad i = 1, \cdots, n. \qquad (2.10)$$

For example, the normal base vector $\boldsymbol{\nabla}\xi^i$ is expanded through the base tangential vectors $\boldsymbol{x}_{\xi^j}$, $j = 1, \cdots, n$, by the following formula:

$$\boldsymbol{\nabla}\xi^i = (\boldsymbol{\nabla}\xi^i \cdot \boldsymbol{\nabla}\xi^k) \boldsymbol{x}_{\xi^k} = \frac{\partial \xi^i}{\partial x^j} \frac{\partial \xi^k}{\partial x^j} \boldsymbol{x}_{\xi^k}, \quad i,j,k = 1, \cdots, n. \qquad (2.11)$$

Analogously, a component $\overline{b}_i$ of the vector $\boldsymbol{b}$ in the basis $\boldsymbol{\nabla}\xi^i$, $i = 1, \cdots, n$, is expressed by the formula

$$\overline{b}_i = b^j \frac{\partial x^j}{\partial \xi^i} = \boldsymbol{b} \cdot \boldsymbol{x}_{\xi^i}, \quad i,j = 1, \cdots, n, \qquad (2.12)$$

and consequently

$$\boldsymbol{b} = \overline{b}_i \boldsymbol{\nabla}\xi^i = (\boldsymbol{b} \cdot \boldsymbol{x}_{\xi^i}) \boldsymbol{\nabla}\xi^i, \quad i = 1, \cdots, n. \qquad (2.13)$$

These components $\overline{b}_i$, $i = 1, \cdots, n$, of the vector $\boldsymbol{b}$ are called covariant. In particular, the base tangential vector $\boldsymbol{x}_{\xi^i}$ is expressed through the base normal vectors $\boldsymbol{\nabla}\xi^j$, $j = 1, \cdots, n$, as follows:

$$\boldsymbol{x}_{\xi^i} = (\boldsymbol{x}_{\xi^i} \cdot \boldsymbol{x}_{\xi^j}) \boldsymbol{\nabla}\xi^j = \frac{\partial x^k}{\partial \xi^i} \frac{\partial x^k}{\partial \xi^j} \boldsymbol{\nabla}\xi^j, \quad i,j,k = 1, \cdots, n. \qquad (2.14)$$

2.5 Metric Tensors

Many grid generation algorithms, in particular those based on the calculus of variations, are typically formulated in terms of fundamental features of coordinate transformations and the corresponding mesh cells. These features are compactly described with the use of the metric notation, which is discussed in this section.

2.5.1 Covariant Metric Tensor

The matrix

$$(g_{ij}), \quad i,j = 1, \cdots, n,$$

whose elements g_{ij} are the dot products of the pairs of the basic tangential vectors $\boldsymbol{x}_{\xi^i}$, $i = 1, \ldots, n$,

$$g_{ij} = \boldsymbol{x}_{\xi^i} \cdot \boldsymbol{x}_{\xi^j} = \frac{\partial x^k}{\partial \xi^i} \frac{\partial x^k}{\partial \xi^j}, \quad i,j,k = 1, \cdots, n, \qquad (2.15)$$

is called a fundamental or covariant metric tensor of the domain X^n in the coordinates $\xi^1, \cdots, \xi^n$.

Geometrically, each diagonal element g_{ii} of the matrix (g_{ij}) is the length of the tangent vector $\boldsymbol{x}_{\xi^i}$ squared:

$$g_{ii} = |\boldsymbol{x}_{\xi^i}|^2, \quad i = 1, \cdots, n, \ i \text{ fixed}.$$

Also,

$$g_{ij} = |\boldsymbol{x}_{\xi^i}||\boldsymbol{x}_{\xi^j}|\cos\theta = \sqrt{g_{ii}}\sqrt{g_{jj}}\cos\theta, \quad i,j = 1, \cdots, n, \ i,j \text{ fixed},$$

where θ is the angle between the tangent vectors $\boldsymbol{x}_{\xi^i}$ and $\boldsymbol{x}_{\xi^j}$. Remind, the notification "fixed" in these expressions for g_{ii} and g_{ij} means that the subscripts ii and jj are fixed, i.e. here the summation over the repeated indices is not carried out.

We designate by g the Jacobian of the covariant matrix (g_{ij}). It is evident that

$$(g_{ij}) = \jmath\jmath^T, \quad i,j = 1, \ldots, n,$$

and hence

$$J^2 = g.$$

The covariant metric tensor is a symmetric matrix, i.e. $g_{ij} = g_{ji}$. If a coordinate system at a point $\boldsymbol{\xi}$ is orthogonal then the tensor (g_{ij}) has a simple diagonal form at this point. Note that these advantageous properties are in general not possessed by the Jacobi matrix $(\partial x^i / \partial \xi^i)$ from which the covariant metric tensor (g_{ij}) is defined.

2.5.2 Line Element

Let P be the point of R^n whose curvilinear coordinates are $\xi^1,\ldots,\xi^n$ and let Q be a neighboring point with the curvilinear coordinates $\xi^1 + \mathrm{d}\xi^1,\ldots,\xi^n + \mathrm{d}\xi^n$. Then the Cartesian coordinates of these points are

$$x^1(\boldsymbol{\xi}),\ldots,x^n(\boldsymbol{\xi}),\quad \boldsymbol{\xi} = (\xi^1,\ldots,\xi^n)$$

and

$$x^1(\boldsymbol{\xi} + \mathrm{d}\boldsymbol{\xi}),\ldots,x^n(\boldsymbol{\xi} + \mathrm{d}\boldsymbol{\xi}),\quad \mathrm{d}\boldsymbol{\xi} = (\mathrm{d}\xi^1,\ldots,\mathrm{d}\xi^n),$$

respectively. The infinitesimal distance PQ denoted by $\mathrm{d}s$ is called the element of length or the line element. In the Cartesian coordinates the line element is the length of the diagonal of the elementary parallelepiped whose edges are $\mathrm{d}x^1,\ldots,\mathrm{d}x^n$, where

$$\mathrm{d}x^i = x^i(\boldsymbol{\xi} + \mathrm{d}\boldsymbol{\xi}) - x^i(\boldsymbol{\xi}) = \frac{\partial x^i}{\partial \xi^j}\mathrm{d}\xi^j + o(|\mathrm{d}\boldsymbol{\xi}|),\quad i,j = 1,\ldots,n,$$

(see Fig. 2.4). Therefore

$$\mathrm{d}s = \sqrt{(\mathrm{d}x^1)^2 + \ldots + (\mathrm{d}x^n)^2} = \sqrt{\mathrm{d}\boldsymbol{x}\cdot\mathrm{d}\boldsymbol{x}},$$

where

$$\mathrm{d}\boldsymbol{x} = \boldsymbol{x}(\boldsymbol{\xi} + \mathrm{d}\boldsymbol{\xi}) - \boldsymbol{x}(\boldsymbol{\xi}) = \boldsymbol{x}_{\xi^i}\mathrm{d}\xi^i + o(|\mathrm{d}\boldsymbol{\xi}|),\quad i = 1,\ldots,n,$$

and we readily find that the expression for $\mathrm{d}s$ in the curvilinear coordinates is as follows:

$$\mathrm{d}s = \sqrt{\boldsymbol{x}_{\xi^i}\mathrm{d}\xi^i \cdot \boldsymbol{x}_{\xi^j}\mathrm{d}\xi^j} + o(|\mathrm{d}\boldsymbol{\xi}|) = \sqrt{g_{ij}\mathrm{d}\xi^i\mathrm{d}\xi^j} + o(|\mathrm{d}\boldsymbol{\xi}|),\quad i,j = 1,\cdots,n.$$

Thus the length s of the curve in X^n, prescribed by the parametrization

$$\boldsymbol{x}[\boldsymbol{\xi}(t)] : [a,b] \to X^n,$$

is computed by the formula

$$s = \int_a^b \sqrt{g_{ij}\frac{\mathrm{d}\xi^i}{\mathrm{d}t}\frac{\mathrm{d}\xi^j}{\mathrm{d}t}}\,\mathrm{d}t,\quad i,j = 1,\ldots,n. \tag{2.16}$$

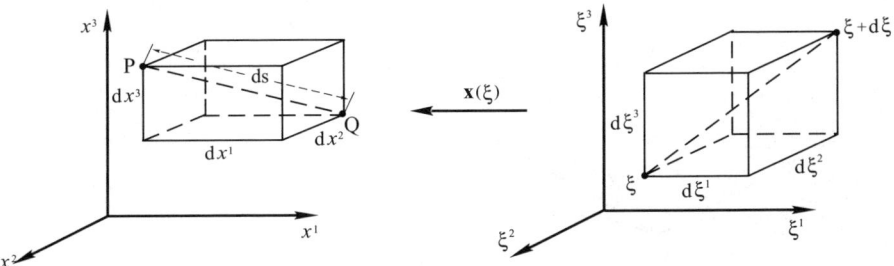

Fig. 2.4. Illustration for the line element

2.5.3 Contravariant Metric Tensor

The contravariant metric tensor of the domain X^n in the coordinates $\xi^1, \cdots, \xi^n$ is the matrix

$$(g^{ij}), \quad i,j = 1, \cdots, n,$$

inverse to (g_{ij}), i.e.

$$g_{ij} g^{jk} = \delta_i^k, \quad i,j,k = 1, \cdots, n. \tag{2.17}$$

Therefore

$$\det(g^{ij}) = \frac{1}{g}, \quad i,j = 1, \ldots, n.$$

It is easily shown that (2.17) is satisfied if and only if

$$g^{ij} = \boldsymbol{\nabla}\xi^i \cdot \boldsymbol{\nabla}\xi^j = \frac{\partial \xi^i}{\partial x^k} \frac{\partial \xi^j}{\partial x^k}, \quad i,j,k = 1, \cdots, n, \tag{2.18}$$

where $\boldsymbol{\nabla}\xi^l$, $l = 1, \ldots, n$, is the base normal vector determined by (2.5). Thus, each diagonal element g^{ii} (where i is fixed) of the matrix (g^{ij}) is the square of the length of the vector $\boldsymbol{\nabla}\xi^i$:

$$g^{ii} = |\boldsymbol{\nabla}\xi^i|^2, \quad i = 1, \ldots, n, \ i \text{ fixed}. \tag{2.19}$$

Geometric Interpretation. Now we discuss the geometric meaning of a diagonal element g^{ii} with a fixed index i, say g^{11}, of the matrix (g^{ij}). Let us consider a three-dimensional coordinate transformation $\boldsymbol{x}(\boldsymbol{\xi}): \varXi^3 \to X^3$. Its tangential vectors $\boldsymbol{x}_{\xi^1}, \boldsymbol{x}_{\xi^2}, \boldsymbol{x}_{\xi^3}$ at some point P form the basic parallelepiped whose edges are these vectors (Fig. 2.5). For the distance d_1 between the opposite faces of the parallelepiped which are defined by the vectors $\boldsymbol{x}_{\xi^2}$ and $\boldsymbol{x}_{\xi^3}$, we have

$$d_1 = \boldsymbol{x}_{\xi^1} \cdot \boldsymbol{n}_1,$$

where $\boldsymbol{n}_1$ is the unit normal to the plane spanned by the vectors $\boldsymbol{x}_{\xi^2}$ and $\boldsymbol{x}_{\xi^3}$. It is clear, that

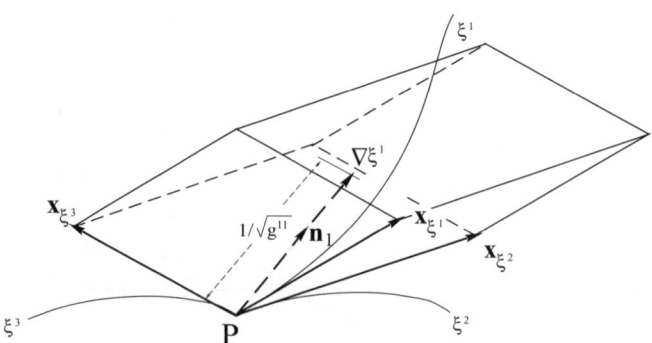

Fig. 2.5. Geometric meaning of the diagonal contravariant metric element g^{11}

$$\nabla_{\xi^1} \cdot \bm{x}_{\xi^j} = 0 \,, \quad j = 2, 3 \,,$$

and hence the unit normal $\bm{n}_1$ is parallel to the normal base vector ∇_{ξ^1}. Thus we obtain

$$\bm{n}_1 = \nabla_{\xi^1}/|\nabla_{\xi^1}| = \nabla_{\xi^1}/\sqrt{g^{11}} \,.$$

Therefore

$$d_1 = \nabla_{\xi^1} \cdot \nabla_{\xi^1}/\sqrt{g^{11}} = 1/\sqrt{g^{11}} \,,$$

and consequently

$$g^{11} = 1/(d_1)^2 \,.$$

Analogous relations are valid for g^{22} and g^{33}, i.e. in three dimensions the diagonal element g^{ii} for a fixed i means the inverse square of the distance d_i between those faces of the basic parallelepiped which are connected by the vector $\bm{x}_{\xi^i}$. In two-dimensional space the element g^{ii} (where i is fixed) is the inverse square of the distance between the edges of the basic parallelogram defined by the tangential vectors $\bm{x}_{\xi^1}$ and $\bm{x}_{\xi^2}$.

The same interpretation of g^{ii} is valid for general multidimensional coordinate transformations:

$$g^{ii} = 1/(d_i)^2, \quad i = 1, \cdots, n \,, \tag{2.20}$$

where the index i is fixed, and d_i is the distance between those $(n-1)$-dimensional faces of the n-dimensional parallelepiped which are linked by the tangential vector $\bm{x}_{\xi^i}$.

2.5.4 Relations Between Covariant and Contravariant Elements

Now, in analogy with (2.1) and (2.2), we write out very convenient formulas for natural relations between the contravariant elements g^{ij} and the covariant ones g_{ij} in two and three dimensions.

For $n = 2$,

$$g^{ij} = (-1)^{i+j} \frac{g_{3-i\ 3-j}}{g} \,,$$

$$g_{ij} = (-1)^{i+j} g g^{3-i\ 3-j} \,, \quad i, j = 1, 2 \,, \tag{2.21}$$

where the indices i, j on the right-hand side of the relations (2.21) are fixed, i.e. summation over the repeated indices is not carried out here. For $n = 3$ we have

$$g^{ij} = \frac{1}{g}(g_{i+1\ j+1}\, g_{i+2\ j+2} - g_{i+1\ j+2}\, g_{i+2\ j+1}) \,,$$

$$g_{ij} = g(g^{i+1\ j+1}\, g^{i+2\ j+2} - g^{i+1\ j+2}\, g^{i+2\ j+1}) \,, \quad i, j = 1, 2, 3 \,, \tag{2.22}$$

with the convention that any index, say l, is identified with $l \pm 3$, so, for instance, $g_{45} = g_{12}$.

We also note that, in accordance with the expressions (2.15), (2.18) for g_{ij} and g^{ij}, respectively, the relations (2.11) and (2.14) between the basic vectors $\boldsymbol{x}_{\xi^i}$ and $\nabla \xi^j$ can be written in the form

$$\boldsymbol{x}_{\xi^i} = g_{ik} \nabla \xi^k \; ,$$

$$\nabla \xi^i = g^{ik} \boldsymbol{x}_{\xi^k} \; , \quad i,k = 1, \cdots, n \; . \tag{2.23}$$

So the first derivatives $\partial x^i / \partial \xi^j$ and $\partial \xi^k / \partial x^m$ of the transformations $\boldsymbol{x}(\boldsymbol{\xi})$ and $\boldsymbol{\xi}(\boldsymbol{x})$, respectively, are connected through the metric elements:

$$\frac{\partial x^i}{\partial \xi^j} = g_{mj} \frac{\partial \xi^m}{\partial x^i} \; ,$$

$$\frac{\partial \xi^i}{\partial x^j} = g^{mi} \frac{\partial x^j}{\partial \xi^m} \; , \quad i,j,m = 1, \cdots, n \; . \tag{2.24}$$

2.6 Cross Product

In addition to the dot product there is another important operation on three-dimensional vectors. This is the cross product, $\times$, which for any two vectors $\boldsymbol{a} = (a^1, a^2, a^3)$, $\boldsymbol{b} = (b^1, b^2, b^3)$ expanded in the Cartesian vector basis $(\boldsymbol{e}_1, \boldsymbol{e}_2, \boldsymbol{e}_3)$ of the Euclidean space R^3 is expressed as the determinant of a matrix:

$$\boldsymbol{a} \times \boldsymbol{b} = \det \begin{pmatrix} \boldsymbol{e}_1 & \boldsymbol{e}_2 & \boldsymbol{e}_3 \\ a^1 & a^2 & a^3 \\ b^1 & b^2 & b^3 \end{pmatrix} \; . \tag{2.25}$$

Thus

$$\boldsymbol{a} \times \boldsymbol{b} = (a^2 b^3 - a^3 b^2, \; a^3 b^1 - a^1 b^3, \; a^1 b^2 - a^2 b^1) \; ,$$

or, with the previously assumed convention in three dimensions of the identification of any index j with $j \pm 3$,

$$\boldsymbol{a} \times \boldsymbol{b} = (a^{i+1} b^{i+2} - a^{i+2} b^{i+1}) \boldsymbol{e}_i \; , \quad i = 1, 2, 3 \; . \tag{2.26}$$

We will now state some facts connected with the cross product operation.

2.6.1 Geometric Meaning

We can readily see that $\boldsymbol{a} \times \boldsymbol{b} = \boldsymbol{0}$ if the vectors $\boldsymbol{a}$ and $\boldsymbol{b}$ are parallel. Also, from (2.26) we find that $\boldsymbol{a} \cdot (\boldsymbol{a} \times \boldsymbol{b}) = 0$ and $\boldsymbol{b} \cdot (\boldsymbol{a} \times \boldsymbol{b}) = 0$, i.e. the vector $\boldsymbol{a} \times \boldsymbol{b}$ is orthogonal to each of the vectors $\boldsymbol{a}$ and $\boldsymbol{b}$. Thus, if these vectors are not parallel then

$$\boldsymbol{a} \times \boldsymbol{b} = \alpha |\boldsymbol{a} \times \boldsymbol{b}| \boldsymbol{n} , \tag{2.27}$$

where $\alpha = 1$ or $\alpha = -1$ and $\boldsymbol{n}$ is a unit normal vector to the plane determined by the vectors $\boldsymbol{a}$ and $\boldsymbol{b}$.

Now we show that the length of the vector $\boldsymbol{a} \times \boldsymbol{b}$ equals the area of the parallelogram formed by the vectors $\boldsymbol{a}$ and $\boldsymbol{b}$, i.e.

$$|\boldsymbol{a} \times \boldsymbol{b}| = |\boldsymbol{a}||\boldsymbol{b}| \sin \theta , \tag{2.28}$$

where θ is the angle between the two vectors $\boldsymbol{a}$ and $\boldsymbol{b}$. To prove (2.28) we first note that

$$|\boldsymbol{a}|^2 |\boldsymbol{b}|^2 \sin^2 \theta = |\boldsymbol{a}|^2 |\boldsymbol{b}|^2 (1 - \cos^2 \theta) = |\boldsymbol{a}|^2 |\boldsymbol{b}|^2 - (\boldsymbol{a} \cdot \boldsymbol{b})^2 .$$

We have, further,

$$\begin{aligned} |\boldsymbol{a}|^2 |\boldsymbol{b}|^2 - |\boldsymbol{a} \cdot \boldsymbol{b}|^2 &= \left(\sum_{i=1}^{3} a^i a^i \right) \left(\sum_{j=1}^{3} b^j b^j \right) - \left(\sum_{k=1}^{3} a^k b^k \right)^2 \\ &= \sum_{k=1}^{3} [(a^l)^2 (b^m)^2 + (a^m)^2 (b^l)^2 - 2 a^l b^l a^m b^m] \\ &= \sum_{k=1}^{3} (a^l b^m - a^m b^l)^2 , \end{aligned}$$

where (k, l, m) are cyclic, i.e. $l = k+1$, $m = k+2$ with the convention that $j+3$ is equivalent to j for any index j. According to (2.26) the quantity $a^l b^m - a^m b^l$ for the cyclic sequence (k, l, m) is the kth component of the vector $\boldsymbol{a} \times \boldsymbol{b}$, so we find that

$$|\boldsymbol{a}||\boldsymbol{b}| \sin^2 \theta = |\boldsymbol{a}|^2 |\boldsymbol{b}|^2 - |\boldsymbol{a} \cdot \boldsymbol{b}|^2 = |\boldsymbol{a} \times \boldsymbol{b}|^2 , \tag{2.29}$$

what proves (2.28). Thus we obtain the result that if the vectors $\boldsymbol{a}$ and $\boldsymbol{b}$ are not parallel then the vector $\boldsymbol{a} \times \boldsymbol{b}$ is orthogonal to the parallelogram formed by these vectors and its length equals the area of the parallelogram. Therefore the three vectors $\boldsymbol{a}$, $\boldsymbol{b}$ and

$\boldsymbol{a} \times \boldsymbol{b}$ are independent in this case and represent a base vector system in the three-dimensional space R^3. Moreover, the vectors $\boldsymbol{a}$, $\boldsymbol{b}$ and $\boldsymbol{a} \times \boldsymbol{b}$ form a right-handed triad since $\boldsymbol{a} \times \boldsymbol{b} \neq \boldsymbol{0}$, and consequently the Jacobian of the matrix determined by $\boldsymbol{a}$, $\boldsymbol{b}$, and $\boldsymbol{a} \times \boldsymbol{b}$ is positive; it equals

$$(\boldsymbol{a} \times \boldsymbol{b}) \cdot (\boldsymbol{a} \times \boldsymbol{b}) = (\boldsymbol{a} \times \boldsymbol{b})^2 .$$

2.6.2 Relation to Volumes

Let $\boldsymbol{c} = (c^1, c^2, c^3)$ be one more vector. The volume V of the parallelepiped whose edges are the vectors $\boldsymbol{a}$, $\boldsymbol{b}$ and $\boldsymbol{c}$ equals the area of the parallelogram formed by the vectors $\boldsymbol{a}$ and $\boldsymbol{b}$ multiplied by the modulas of the dot product of the vector $\boldsymbol{c}$ and the unit normal $\boldsymbol{n}$ to the parallelogram. Thus

46   2. General Coordinate Systems in Domains

$$V = |\boldsymbol{a} \times \boldsymbol{b}||\boldsymbol{n} \cdot \boldsymbol{c}|$$

and from (2.27) we obtain

$$V = |(\boldsymbol{a} \times \boldsymbol{b}) \cdot \boldsymbol{c}| \ . \tag{2.30}$$

Taking into account (2.26), we obtain

$$(\boldsymbol{a} \times \boldsymbol{b}) \cdot \boldsymbol{c} = c^1(a^2 b^3 - a^3 b^2) + c^2(a^3 b^1 - a^1 b^3) + c^3(a^1 b^2 - a^2 b^1) \ .$$

The right-hand side of this equation is the Jacobian of the matrix whose rows are formed by the vectors $\boldsymbol{a}$, $\boldsymbol{b}$, and $\boldsymbol{c}$, i.e.

$$(\boldsymbol{a} \times \boldsymbol{b}) \cdot \boldsymbol{c} = \det \begin{pmatrix} a^1 & a^2 & a^3 \\ b^1 & b^2 & b^3 \\ c^1 & c^2 & c^3 \end{pmatrix} \ . \tag{2.31}$$

From this equation we readily obtain

$$(\boldsymbol{a} \times \boldsymbol{b}) \cdot \boldsymbol{c} = \boldsymbol{a} \cdot (\boldsymbol{b} \times \boldsymbol{c}) = (\boldsymbol{c} \times \boldsymbol{a}) \cdot \boldsymbol{b} \ .$$

Thus the volume of the parallelepiped determined by the vectors $\boldsymbol{a}$, $\boldsymbol{b}$, and $\boldsymbol{c}$ equals the Jacobian of the matrix formed by the components of these vectors. In particular, we obtain that the Jacobian of a three-dimensional coordinate transformation $\boldsymbol{x}(\boldsymbol{\xi})$ is expressed as follows:

$$J = \boldsymbol{x}_{\xi^1} \cdot (\boldsymbol{x}_{\xi^2} \times \boldsymbol{x}_{\xi^3}) \ . \tag{2.32}$$

2.6.3 Relation to Base Vectors

Applying the operation of the cross product to two base tangential vectors $\boldsymbol{x}_{\xi^l}$ and $\boldsymbol{x}_{\xi^m}$, we find that the vector $\boldsymbol{x}_{\xi^l} \times \boldsymbol{x}_{\xi^m}$ is a normal to the coordinate surface $\xi^i = \xi_0^i$ with (i,l,m) cyclic. The base normal vector $\boldsymbol{\nabla}\xi^i$ is also orthogonal to the surface and therefore it is a scalar multiple of $\boldsymbol{x}_{\xi^l} \times \boldsymbol{x}_{\xi^m}$, i.e.

$$\boldsymbol{\nabla}\xi^i = c(\boldsymbol{x}_{\xi^l} \times \boldsymbol{x}_{\xi^m}) \ .$$

Multiplying this equation for a fixed i by $\boldsymbol{x}_{\xi^i}$, using the operation of the dot product, we obtain, using (2.32),

$$1 = cJ \ ,$$

and therefore

$$\boldsymbol{\nabla}\xi^i = \frac{1}{J}(\boldsymbol{x}_{\xi^l} \times \boldsymbol{x}_{\xi^m}) \ . \tag{2.33}$$

Thus the elements of the three-dimensional contravariant metric tensor (g^{ij}) which are computed by the dot product of the normal base vectors (formula (2.18)) can also be found through the tangential vectors $\boldsymbol{x}_{\xi^i}$ by the formula

$$g^{ij} = \boldsymbol{\nabla}\xi^i \cdot \boldsymbol{\nabla}\xi^j = \frac{1}{g}(\boldsymbol{x}_{\xi^{i+1}} \times \boldsymbol{x}_{\xi^{i+2}}) \cdot (\boldsymbol{x}_{\xi^{j+1}} \times \boldsymbol{x}_{\xi^{j+2}}) \ , \quad i,j = 1,2,3 \ .$$

Analogously, every base vector $\boldsymbol{x}_{\xi^i}$, $i = 1, 2, 3$, is expressed by the tensor product of the normal base vectors $\boldsymbol{\nabla}\xi^j$, $j = 1, 2, 3$:

$$\boldsymbol{x}_{\xi^i} = J(\boldsymbol{\nabla}\xi^l \times \boldsymbol{\nabla}\xi^k), \quad i = 1, 2, 3 , \tag{2.34}$$

where $l = i + 1$, $k = i + 2$, and m is equivalent to $m + 3$ for any index m. Accordingly we have, taking into account (2.15),

$$g_{ij} = g(\boldsymbol{\nabla}\xi^{i+1} \times \boldsymbol{\nabla}\xi^{i+2}) \cdot (\boldsymbol{\nabla}\xi^{j+1} \times \boldsymbol{\nabla}\xi^{j+2}) , \quad i, j = 1, 2, 3 .$$

Using the relations (2.33) and (2.34) in (2.30), we also obtain

$$\frac{1}{J} = \boldsymbol{\nabla}\xi^1 \cdot \boldsymbol{\nabla}\xi^2 \times \boldsymbol{\nabla}\xi^3 . \tag{2.35}$$

Thus the volume of the parallelepiped formed by the base normal vectors $\boldsymbol{\nabla}\xi^1$, $\boldsymbol{\nabla}\xi^2$, and $\boldsymbol{\nabla}\xi^3$ is the modulus of the inverse of the Jacobian J of the transformation $\boldsymbol{x}(\boldsymbol{\xi})$.

2.7 Relations Concerning Second Derivatives

The elements of the covariant and contravariant metric tensors are defined by the dot products of the base tangential and normal vectors, respectively. These elements are suitable for describing the internal features of the cells such as the lengths of the edges, the areas of the faces, their volumes, and the angles between the edges and the faces. However, as they are derived from the first derivatives of the coordinate transformation $\boldsymbol{x}(\boldsymbol{\xi})$, the direct use of the metric elements is not sufficient for the description of the dynamic features of the grid (e.g. curvature), which reflect changes between adjacent cells. This is because the formulation of these grid features relies not only on the first derivatives but also on the second derivatives of $\boldsymbol{x}(\boldsymbol{\xi})$. Therefore there is a need to study relations connected with the second derivatives of the coordinate parametrizations.

This section presents some notations and formulas which are concerned with the second derivatives of the components of the coordinate transformations. These notations and relations will be used to describe the curvatures of the coordinate lines and surfaces.

2.7.1 Christoffel Symbols of Domains

The edge of a grid cell in the ξ^i direction can be represented with high accuracy by the base vector $\boldsymbol{x}_{\xi^i}$ contracted by the factor h_i, which represents the step size of a uniform grid in $\varXi^n$ in the ξ^i direction. Therefore the local change of the edge in the ξ^j direction is characterized by the derivative of $\boldsymbol{x}_{\xi^i}$ with respect to ξ^j, i.e. by $\boldsymbol{x}_{\xi^i \xi^j}$.

Since the second derivatives may be used to formulate quantitative measures of the grid, we describe these vectors $\boldsymbol{x}_{\xi^i \xi^j}$ through the base tangential and normal vectors using certain three-index quantities known as Christoffel symbols. The Christoffel symbols are commonly used in formulating measures of the mutual interaction of the cells and in formulas for differential equations.

Let us denote by Γ_{ij}^k the kth contravariant component of the vector $\boldsymbol{x}_{\xi^i \xi^j}$ in the base tangential vectors $\boldsymbol{x}_{\xi^k}$, $k = 1, \cdots, n$. The superscript k in this designation relates to the base vector $\boldsymbol{x}_{\xi^k}$ and the subscript ij corresponds to the mixed derivative with respect to ξ^i and ξ^j. Thus

$$\boldsymbol{x}_{\xi^i \xi^j} = \Gamma_{ij}^k \boldsymbol{x}_{\xi^k}, \quad i, j, k = 1, \cdots, n, \tag{2.36}$$

and consequently

$$\frac{\partial^2 x^p}{\partial \xi^j \partial \xi^k} = \Gamma_{kj}^m \frac{\partial x^p}{\partial \xi^m}, \quad j, k, m, p = 1, \cdots, n. \tag{2.37}$$

Multiplying these equations by $\partial \xi^i / \partial x^p$ gives

$$\Gamma_{kj}^i = \frac{\partial^2 x^p}{\partial \xi^k \partial \xi^j} \frac{\partial \xi^i}{\partial x^p}, \quad i, j, k, p = 1, \cdots, n, \tag{2.38}$$

or in vector form,

$$\Gamma_{kj}^i = \boldsymbol{x}_{\xi^k \xi^j} \cdot \boldsymbol{\nabla} \xi^i. \tag{2.39}$$

The quantities Γ_{kj}^i are called the space Christoffel symbols of the second kind and the expression (2.36) is a form of the Gauss relation representing the second derivatives of the position vector $\boldsymbol{x}(\boldsymbol{\xi})$ through the tangential vectors $\boldsymbol{x}_{\xi^i}$.

Analogously, the components of the second derivatives of the position vector $\boldsymbol{x}(\boldsymbol{\xi})$ expanded in the base normal vectors $\boldsymbol{\nabla} \xi^i$, $i = 1, \cdots, n$, are referred to as the space Christoffel symbols of the first kind. The mth component of the vector $\boldsymbol{x}_{\xi^k \xi^j}$ in the base vectors $\boldsymbol{\nabla} \xi^i$, $i = 1, \cdots, n$, is denoted by $[kj, m]$. Thus, according to (2.12),

$$[kj, m] = \boldsymbol{x}_{\xi^k \xi^j} \cdot \boldsymbol{x}_{\xi^m} = \frac{\partial^2 x^l}{\partial \xi^k \partial \xi^j} \frac{\partial x^l}{\partial \xi^m}, \quad j, k, l, m = 1, \cdots, n, \tag{2.40}$$

and consequently

$$\boldsymbol{x}_{\xi^k \xi^j} = [kj, m] \boldsymbol{\nabla} \xi^m. \tag{2.41}$$

So, in analogy with (2.37), we obtain

$$\frac{\partial^2 x^l}{\partial \xi^j \partial \xi^k} = [kj, m] \frac{\partial \xi^m}{\partial x^i}, \quad i, j, k, m = 1, \cdots, n. \tag{2.42}$$

Multiplying (2.40) by g^{im} and summing over m we find that the space Christoffel symbols of the first and second kind are connected by the following relation:

$$\Gamma_{kj}^i = g^{im} [kj, m], \quad i, j, k, m = 1, \cdots, n. \tag{2.43}$$

Conversely, from (2.38) and (2.40),

$$[kj, m] = g_{ml}\Gamma^l_{kj}, \quad j, k, l, m = 1, \cdots, n. \tag{2.44}$$

The space Christoffel symbols of the first kind $[kj, m]$ can be expressed through the first derivatives of the covariant elements g_{ij} of the metric tensor (g_{ij}) by the following readily verified formula:

$$[kj, m] = \frac{1}{2}\left(\frac{\partial g_{jm}}{\partial \xi^k} + \frac{\partial g_{km}}{\partial \xi^j} - \frac{\partial g_{kj}}{\partial \xi^m}\right), \quad i, j, k, m = 1, \cdots, n. \tag{2.45}$$

Thus, taking into account (2.43), we see that the space Christoffel symbols of the second kind Γ^i_{kj} can be written in terms of metric elements and their first derivatives. In particular, in the case of an orthogonal coordinate system ξ^i, we obtain from (2.43, 2.45)

$$\Gamma^i_{kj} = \frac{1}{g}g^{ii}\left(\frac{\partial g_{ii}}{\partial \xi^k} + \frac{\partial g_{ii}}{\partial \xi^j} - \frac{\partial g_{kj}}{\partial \xi^i}\right), \quad i, j, k = 1, \cdots, n.$$

Here the index i is fixed, i.e. the summation over i is not carried out.

2.7.2 Differentiation of the Jacobian

Of critical importance in establishing relations between geometric characteristics is the formula for differentiation of the Jacobian of a coordinate transformation $\boldsymbol{x}(\boldsymbol{\xi})$

$$\frac{\partial J}{\partial \xi^k} \equiv J\frac{\partial^2 x^i}{\partial \xi^k \partial \xi^m}\frac{\partial \xi^m}{\partial x^i} \equiv J\frac{\partial}{\partial x^i}\left(\frac{\partial x^i}{\partial \xi^k}\right) \equiv J\mathrm{div}_x \frac{\partial \boldsymbol{x}}{\partial \xi^k},$$

$$i, k, m = 1, \cdots, n. \tag{2.46}$$

In accordance with (2.38), this identity can also be expressed through the space Christoffel symbols of the second kind Γ^i_{kj} by

$$\frac{\partial J}{\partial \xi^k} = J\Gamma^i_{ik}, \quad i, k = 1, \cdots, n,$$

with the summation convention over the repeated index i.

In order to prove the identity (2.46) we note that in the case of an arbitrary matrix (a_{ij}) the first derivative of its Jacobian is obtained by the process of differentiating the first row (the others are left unchanged), then performing the same operation on the second row, and so on with all of the rows of the matrix. The summation of the Jacobians of the matrices derived in such a manner gives the first derivative with respect to ξ^k of the Jacobian of the original matrix (a_{ij}). Thus

$$\frac{\partial}{\partial \xi^k}\det(a_{ij}) = \frac{\partial a_{im}}{\partial \xi^k}G^{im}, \quad i, j, k, m = 1, \cdots, n, \tag{2.47}$$

where G^{im} is the cofactor of the element a_{im}. For the Jacobi matrix $(\partial x^i/\partial \xi^j)$ of the coordinate transformation $\boldsymbol{x}(\boldsymbol{\xi})$ we have

$$G^{im} = J\frac{\partial \xi^m}{\partial x^i}, \quad i,j = 1,\cdots,n \ .$$

Therefore, applying (2.47) to the Jacobi matrix, we obtain (2.46).

2.7.3 Basic Identity

The identity (2.46) implies the extremely important relation

$$\frac{\partial}{\partial \xi^j}\left(J\frac{\partial \xi^j}{\partial x^i}\right) \equiv 0, \quad i,j = 1,\cdots,n \ , \tag{2.48}$$

which leads to specific forms of new dependent variables for conservation-law equations. To prove (2.48) we first note that

$$\frac{\partial^2 \xi^p}{\partial x^k \partial x^j}\frac{\partial x^l}{\partial \xi^p} = -\frac{\partial^2 x^l}{\partial \xi^p \partial \xi^m}\frac{\partial \xi^m}{\partial x^k}\frac{\partial \xi^p}{\partial x^j} \ .$$

Multiplying this equation by $\partial \xi^i/\partial x^l$ and summing over l, we obtain a formula representing the second derivative $\partial^2 \xi^i/\partial x^k \partial x^m$ of the functions $\xi^i(\boldsymbol{x})$ through the second derivatives $\partial^2 x^m/\partial \xi^l \partial \xi^p$ of the functions $x^m(\boldsymbol{\xi})$, $m = 1,\cdots,n$:

$$\begin{aligned}\frac{\partial^2 \xi^i}{\partial x^k \partial x^m} &= -\frac{\partial^2 x^p}{\partial \xi^l \partial \xi^j}\frac{\partial \xi^j}{\partial x^k}\frac{\partial \xi^l}{\partial x^m}\frac{\partial \xi^i}{\partial x^p} \\ &= -\Gamma^i_{lj}\frac{\partial \xi^l}{\partial x^m}\frac{\partial \xi^j}{\partial x^k}, \quad i,j,k,l,m,p = 1,\cdots,n \ .\end{aligned} \tag{2.49}$$

Now, using this relation and the formula (2.46) for differentiation of the Jacobian in the identity

$$\frac{\partial}{\partial \xi^j}\left(J\frac{\partial \xi^j}{\partial x^i}\right) = \frac{\partial J}{\partial \xi^j}\frac{\partial \xi^j}{\partial x^i} + J\frac{\partial^2 \xi^j}{\partial x^i \partial x^k}\frac{\partial x^k}{\partial \xi^j} \ ,$$

we obtain

$$\begin{aligned}\frac{\partial}{\partial \xi^j}\left(J\frac{\partial \xi^j}{\partial x^i}\right) &= J\frac{\partial^2 x^k}{\partial \xi^p \partial \xi^j}\frac{\partial \xi^p}{\partial x^k}\frac{\partial \xi^j}{\partial x^i} - J\frac{\partial^2 x^p}{\partial \xi^l \partial \xi^m}\frac{\partial \xi^m}{\partial x^i}\frac{\partial \xi^l}{\partial x^k}\frac{\partial \xi^j}{\partial x^p}\frac{\partial x^k}{\partial \xi^j} \\ &= J\frac{\partial^2 x^k}{\partial \xi^p \partial \xi^j}\frac{\partial \xi^p}{\partial x^k}\frac{\partial \xi^j}{\partial x^i} - J\frac{\partial^2 x^p}{\partial \xi^l \partial \xi^m}\frac{\partial \xi^l}{\partial x^p}\frac{\partial \xi^m}{\partial x^i} = 0 \ ,\end{aligned}$$

$$i,j,k,l,m,p = 1,\cdots,n \ ,$$

i.e. (2.48) has been proved.

The identity (2.48) is obvious when $n = 1$ or $n = 2$. For example, for $n = 2$ we have from (2.1)

$$J\frac{\partial \xi^j}{\partial x^i} = (-1)^{i+j}\frac{\partial x^{3-i}}{\partial \xi^{3-j}}, \quad i,j = 1,2 \ ,$$

2.7 Relations Concerning Second Derivatives

with fixed indices i and j, and therefore

$$\frac{\partial}{\partial \xi^j}\left(J\frac{\partial \xi^j}{\partial x^i}\right) = (-1)^{i+1}\left(\frac{\partial}{\partial \xi^1}\frac{\partial x^{3-i}}{\partial \xi^2} - \frac{\partial}{\partial \xi^2}\frac{\partial x^{3-i}}{\partial \xi^1}\right) = 0, \quad i,j = 1,2.$$

An inference of (2.48) for $n = 3$ also follows from the differentiation of the cross product of the base tangential vectors $\boldsymbol{r}_{\xi^i}$, $i = 1, 2, 3$. Taking into account (2.26), we readily obtain the following formula for the differentiation of the cross product of two three-dimensional vector-valued functions $\boldsymbol{a}$ and $\boldsymbol{b}$:

$$\frac{\partial}{\partial \xi^i}(\boldsymbol{a} \times \boldsymbol{b}) = \frac{\partial}{\partial \xi^i}\boldsymbol{a} \times \boldsymbol{b} + \boldsymbol{a} \times \frac{\partial}{\partial \xi^i}\boldsymbol{b}, \quad i = 1, 2, 3.$$

With this formula we obtain

$$\sum_{i=1}^{3}\frac{\partial}{\partial \xi^i}(\boldsymbol{x}_{\xi^j} \times \boldsymbol{x}_{\xi^k}) = \sum_{i=1}^{3}\boldsymbol{x}_{\xi^j\xi^i} \times \boldsymbol{x}_{\xi^k} + \sum_{i=1}^{3}\boldsymbol{x}_{\xi^j} \times \boldsymbol{x}_{\xi^k\xi^i}, \quad (2.50)$$

where the indices (i, j, k) are cyclic, i.e. $j = i+1$, $k = i+2$, m is equivalent to $m + 3$. For the last summation of the above formula, we obtain

$$\sum_{i=1}^{3}\boldsymbol{x}_{\xi^j} \times \boldsymbol{x}_{\xi^k\xi^i} = \sum_{i=1}^{3}\boldsymbol{x}_{\xi^k} \times \boldsymbol{x}_{\xi^i\xi^j}.$$

Therefore, from (2.50),

$$\sum_{i=1}^{3}\frac{\partial}{\partial \xi^i}(\boldsymbol{x}_{\xi^j} \times \boldsymbol{x}_{\xi^k}) = 0,$$

since

$$\boldsymbol{x}_{\xi^i} \times \boldsymbol{x}_{\xi^j\xi^k} = -\boldsymbol{x}_{\xi^j\xi^k} \times \boldsymbol{x}_{\xi^i}$$

and (2.33) implies (2.48) for $n = 3$.

3. Geometry of Curves

3.1 Curves in Multidimensional Space

3.1.1 Definition

Commonly, a curve in the n-dimensional Euclidean space R^n is the locus of points of R^n whose positions are represented by a vector-valued position function $\boldsymbol{r}$ of a single parameter, say φ,

$$\boldsymbol{r}(\varphi) : [a,b] \to R^n, \quad \boldsymbol{r}(\varphi) = [x^1(\varphi), \ldots, x^n(\varphi)], \tag{3.1}$$

(see Fig. 3.1). The position function $\boldsymbol{r}(\varphi)$ is referred to as a parametrization of the curve. Note each curve can be given parametric representation in an infinity of ways, namely, as

$$\boldsymbol{r}[\varphi(\psi)] : [a_1, b_1] \to [a, b],$$

where

$$\psi : [a_1, b_1] \to [a, b]$$

is an arbitrary univariate one-to-one monotone function.

It is assumed that the parametrization (3.1) is $p \geq 1$ times continuously differentiable with respect to φ and $\boldsymbol{r}_\varphi \neq \boldsymbol{0}$ for all $\varphi \in [a,b]$. In our considerations we will use the designation S^{r1} for the curve with the parametrization

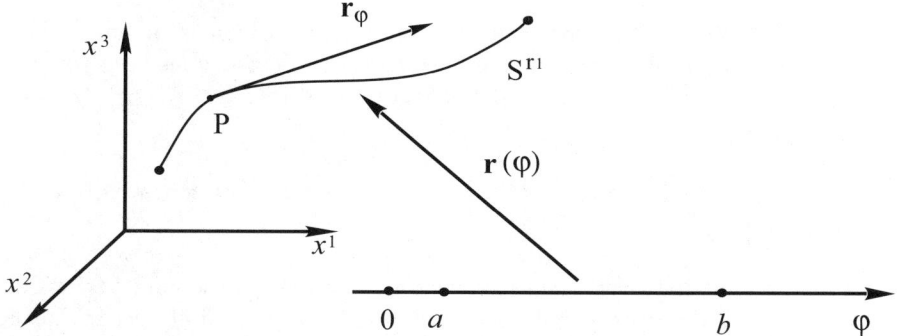

Fig. 3.1. Scheme of a curve representation in R^3

$r(\varphi)$. In this chapter we discuss the important measures of the local curve quality known as curvature and torsion. These measures are derived by some manipulations of basic curve vectors using the operations of the dot and cross products.

3.1.2 Basic Curve Vectors

Tangent Vector. The first derivative of the parametrization $r(\varphi)$ is a tangential vector

$$r_\varphi = \left(\frac{dx^1}{d\varphi}, \ldots, \frac{dx^n}{d\varphi} \right)$$

to the curve S^{r1} (Fig. 3.1). The quantity

$$g^{r\varphi} = r_\varphi \cdot r_\varphi = \frac{dx^i}{d\varphi} \frac{dx^i}{d\varphi}, \quad i = 1, \ldots, n, \tag{3.2}$$

is the metric tensor of the curve and its square root is the length of the tangent vector r_φ. Accordingly the length l of the curve S^{r1} is computed from the integral

$$l = \int_a^b \sqrt{g^{r\varphi}} d\varphi.$$

The most important notions related to curves are connected with the arc length parameter s defined by the equation

$$s(\varphi) = \int_a^\varphi \sqrt{g^{r\varphi}} d\varphi. \tag{3.3}$$

The vector $dr[\varphi(s)]/ds$, where $\varphi(s)$ is the inverse of $s(\varphi)$, is a tangent vector designated by t. Using (3.3) we obtain

$$t = \frac{d}{ds} r[\varphi(s)] = \frac{d\varphi}{ds} r_\varphi = \frac{1}{\sqrt{g^{r\varphi}}} r_\varphi. \tag{3.4}$$

Therefore t is the unit tangent vector.

Principal Normal Vector. Any nonzero vector, which is orthogonal to the tangent vector, is called a normal vector to the curve. Thus a vector v is normal to S^{r1} if $v \cdot t = 0$. Since (3.4) t is a unit vector, i.e.

$$t \cdot t = 1,$$

and if we take the derivative with respect to s of this equation, we obtain

$$t_s \cdot t = 0.$$

This means that if $t_s \neq 0$ then this vector is normal to the curve S^{r1}. The vector t_s is called the principal normal of S^{r1}. Let n be a unit vector that is co-directional with t_s. Then

$$\boldsymbol{n} = \frac{1}{k}\boldsymbol{t}_s, \quad k = (\boldsymbol{t}_s \cdot \boldsymbol{t}_s)^{1/2}. \tag{3.5}$$

The magnitude k is called the curvature at the point in question, while the quantity $\rho = 1/k$ is called the radius of curvature of the curve.

Using the identity $\boldsymbol{r}_\varphi = \sqrt{g^{r\varphi}}\boldsymbol{t}$ obtained from (3.4) we find by virtue of (3.2) and (3.3),

$$\begin{aligned}
\boldsymbol{r}_{\varphi\varphi} &= \frac{1}{\sqrt{g^{r\varphi}}}(\boldsymbol{r}_{\varphi\varphi} \cdot \boldsymbol{r}_\varphi)\boldsymbol{t} + g^{r\varphi}\boldsymbol{t}_s \\
&= \frac{1}{\sqrt{g^{r\varphi}}}(\boldsymbol{r}_{\varphi\varphi} \cdot \boldsymbol{r}_\varphi)\boldsymbol{t} + g^{r\varphi}k\boldsymbol{n}.
\end{aligned} \tag{3.6}$$

The identity (3.6) is an analog of the Gauss relations (2.36). This identity shows that the vector $\boldsymbol{r}_{\varphi\varphi}$ lies in the $\boldsymbol{t} - \boldsymbol{n}$ plane.

3.2 Curves in Three-Dimensional Space

3.2.1 Basic Vectors

In three dimensions we can apply the operation of the cross product to the basic tangential and normal vectors of a curve. The vector $\boldsymbol{b} = \boldsymbol{t} \times \boldsymbol{n}$ is a unit vector which is orthogonal to both $\boldsymbol{t}$ and $\boldsymbol{n}$. It is called the binormal vector. From (3.6) we find that $\boldsymbol{b}$ is orthogonal to $\boldsymbol{r}_{\varphi\varphi}$.

The three vectors $(\boldsymbol{t},\ \boldsymbol{n},\ \boldsymbol{b})$ form a right-handed triad (Fig. 3.2). Note that if the curve lies in a plane, then the vectors $\boldsymbol{t}$ and $\boldsymbol{n}$ lie in the plane as well and $\boldsymbol{b}$ is a constant unit vector normal to the plane.

The vectors $\boldsymbol{t}$, $\boldsymbol{n}$, and $\boldsymbol{b}$ are connected by the Serret–Frenet equations

$$\frac{\mathrm{d}\boldsymbol{t}}{\mathrm{d}s} = k\boldsymbol{n},$$

$$\frac{\mathrm{d}\boldsymbol{n}}{\mathrm{d}s} = -k\boldsymbol{t} + \tau\boldsymbol{b},$$

$$\frac{\mathrm{d}\boldsymbol{b}}{\mathrm{d}s} = -\tau\boldsymbol{n}, \tag{3.7}$$

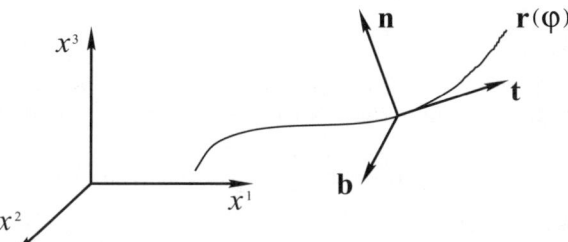

Fig. 3.2. Base curve vectors

where the coefficient τ is called the second curvature or torsion of the curve. The first equation of the system (3.7) is taken from (3.5). The second and third equations are readily obtained from the formula (2.6) by replacing the $\boldsymbol{b}$ in (2.6) by the vectors on the left-hand side of (3.7), while the vectors $\boldsymbol{t}$, $\boldsymbol{n}$, and $\boldsymbol{b}$ substitute for $\boldsymbol{e}_1$, $\boldsymbol{e}_2$, and $\boldsymbol{e}_3$, respectively. The vectors $\boldsymbol{t}$, $\boldsymbol{n}$, and $\boldsymbol{b}$ constitute an orthonormal basis, i.e.

$$a_{ij} = a^{ij} = \delta^i_j, \quad i,j = 1,2,3,$$

where, in accordance with Sect. 2.4 $a_{ij} = \boldsymbol{e}_i \cdot \boldsymbol{e}_j$, and (a^{ij}) is the inverse of the tensor (a_{ij}). Now, using (2.6) we obtain

$$\frac{d\boldsymbol{n}}{ds} = \left(\frac{d\boldsymbol{n}}{ds} \cdot \boldsymbol{t}\right)\boldsymbol{t} + \left(\frac{d\boldsymbol{n}}{ds} \cdot \boldsymbol{n}\right)\boldsymbol{n} + \left(\frac{d\boldsymbol{n}}{ds} \cdot \boldsymbol{b}\right)\boldsymbol{b} = -k\boldsymbol{t} + \left(\frac{d\boldsymbol{n}}{ds} \cdot \boldsymbol{b}\right)\boldsymbol{b},$$

since $\boldsymbol{n}_s \cdot \boldsymbol{t} = -\boldsymbol{n} \cdot \boldsymbol{t}_s$, $\boldsymbol{n}_s \cdot \boldsymbol{n} = 0$. Thus we obtain the second equation of (3.7) with $\tau = \boldsymbol{n}_s \cdot \boldsymbol{b}$. Analogously we obtain the last equation of (3.7) by expanding the vector $\boldsymbol{b}_s$ through $\boldsymbol{t}$, $\boldsymbol{n}$, and $\boldsymbol{b}$ using the relation (2.6):

$$\frac{d\boldsymbol{b}}{ds} = \left(\frac{d\boldsymbol{b}}{ds} \cdot \boldsymbol{t}\right)\boldsymbol{t} + \left(\frac{d\boldsymbol{b}}{ds} \cdot \boldsymbol{n}\right)\boldsymbol{n} + \left(\frac{d\boldsymbol{b}}{ds} \cdot \boldsymbol{b}\right)\boldsymbol{b} = -\left(\frac{d\boldsymbol{n}}{ds} \cdot \boldsymbol{b}\right)\boldsymbol{n} = -\tau\boldsymbol{n},$$

as $\boldsymbol{b}_s \cdot \boldsymbol{t} = -\boldsymbol{b} \cdot \boldsymbol{t}_s = 0$, $\boldsymbol{b}_s \cdot \boldsymbol{b} = 0$.

Note the formula (3.7) also has the following form

$$\frac{d\boldsymbol{p}}{ds} = \begin{pmatrix} 0 & k & 0 \\ -k & 0 & \tau \\ 0 & -\tau & 0 \end{pmatrix} \boldsymbol{p},$$

where $\boldsymbol{p} = (\boldsymbol{t}, \boldsymbol{n}, \boldsymbol{b})^T$. We also find from this expression

$$\begin{pmatrix} \dfrac{d\boldsymbol{r}}{ds} \\ \dfrac{d^2\boldsymbol{r}}{ds} \\ \dfrac{d^3\boldsymbol{r}}{ds} \end{pmatrix} = \begin{pmatrix} 1 & 0 & 0 \\ 0 & k & 0 \\ -k^2 & k' & k\tau \end{pmatrix} \boldsymbol{p},$$

where $'$ is a sign of the first derivative with respect to s.

3.2.2 Curvature

A very important characteristic of a curve which is related to grid generation is the curvature k. This quantity is used as a measure of coordinate line bending.

One way to compute the curvature is to multiply (3.5) by $\boldsymbol{n}$ using the dot product operation. As from (3.3–3.5),

$$kn = \frac{d\boldsymbol{t}}{ds} = \frac{1}{\sqrt{g^{r\varphi}}}\frac{d}{d\varphi}\left(\frac{1}{\sqrt{g^{r\varphi}}}\boldsymbol{r}_\varphi\right) = \frac{1}{g^{r\varphi}}\boldsymbol{r}_{\varphi\varphi} - \frac{1}{(g^{r\varphi})^2}(\boldsymbol{r}_\varphi \cdot \boldsymbol{r}_{\varphi\varphi})\boldsymbol{r}_\varphi ,$$

the result is

$$k = \frac{1}{g^{r\varphi}}\boldsymbol{r}_{\varphi\varphi} \cdot \boldsymbol{n} . \tag{3.8}$$

The vector $\boldsymbol{n}$ is independent of the curve parametrization, and therefore we find from (3.6) and (3.8) that k is an invariant of parametrizations of the curve.

In two dimensions,

$$\boldsymbol{n} = \frac{1}{\sqrt{g^{r\varphi}}}(-x_\varphi^2, x_\varphi^1) ,$$

therefore in this case we obtain from (3.8),

$$k^2 = \frac{(x_\varphi y_{\varphi\varphi} - y_\varphi x_{\varphi\varphi})^2}{[(x_\varphi)^2 + (y_\varphi)^2]^3} \tag{3.9}$$

with the convention $x = x^1$, $y = x^2$. In particular, when the curve in R^2 is defined by a function $u = u(x)$, we obtain from (3.9), assuming $\boldsymbol{r}(\varphi) = [\varphi, u(\varphi)]$, $\varphi = x$,

$$k^2 = (u_{xx})^2/[1 + (u_x)^2]^3 . \tag{3.10}$$

In the case of three-dimensional space the curvature k can also be computed from the relation obtained by multiplying (3.6) by $\boldsymbol{r}_\varphi$ with the cross product operation:

$$\boldsymbol{r}_\varphi \times \boldsymbol{r}_{\varphi\varphi} = g^{r\varphi}k(\boldsymbol{r}_\varphi \times \boldsymbol{n}) = (g^{r\varphi})^{3/2}k\boldsymbol{b} .$$

Thus we obtain

$$k^2 = \frac{|\boldsymbol{r}_\varphi \times \boldsymbol{r}_{\varphi\varphi}|^2}{(g^{r\varphi})^3} \tag{3.11}$$

and, consequently, from (2.26)

$$k^2 = \frac{(x_\varphi^1 x_{\varphi\varphi}^2 - x_\varphi^2 x_{\varphi\varphi}^1)^2 + (x_\varphi^2 x_{\varphi\varphi}^3 - x_\varphi^3 x_{\varphi\varphi}^2)^2 + (x_\varphi^3 x_{\varphi\varphi}^1 - x_\varphi^1 x_{\varphi\varphi}^3)^2}{[(x_\varphi^1)^2 + (x_\varphi^2)^2 + (x_\varphi^3)^2]^3} .$$

$$\tag{3.12}$$

3.2.3 Torsion

Another important quality measure of curves in three-dimensional space is the torsion τ. This quantity is suitable for measuring the rate of twisting of the lines of coordinate grids.

58   3. Geometry of Curves

In order to figure out the value of τ for a curve in R^3 we use the last relation in (3.7), which yields

$$\tau = -\frac{d\boldsymbol{b}}{ds} \cdot \boldsymbol{n}.$$

As $\boldsymbol{b} = \boldsymbol{t} \times \boldsymbol{n}$, we obtain

$$\frac{d\boldsymbol{b}}{ds} = \frac{d\boldsymbol{t}}{ds} \times \boldsymbol{n} + \boldsymbol{t} \times \frac{d\boldsymbol{n}}{ds} = \boldsymbol{t} \times \frac{d\boldsymbol{n}}{ds},$$

since $d\boldsymbol{t}/ds = k\boldsymbol{n}$. Thus

$$\tau = \left(-\boldsymbol{t} \times \frac{d\boldsymbol{n}}{ds}\right) \cdot \boldsymbol{n}. \tag{3.13}$$

From (3.3–3.5) we have the following obvious relations for the basic vectors $\boldsymbol{t}$ and $\boldsymbol{n}$ in terms of the parametrization $\boldsymbol{r}(\varphi)$ and its derivatives:

$$\boldsymbol{t} = \frac{1}{\sqrt{g^{r\varphi}}}\boldsymbol{r}_\varphi,$$

$$\boldsymbol{n} = \frac{1}{k}\frac{d\boldsymbol{t}}{ds} = \frac{1}{k}\left(\frac{1}{g^{r\varphi}}\boldsymbol{r}_{\varphi\varphi} - \frac{\boldsymbol{r}_\varphi \cdot \boldsymbol{r}_{\varphi\varphi}}{(g^{r\varphi})^2}\boldsymbol{r}_\varphi\right),$$

$$\frac{d\boldsymbol{n}}{ds} = \frac{1}{k}\left(\frac{1}{(g^{r\varphi})^{3/2}}\boldsymbol{r}_{\varphi\varphi\varphi} - 2\frac{\boldsymbol{r}_\varphi \cdot \boldsymbol{r}_{\varphi\varphi}}{(g^{r\varphi})^2}\boldsymbol{r}_{\varphi\varphi}\right.$$

$$\left. - \frac{d}{d\varphi}\left(\frac{\boldsymbol{r}_\varphi \cdot \boldsymbol{r}_{\varphi\varphi}}{(g^{r\varphi})^2}\right)\boldsymbol{r}_\varphi - \frac{1}{k}\frac{dk}{ds}\boldsymbol{n}\right). \tag{3.14}$$

Thus

$$\boldsymbol{t} \times \frac{d\boldsymbol{n}}{ds} = \frac{1}{k(g^{r\varphi})^2}\boldsymbol{r}_\varphi \times \boldsymbol{r}_{\varphi\varphi\varphi} - 2\frac{\boldsymbol{r}_\varphi \cdot \boldsymbol{r}_{\varphi\varphi}}{k(g^{r\varphi})^{5/2}}\boldsymbol{r}_\varphi \times \boldsymbol{r}_{\varphi\varphi} - \frac{1}{k^2\sqrt{g^{r\varphi}}}\frac{dk}{ds}\boldsymbol{r}_\varphi \times \boldsymbol{n}.$$

As $(\boldsymbol{a} \times \boldsymbol{b}) \cdot \boldsymbol{a} = (\boldsymbol{a} \times \boldsymbol{b}) \cdot \boldsymbol{b} = 0$ for arbitrary vectors $\boldsymbol{a}$ and $\boldsymbol{b}$, we obtain from (3.13, 3.14)

$$\tau = -\frac{1}{k^2(g^{r\varphi})^3}(\boldsymbol{r}_\varphi \times \boldsymbol{r}_{\varphi\varphi\varphi}) \cdot \boldsymbol{r}_{\varphi\varphi} = \frac{1}{k^2(g^{r\varphi})^3}(\boldsymbol{r}_\varphi \times \boldsymbol{r}_{\varphi\varphi}) \cdot \boldsymbol{r}_{\varphi\varphi\varphi}. \tag{3.15}$$

And, using (2.31), we also find

$$\tau = \frac{1}{k^2(g^{r\varphi})^3}\det\begin{pmatrix} \frac{dx^1}{d\varphi} & \frac{dx^2}{d\varphi} & \frac{dx^1}{d\varphi} \\ \frac{d^2x^1}{d\varphi^2} & \frac{d^2x^2}{d\varphi^2} & \frac{d^2x^3}{d\varphi^2} \\ \frac{d^3x^1}{d\varphi^3} & \frac{d^3x^2}{d\varphi^3} & \frac{d^3x^3}{d\varphi^3} \end{pmatrix}. \tag{3.16}$$

4. Multidimensional Geometry

The notion of a curve in R^n is readily extended to a notion of an n-dimensional surface in R^{n+l}, $l \geq 0$.

A regular n-dimensional surface of class $C^m (m \geq 1)$ is the point set in real $(n+l)$-dimensional space R^{n+l} locally represented by some parametric n-dimensional domain S^n and a parametrization

$$\boldsymbol{r}(\boldsymbol{s}) : S^n \to R^{n+l} , \quad \boldsymbol{r}(\boldsymbol{s}) = [r^1(\boldsymbol{s}), \ldots, r^{n+l}(\boldsymbol{s})] , \quad \boldsymbol{s} = (s^1, \ldots, s^n) , \quad (4.1)$$

such that all partial derivatives of $\boldsymbol{r}(\boldsymbol{s})$ of order m are continuous in S^n and the rank of the matrix $(\partial r^i / \partial s^j)$, $i = 1, \ldots, n+l$, $j = 1, \ldots, n$, equals n at each point of S^n. The vector equation (4.1) is called a parametric equation of the n-dimensional surface while the variables s^i, $i = 1, \ldots, n$, are referred to as curvilinear coordinates on the surface. We shall use the designation S^{rn} for the surface represented by (4.1). In accordance with the definition a curve is meant as a one-dimensional surface.

In grid generation methods regular n-dimensional surfaces are typical objects as boundaries of the domains under consideration, coordinate hypersurfaces, and monitor surfaces specified to generate adaptive meshes. The advanced grid technology also requires the application of the theories of more sophisticated geometries, namely, Riemannian manifolds which generalize regular surfaces. These geometries have real potential to provide efficient means to control the qualitative properties of grids and develop advanced grid technologies.

This chapter gives an introduction to the theory of multidimensional surfaces and Riemannian manifolds.

4.1 Tangent and Normal Vectors and Tangent Plane

Let $s(t) : [a, b] \to S^n$ be a representation of a curve in S^n. Then the parametrization

$$\boldsymbol{r}[\boldsymbol{s}(t)] : [a, b] \to R^{n+l} \qquad (4.2)$$

represents a curve in the n-dimensional surface S^{rn} specified by (4.1). The tangent vector to this curve forms a tangent vector to the surface S^{rn}. The

4. Multidimensional Geometry

set of all tangent vectors at a point on the surface S^{rn} forms the tangent n-dimensional plane to the surface at this point.

Analogously to the definition of the coordinate line in space (Sect. 2.2) there is defined the s^ith coordinate line in the surface S^{rn} as a curve represented by the following vector-valued function dependent upon a variable φ in the capacity of s^i:

$$\boldsymbol{r}[\boldsymbol{s}^i(\varphi)] : [a,b] \to R^{n+l} ,$$
$$\boldsymbol{s}^i(\varphi) = (s_0^1, \ldots, s_0^{i-1}, \varphi, s_0^{i+1}, \ldots, s_0^n) , \quad i \text{ fixed} ,$$
(4.3)

here $\boldsymbol{r}(\boldsymbol{s})$ is the function from (4.1), $\boldsymbol{s}^i(\varphi) \in S^n$, $\varphi \in [a,b]$, the constants s_0^j, $j \neq i$ are fixed.

Each s^ith coordinate line defines one basic tangent vector along this curve

$$\boldsymbol{r}_{s^i} = \frac{\partial \boldsymbol{r}}{\partial s^i} , \quad i = 1, \ldots, n ,$$

assuming in (4.3) $\varphi = s^i$. The transformation $\boldsymbol{r}(\boldsymbol{s})$ is of rank n hence the basic tangent vectors $\boldsymbol{r}_{s^i}$, $i = 1, \ldots, n$, at a point P are independent and therefore form the tangent plane at this point (Fig. 4.1 for $n = 2$).

Similarly to the coordinate hypersurface in space (section 2.3) we define a coordinate hypersurface in S^{rn} as an $(n-1)$-dimensional surface lying in S^{rn} along which all of the coordinates $s^1, \ldots, s^n$ except one, say s^i, are varied. Thus the s^ith coordinate hypersurface is specified by the parametrization

$$\boldsymbol{r}[\boldsymbol{s}_i(s^1, \ldots, s^{i-1}, s^{i+1}, \ldots, s^n)] : S^{n-1} \to R^{n+l} ,$$
$$\boldsymbol{s}_i(s^1, \ldots, s^{i-1}, s^{i+1}, \ldots, s^n) = (s^1, \ldots, s^{i-1}, s_0^i, s^{i+1}, \ldots, s^n) ,$$
(4.4)

where s_0^i fixed, while $\boldsymbol{r}(\boldsymbol{s})$ is the function from (4.1). We personify this coordinate hypersurface with the equation $s^i = s_0^i$. Equation (4.4) readily yields that the basic tangent vectors to the hypersurface $s^i = s_0^i$ are the vectors $\boldsymbol{r}_{s^j}$, $j \neq i$.

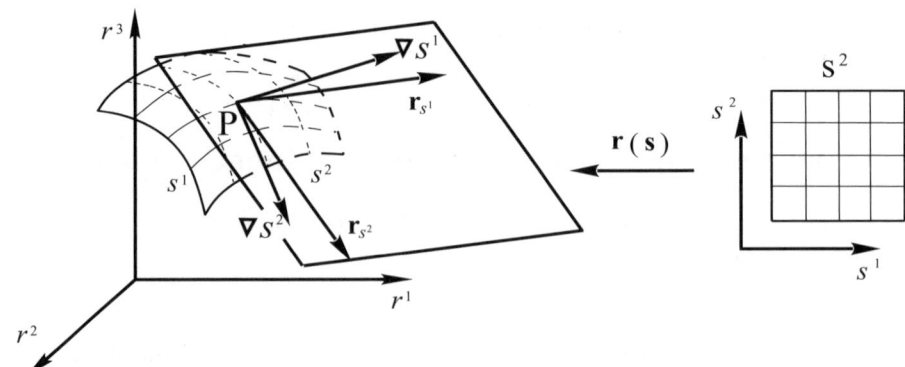

Fig. 4.1. Base tangent and normal vectors to coordinate lines

A vector lying in the tangent plane to S^{rn} and orthogonal to the coordinate hypersurface $s^i = s_0^i$ in S^{rn} (consequently to its basic tangent vectors r_{s^j}, $j \neq i$) is called a vector orthogonal or a normal vector to this hypersurface in S^{rn}. Let us introduce, analogously to the normal vectors in space, as a basic normal vector to the coordinate hypersurface $s^i = s_0^i$ in S^{rn} the vector designated by ∇s^i for which

$$\nabla s^i \cdot r_{s^j} = \delta_j^i, \quad i,j = 1,\ldots,n, \quad i - \text{ fixed}, \tag{4.5}$$

(Fig. 4.1).

The basic normal vectors to the coordinate hypersurface in S^{rn} are described inambiguously in the following section through the basic tangent vectors and elements of the metric tensors of S^{rn}.

4.2 First Groundform

All the properties of an n-dimensional surface S^{rn} which can be described without referring to the surrounding space are called intrinsic properties of the surface and their description constitutes the intrinsic geometry of the surface. Any characteristic of the intrinsic geometry is defined by the surface metric tensor whose elements are specified through the operation of the dot product on the basic tangent vectors.

4.2.1 Covariant Metric Tensor

Definition. The covariant metric tensor of any regular n-dimensional surface S^{rn} represented in the coordinates $s^1,\ldots,s^n$ by (4.1) is the matrix (g_{ij}^{rs}), $i,j = 1,\ldots,n$, where

$$g_{ij}^{rs} = r_{s^i} \cdot r_{s^j}, \quad i,j = 1,\ldots,n. \tag{4.6}$$

In particular, when the surface is a monitor surface i.e. it is identified with the graph of the values of some vector-valued function $u(s)$ over a domain S^n then this surface is parametrized by the equation $r(s) = [s, r(s)]$ and consequently

$$g_{ij}^{rs} = \delta_i^j + \frac{\partial u}{\partial s^i} \cdot \frac{\partial u}{\partial s^j}, \quad i,j = 1,\ldots,n.$$

Quadratic Form. The differential quadratic form

$$g_{ij}^{rs} \mathrm{d}s^i \mathrm{d}s^j, \quad i,j = 1,\ldots,n,$$

relating to the line elements in space, is called the first groundform or fundamental form of the surface. It represents the value of the square of the length of an elementary displacement $\mathrm{d}r$ (see Fig. 2.4) on the surface. Therefore the length of the curve (4.2) in the surface S^{rn} is computed by the formula

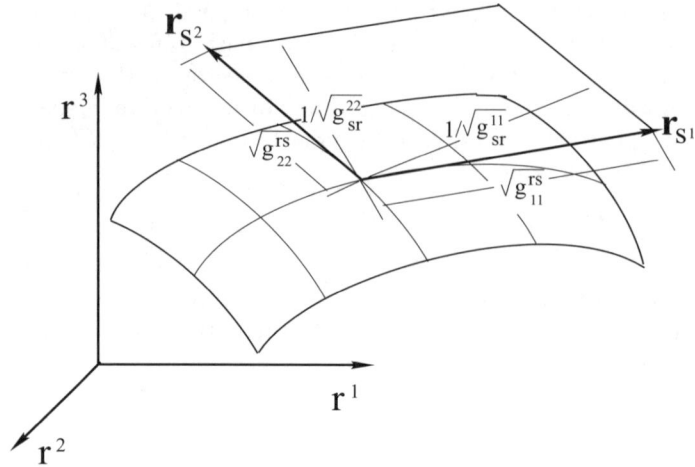

Fig. 4.2. Geometric meaning of the metric elements

$$l = \int_a^b \sqrt{g_{ij} \frac{ds^i}{dt} \frac{ds^j}{dt}} dt, \quad i,j = 1,\ldots,n,$$

which is similar to (2.16).

Let the Jacobian of (g_{ij}^{rs}) be designated by g^{rs}. Then, analogously to the length of the line (4.2), the n-dimensional area of the surface S^{rn} is computed from the formula

$$S = \int_{S^n} \sqrt{g^{rs}} d\mathbf{s}.$$

Basic Parallelepiped. The basic parallelepiped in S^{rn} with respect to the coordinates $s^1, \ldots, s^n$ is an n-dimensional parallelepiped whose edges are the basic tangent vectors $\mathbf{r}_{s^i}$, $i = 1, \ldots, n$. So the quantity $\sqrt{g_{ii}^{rs}}$ for a fixed index i has the geometrical meaning of the length of the ith edge of the basic parallelepiped (see Fig. 4.2 in the case $n = 2$). Note the uniformly contracted basic parallelepiped represents to a high order of accuracy the cell of the coordinate grid in S^{rn} in the case when the parametric domain S^n is a logical domain (see also Sect. 2.2).

4.2.2 Contravariant Metric Tensor

Definition. The contravariant metric tensor of the surface S^{rn} in the coordinates $s^1, \ldots, s^n$ is the matrix (g_{sr}^{ij}), $i,j = 1, \ldots, n$, inverse to (g_{ij}^{rs}), i.e. the folowing relations are valid

$$g_{ij}^{rs} g_{sr}^{jk} = \delta_k^i, \quad i,j,k = 1,\ldots,n. \tag{4.7}$$

Thus analogously to (2.21) we obtain, in the case $n = 2$,

$$g_{sr}^{ij} = (-1)^{i+j} g_{3-i\ 3-j}^{rs}/g^{rs},$$

$$g_{ij}^{rs} = (-1)^{i+j} g^{rs} g_{sr}^{3-i\ 3-j}, \quad i,j = 1,2, \tag{4.8}$$

with fixed indices i and j. Similarly to (2.22) we find, in the case $n = 3$,

$$g_{sr}^{ij} = \frac{1}{g^{rs}}(g_{i+1\ j+1}^{rs} g_{i+2\ j+2}^{rs} - g_{i+1\ j+2}^{rs} g_{i+2\ j+1}^{rs}),$$

$$g_{ij}^{rs} = g^{rs}(g_{sr}^{i+1\ j+1} g_{sr}^{i+2\ j+2} - g_{sr}^{i+1\ j+2} g_{sr}^{i+2\ j+1}), \tag{4.9}$$

$i,j = 1,2,3, \quad i,j$ fixed,

with the convention that any index, say l, is identified with $l \pm 3$.

Computation of Basic Normal Vectors. Using the elements of the contravariant metric tensor we can readily find the expression for the basic normal vector ∇s^i to the coordinate hypersurface $s^i = s_0^i$, satisfying (4.5), through the basic tangent vectors $\boldsymbol{r}_{s^j}$, $j = 1, \ldots, n$. Namely

$$\nabla s^i = g_{sr}^{ij} \boldsymbol{r}_{s^j}, \quad i,j = 1, \ldots, n. \tag{4.10}$$

Indeed, the condition (4.5) is observed since

$$\nabla s^i \cdot \boldsymbol{r}_{s^k} = g_{sr}^{ij} \boldsymbol{r}_{s^j} \cdot \boldsymbol{r}_{s^k} = g_{sr}^{ij} g_{jk}^{rs} = \delta_k^i, \quad i,j,k = 1, \ldots, n,$$

i.e. the vector ∇s^i is orthogonal to the coordinate hypersurface $s^i = const$ in S^{rn} at the point of consideration. As $\nabla s^i \cdot \boldsymbol{r}_{s^i} = 1$, i fixed, we conclude that the both basic vectors ∇s^i and $\boldsymbol{r}_{s^i}$ have the same direction with respect to the hypersurface $s^i = const$ in S^{rn}.

From (4.10) we readily find that

$$\nabla s^i \cdot \nabla s^j = g_{sr}^{ij}, \quad i,j = 1, \ldots, n. \tag{4.11}$$

Note also that for the length of ∇s^i we obtain the following expression

$$|\nabla s^i| = \sqrt{\nabla s^i \cdot \nabla s^i} = \sqrt{g_{sr}^{ii}}, \quad i = 1, \ldots, n, \quad i \text{ fixed}.$$

Since the normal vector ∇s^i is orthogonal to the vectors $\boldsymbol{r}_{s^j}$, $j \neq i$, we found that the distance d_i between the ith $(n-1)$-dimensional faces of the basic parallelepiped formed by the tangent vectors $\boldsymbol{r}_{s^j}$, $j = 1, \ldots, n$, is computed as follows:

$$d_i = \boldsymbol{r}_{s^i} \cdot \frac{\nabla s^i}{\|\nabla s^i\|} = \frac{1}{\sqrt{g_{sr}^{ii}}}, \quad i = 1, \ldots, n, \quad i \text{ fixed}.$$

Thus with respect to the basic parallelepiped the quantity $\sqrt{g_{ii}^{rs}}$ (i fixed) is the length of its ith edge, while the quantity $1/\sqrt{g_{sr}^{ii}}$ (i fixed) is the distance between the parallel $(n-1)$-dimensional faces of the parallelepiped, which are formed by the vectors $\boldsymbol{r}_{s^j}$, $j \neq i$ (see Fig. 4.2 for $n = 2$).

4. Multidimensional Geometry

Normal Vector to a Hypersurface. Formula (4.10) is readily extended to the case of the hypersurface in S^{rn} defined by an equation $\varphi(s) = 0$. Namely, a normal n to this hypersurface in S^{rn} is specified by the following formula

$$n = \varphi_{s^i} g^{ik}_{sr} r_{s^k}, \quad i, k = 1, \ldots, n, \tag{4.12}$$

that is a generalization of (4.10) obtained for the equation $\varphi(s) \equiv s^i - const$. The validity of (4.12) will be proved if we show that

$$n \cdot t = 0,$$

where t is an arbitrary tangent vector to the hypersurface $\varphi(s) = 0$. Without loss of generality we can assume that $\varphi_{s^n} \neq 0$ at a point $s \in S^n$ under consideration. Then the equation $\varphi(s) = 0$ is resolved with respect to s^n, i.e. there exists a function $s^n(s^1, \ldots, s^{n-1})$, $(s^1, \ldots, s^{n-1}) \in S^{n-1}$ such that

$$\varphi[s^1, \ldots, s^{n-1}, s^n(s^1, \ldots, s^{n-1})] \equiv 0, \quad (s^1, \ldots, s^{n-1}) \in S^{n-1}.$$

Therefore, using (4.1), we can locally specify the hypersurface $\varphi(s) = 0$ in the coordinates $s^1, \ldots, s^{n-1}$ by the following parametrization

$$r^\varphi(s^1, \ldots, s^{n-1}) = r[s^1, \ldots, s^{n-1}, s^n(s^1, \ldots, s^{n-1})] : S^{n-1} \to R^{n+l}.$$

Consequently the basic tangent vectors to this hypersurface with respect to the coordinates $s^1, \ldots, s^{n-1}$ are computed by

$$r^\varphi_{s^i} = r_{s^i} + \frac{\partial s^n}{\partial s^i} r_{s^n} = r_{s^i} - \frac{\varphi_{s^i}}{\varphi_{s^n}} r_{s^n}, \quad i = 1, \ldots, n-1,$$

since $\partial s^n / \partial s^i = -\varphi_{s^i} / \varphi_{s^n}$, $i = 1, \ldots, n-1$. Using these relations and (4.12) gives

$$\begin{aligned}
n \cdot r^\varphi_{s^i} &= \varphi_{s^m} g^{mk}_{sr} r_{s^k} \left(r_{s^i} - \frac{\varphi_{s^i}}{\varphi_{s^n}} r_{s^n} \right) \\
&= \varphi_{s^m} g^{mk}_{sr} \left(g^{rs}_{ki} - \frac{\varphi_{s^i}}{\varphi_{s^n}} g^{rs}_{kn} \right) \\
&= \varphi_{s^i} - \frac{\varphi_{s^i} \varphi_{s^n}}{\varphi_{s^n}} = 0, \quad i = 1, \ldots, n-1.
\end{aligned}$$

As an arbitrary tangent vector t to the hypersurface is expanded by $r^\varphi_{s^i}$, $i = 1, \ldots, n-1$, we obtain that $n \cdot t = 0$, i.e. formula (4.12) gives a real expression for the vector n normal to the hypersurface $\varphi(s) = 0$.

We have, by virtue of (4.12), the following formula for the length of n

$$\begin{aligned}
|n| &= \sqrt{n \cdot n} = \sqrt{(\varphi_{s^i} g^{ik}_{sr} r_{s^k}) \cdot (\varphi_{s^m} g^{ml}_{sr} r_{s^l})} \\
&= \sqrt{\varphi_{s^i} \varphi_{s^m} g^{ik}_{sr} g^{ml}_{sr} g^{rs}_{kl}} \\
&= \sqrt{\varphi_{s^m} \varphi_{s^l} g^{ml}_{sr}}, \quad i, k, l, m = 1, \ldots, n.
\end{aligned} \tag{4.13}$$

4.3 Generalization to Riemannian Manifolds

4.3.1 Definition of the Manifolds

Formula (4.6) readily yields the result that the elements of the covariant metric tensor (g_{ij}^{rs}) and (g_{ij}^{rv}) of the regular surface S^{rn} in arbitrary coordinates $s^1, \ldots, s^n$ and $v^1, \ldots, v^n$, respectively, are connected by the following relations

$$g_{ij}^{rs} = g_{kl}^{rv} \frac{\partial v^k}{\partial s^i} \frac{\partial v^l}{\partial s^j}, \quad i,j,k,l = 1, \ldots, n. \tag{4.14}$$

Indeed,

$$g_{ij}^{rs} = \boldsymbol{r}_{s^i} \cdot \boldsymbol{r}_{s^j} = \boldsymbol{r}_{v^k} \frac{\partial v^k}{\partial s^i} \cdot \boldsymbol{r}_{v^m} \frac{\partial v^m}{\partial s^j}$$

$$= \boldsymbol{r}_{v^k} \cdot \boldsymbol{r}_{v^m} \frac{\partial v^k}{\partial s^i} \frac{\partial v^m}{\partial s^j} = g_{km}^{rv} \frac{\partial v^k}{\partial s^i} \frac{\partial v^m}{\partial s^j}, \quad i,j,k,m = 1, \ldots, n,$$

i.e. equations (4.14) are valid.

Analogously, the elements of the contravariant metric tensor (g_{sr}^{ij}) and (g_{vr}^{ij}) of the regular surface S^{rn} in the coordinates $s^1, \ldots, s^n$ and $v^1, \ldots, v^n$, respectively, are connected by

$$g_{vr}^{ij} = g_{sr}^{kl} \frac{\partial v^i}{\partial s^k} \frac{\partial v^j}{\partial s^l}, \quad i,j,k,l = 1, \ldots, n. \tag{4.15}$$

For showing that the components g_{vr}^{ij} are subject to (4.15), it is sufficient to demonstrate that the matrix

$$\left(g_{sr}^{kl} \frac{\partial v^i}{\partial s^k} \frac{\partial v^j}{\partial s^l} \right), \quad i,j,k,l = 1, \ldots, n,$$

coincides with (g_{vr}^{ij}), i.e. it is the inverse of (g_{ij}^{rv}) provided the matrix (g_{sr}^{ij}) is the inverse of (g_{ij}^{rs}). Since (4.14)

$$g_{ij}^{rv} g_{sr}^{kl} \frac{\partial v^j}{\partial s^k} \frac{\partial v^t}{\partial s^l} = g_{mp}^{rs} \frac{\partial s^m}{\partial v^i} \frac{\partial s^p}{\partial v^j} \frac{\partial v^j}{\partial s^k} g_{sr}^{kl} \frac{\partial v^t}{\partial s^l} = \delta_i^t,$$

$$i,j,k,l,m,p,t = 1, \ldots, n,$$

i.e. equations (4.15) are valid.

The relations (4.14) and (4.15) valid for the metrics of arbitrary regular n-dimensional surfaces give rise to the definition of the geometrical objects called Riemannian manifolds which generalize the regular surfaces.

Namely, a point set M^n of R^{n+l}, $l \geq 1$ is called a C^m-differential Riemannian manifold (Fig. 4.3) of dimension n if there is a collection (atlas) Φ of local parametrizations of M (local maps)

$$\begin{aligned} &\boldsymbol{r}_\varphi(\boldsymbol{\varphi}) : S_\varphi^n \to M, \quad \varphi \in \Phi, \quad \boldsymbol{\varphi} \in S_\varphi^n, \\ &\boldsymbol{\varphi} = (\varphi^1, \ldots, \varphi^n), \quad \boldsymbol{r}_\varphi = (r_\varphi^1, \ldots, r_\varphi^{n+l}), \end{aligned} \tag{4.16}$$

66   4. Multidimensional Geometry

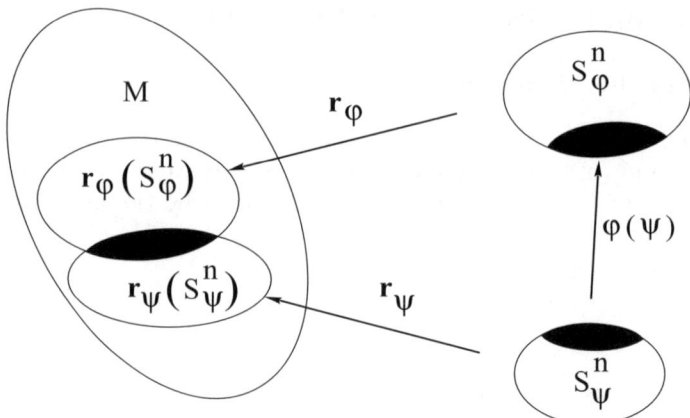

Fig. 4.3. Illustration for Riemannian manifold maps

where $S_\varphi^n \subset R^n$ is an n-dimensional parametric domain, and the same collection of local variable matrices (local covariant metric tensors)

$$(g_{ij}^{r\varphi}(\boldsymbol{\varphi})), \quad i,j = 1,\ldots,n, \quad \boldsymbol{\varphi} \in S_\varphi^n, \quad \varphi \in \Phi, \tag{4.17}$$

such that

1) $\cup_{\varphi \in \Phi} \boldsymbol{r}_\varphi(S_\varphi^n) = M^n$, where $\boldsymbol{r}_\varphi(S_\varphi^n)$ is the image of S_φ^n in M^n built by $\boldsymbol{r}_\varphi$,
2) for each $\varphi \in \Phi$, $\boldsymbol{r}_\varphi(S_\varphi^n)$ is an open space of M^n and

$$\boldsymbol{r}_\varphi : S_\varphi^n \to \boldsymbol{r}_\varphi(S_\varphi^n), \quad \varphi \in \Phi,$$

is a one-to-one continious mapping,
3) for each $\varphi, \psi \in \Phi$ with $B_{\varphi\psi} \neq \emptyset$, where

$$B_{\varphi\psi} = \boldsymbol{r}_\varphi(S_\varphi^n) \cap \boldsymbol{r}_\psi(S_\psi^n),$$

the map

$$\boldsymbol{\varphi}(\boldsymbol{\psi}) : S_\psi^n \cap \boldsymbol{r}_\psi^{-1}(B_{\varphi\psi}) \to R^n,$$

where $\boldsymbol{\varphi}(\boldsymbol{\psi}) = \boldsymbol{r}_\varphi^{-1}[\boldsymbol{r}_\psi(\boldsymbol{\psi})]$, is a C^m – map,
4) for each $\varphi \in \Phi$, $\boldsymbol{\varphi} \in S_\varphi^n$ the matrix $(g_{ij}^{r\varphi}(\boldsymbol{\varphi}))$ is positive, symmetric, and nondegenerate and the functions

$$g_{ij}^{r\varphi}(\boldsymbol{\varphi}) : S_\varphi^n \to R, \quad i,j = 1,\ldots,n,$$

are C^m-functions,
5) for each $\varphi, \psi \in \Phi$, $B_{\varphi\psi} \neq \emptyset$,

$$g_{ij}^{r\psi}(\boldsymbol{\psi}) = g_{kl}^{r\varphi}[\boldsymbol{\varphi}(\boldsymbol{\psi})]\frac{\partial \varphi^k}{\partial \psi^i}(\boldsymbol{\psi})\frac{\partial \varphi^l}{\partial \psi^j}(\boldsymbol{\psi}), \quad i,j,k,l = 1,\ldots,n, \tag{4.18}$$

where

$$\boldsymbol{\varphi} \in S_\varphi^n \cap \boldsymbol{r}_\varphi^{-1}(B_{\varphi\psi}), \quad \boldsymbol{\psi} \in S_\psi^n \cap \boldsymbol{r}_\psi^{-1}(B_{\varphi\psi}),$$

$\varphi^i(\boldsymbol{\psi})$, $i = 1, \ldots, n$, is the ith component of the function $\boldsymbol{\varphi}(\boldsymbol{\psi}) = \boldsymbol{r}_\varphi^{-1}(\boldsymbol{r}_\psi)$,
6) Φ is maximal relative to (2-5) that if there is one more local parametrization

$$\boldsymbol{r}_\theta(\boldsymbol{\theta}) : S_\theta^n \to X, \quad S_\theta^n \subset R^n, \quad \boldsymbol{\theta} \in S_\theta^n,$$

and a nondegenerate, positive, symmetric C^m – matrix function

$$(g_{ij}^{r\theta}(\boldsymbol{\theta})), \quad \boldsymbol{\theta} \in S_\theta^n,$$

and their inclusion into local parametrizations and local matrices, respectively, does not violate the requirements (2-5) – they are in the corresponding collections of Φ.

Here m may be $1, 2, \ldots, \infty$. C^m for m finite means all partial derivatives of order less than or equal to m exist and are continuous.

The variables $\varphi^1, \ldots, \varphi^n$ of the parametric domain S_φ^n, i.e.

$$\boldsymbol{\varphi} = (\varphi^1, \ldots, \varphi^n) \in S_\varphi^n, \quad \boldsymbol{\varphi} \in \Phi,$$

together with the corresponding transformation

$$\boldsymbol{r}_\varphi(\boldsymbol{\varphi}) : S_\varphi^n \to M^n$$

are called a local coordinate system or local coordinates of M^n.

Each local covariant metric tensor $(g_{ij}^{r\varphi}(\boldsymbol{\varphi}))$ derives the local contravariant metric tensor $(g_{\varphi r}^{ij}(\boldsymbol{\varphi}))$ as the inverse matrix of $(g_{ij}^{r\varphi}(\boldsymbol{\varphi}))$, i.e.

$$g_{\varphi r}^{ij}(\boldsymbol{\varphi}) = G_{ij}^\varphi / \det(g_{ij}^{r\varphi}(\boldsymbol{\varphi})), \quad i, j = 1, \ldots, n,$$

where G_{ij}^φ is the (ij)th cofactor of $(g_{ij}^{r\varphi}(\boldsymbol{\varphi}))$. Analogously to (4.15) it is readily verified that the equations (4.18) yield

$$g_{\psi r}^{ij}(\boldsymbol{\psi}) = g_{\varphi r}^{kl}[\boldsymbol{\varphi}(\boldsymbol{\psi})] \frac{\partial \psi^i}{\partial \varphi^k} \frac{\partial \psi^j}{\partial \varphi^l}, \quad i, j, k, l = 1, \ldots, n, \qquad (4.19)$$

where $\psi^k(\boldsymbol{\varphi})$, $k = 1, \ldots, n$, is the kth component of the function $\boldsymbol{\psi}(\boldsymbol{\varphi}) = \boldsymbol{r}_\psi^{-1}[\boldsymbol{r}_\varphi(\boldsymbol{\varphi})]$.

Note in some specific theories M^n is often assumed as a Hausdorff topological space, but for our purpose we consider M^n as a point set of R^{n+l} with a topology imposed by the topology of R^{n+l} (however the metric may not be taken from R^{n+l}).

4.3.2 Example of a Riemannian Manifold

In the grid generation theory the Riemannian manifolds appear as tools to control the grids derived by the generalized Laplace equations. Typically these manifolds are formulated as a generalization of monitor surfaces. Namely, let S^{xn} be a regular n-dimensional surface lying in R^{n+k} and represented locally by a mapping

$$\boldsymbol{x}(\boldsymbol{s}) : S^n \to R^{n+k} \,, \quad \boldsymbol{s} = (s^1, \ldots, s^n) \,, \quad \boldsymbol{x} = (x^1, \ldots, x^{n+k}) \,,$$

with a covariant metric tensor (g_{ij}^{xs}) in the coordinates $s^1, \ldots, s^n$

$$g_{ij}^{xs} = \boldsymbol{x}_{s^i} \cdot \boldsymbol{x}_{s^j} \,, \quad i,j = 1, \ldots, n \,. \tag{4.20}$$

By a monitor surface over S^{xn} there is meant a regular surface whose points are

$$[\boldsymbol{x}, \boldsymbol{f}(\boldsymbol{x})] \in R^{n+l+k} \,,$$

where $\boldsymbol{x} \in S^{xn}$, while

$$\boldsymbol{f}(\boldsymbol{x}) : S^{xn} \to R^l \,, \quad \boldsymbol{f} = (f^1, \ldots, f^l) \,,$$

is some vector-valued function called a monitor function. The monitor surface is represented in the coordinates $s^1, \ldots, s^n$ by the following local parametrization

$$\boldsymbol{r}(\boldsymbol{s}) : S^n \to R^{n+k+l} \,, \quad \boldsymbol{r}(\boldsymbol{s}) = \{\boldsymbol{x}(\boldsymbol{s}), \boldsymbol{f}[\boldsymbol{x}(\boldsymbol{s})]\} \,.$$

An extension of the monitor surface over S^{xn} is produced, for example, by two scalar-valued weight functions $z(\boldsymbol{s}) > 0$ and $v(\boldsymbol{s}) \geq 0$ and one vector-valued smooth function $\boldsymbol{f}(\boldsymbol{s}) = [f^1(\boldsymbol{s}), \ldots, f^l(\boldsymbol{s})]$ which form a Riemannian manifold M^n whose points and parametrizations are the same as of S^{xn}, while the elements of the covariant metric tensor in the coordinates $s^1, \ldots, s^n$, designated as g_{ij}^s, are defined as follows:

$$g_{ij}^s = z(\boldsymbol{s})g_{ij}^{xs} + v(\boldsymbol{s})\frac{\partial \boldsymbol{f}}{\partial s_i}(\boldsymbol{s}) \cdot \frac{\partial \boldsymbol{f}}{\partial s_j}(\boldsymbol{s}) \,, \quad i,j = 1, \ldots, n \,. \tag{4.21}$$

We shall call the Riemannian manifold with the metric (4.21) imposed by the functions $z(\boldsymbol{s})$, $v(\boldsymbol{s})$, and $\boldsymbol{f}(\boldsymbol{s})$ a monitor manifold. The function $\boldsymbol{f}(\boldsymbol{s})$ will be referred to as a monitor function, while $z(\boldsymbol{s})$ and $v(\boldsymbol{s})$ will be called weight functions.

4.3.3 Christoffel Symbols of Manifolds

Definition of the Symbols. Any covariant metric tensor of a Riemannian manifold M^n, designated by (g_{ij}^s) in the local coordinates s^i, $i = 1, \ldots, n$, derives the quantities called the Christoffel symbols of the first and second kinds, which are also referred to as the three-index symbols. The symbols of the first kind, designated in the coordinates s^i, $i = 1, \ldots, n$, by $[ij, k]^s$, are defined as follows:

4.3 Generalization to Riemannian Manifolds

$$[ij,k]^s = \frac{1}{2}\left(\frac{\partial g^s_{jk}}{\partial s^i} + \frac{\partial g^s_{ik}}{\partial s^j} - \frac{\partial g^s_{ij}}{\partial s^k}\right), \quad i,j,k = 1,\ldots,n. \tag{4.22}$$

Equations (4.22) easily yield the following relations for the first derivatives of the elements of the covariant metric tensor

$$\frac{\partial g^s_{ik}}{\partial s^j} = [ij,k]^s + [kj,i]^s, \quad i,j,k = 1,\ldots,n. \tag{4.23}$$

The Christoffel symbols of the second kind, designated in the coordinates s^i, $i = 1,\ldots,n$, by ${}^s\Upsilon^l_{ij}$, are defined by the equations

$${}^s\Upsilon^l_{ij} = g^{lm}_s [ij,m]^s, \quad i,j,l,m = 1,\ldots,n. \tag{4.24}$$

It is seen at once from (4.22) and (4.24) that the Christoffel symbols are symmetrical in i,j.

Further, when a coordinate system in a formula is fixed we, for simplicity, shall omit the superscript personifying a coordinate system in the Christoffel symbols of the first and second kinds thus designating them merely by $[ij,k]$ and Υ^l_{ij}, respectively.

Analogously to (4.23) we find an expression for the first derivatives of the elements of the contravariant metric tensor through the Christoffel symbols of the second kind

$$\begin{aligned}\frac{\partial g^{ij}_s}{\partial s^k} &= g^{lj}_s g^s_{ml} \frac{\partial g^{im}_s}{\partial s^k} = -g^{lj}_s g^{im}_s \frac{\partial g^s_{ml}}{\partial s^k}\\ &= -g^{lj}_s g^{im}_s ([mk,l] + [lk,m]) = -g^{im}_s \Upsilon^j_{km} - g^{lj}_s \Upsilon^i_{lk},\end{aligned} \tag{4.25}$$

$$i,j,k,l,m = 1,\ldots,n.$$

The Christoffel symbols of the second kind also have some relation to the formula of the differentiation of the Jacobian g^s of the metric tensor (g^s_{ij}). Indeed the rule of the differentiation of the Jacobian g^s gives

$$\frac{\partial g^s}{\partial s^i} = g^s g^{jm} \frac{\partial g^s_{jm}}{\partial s^i}, \quad i,j,m = 1,\ldots,n,$$

and the application of (4.23) to this formula yields

$$\frac{\partial g^s}{\partial s^i} = g^s g^{jm}_s ([ji,m] + [mi,j]) = g^s [\Upsilon^j_{ji} + \Upsilon^m_{mi}] = 2g^s \Upsilon^j_{ji},$$

$$i,j,m = 1,\ldots,n.$$

Remind repeated indices in a single term mean a summation over them so we imply in the above equations

$$\frac{\partial g^s}{\partial s_i} = 2g^s \Upsilon^j_{ji} = 2g^s \sum_{j=1}^n \Upsilon^j_{ji}, \quad i,j = 1,\ldots,n. \tag{4.26}$$

4. Multidimensional Geometry

Symbols for Regular Surfaces. In the case of the regular surface S^{rn} represented by (4.1) whose intrinsic metric is expressed as

$$g_{ij}^s = \mathbf{r}_{s^i} \cdot \mathbf{r}_{s^j}, \quad i,j = 1,\ldots,n,$$

it is readily found, from (4.22), that

$$[ij,k] = \mathbf{r}_{s^i s^j} \cdot \mathbf{r}_{s^k}, \quad i,j,k = 1,\ldots,n. \tag{4.27}$$

Using (4.27), (4.24), and (4.10) yields the following formula for the Christoffel symbols of the second kind of the regular surface S^{rn} in the coordinates $s^1,\ldots,s^n$:

$$\Upsilon_{ij}^l = g_s^{lm}(\mathbf{r}_{s^i s^j} \cdot \mathbf{r}_{s^m}) = \mathbf{r}_{s^i s^j} \cdot \boldsymbol{\nabla} s^l, \quad i,j,l,m = 1,\ldots,n, \tag{4.28}$$

where $\boldsymbol{\nabla} s^l$ is the basic normal vector to the coordinate hypersurface $s^l = c_0$ in S^{rn} (see (4.10)).

Transformation of the Christoffel Symbols. Now we shall establish how the Christoffel symbols of two coordinate systems are related. Let us designate by (g_{ij}^s) and (g_{ij}^v) the elements of the covariant metric tensor of a manifold M^n in the coordinates $s^1,\ldots,s^n$ and $v^1,\ldots,v^n$, respectively. Then, according to (4.18), we have

$$g_{ij}^v = g_{kl}^s \frac{\partial s^k}{\partial v^i} \frac{\partial s^l}{\partial v^j}, \quad i,j,k,l = 1,\ldots,n.$$

Differentiating these equations with respect to v^p gives

$$\frac{\partial g_{ij}^v}{\partial v^p} = \frac{\partial g_{kl}^s}{\partial s^m} \frac{\partial s^m}{\partial v^p} \frac{\partial s^k}{\partial v^i} \frac{\partial s^l}{\partial v^j} + 2 g_{kl}^s \frac{\partial^2 s^k}{\partial v^i \partial v^p} \frac{\partial s^l}{\partial v^j}, \quad i,j,k,l,m,p = 1,\ldots,n,$$

and consequently, applying (4.23),

$$\frac{\partial g_{ij}^v}{\partial v^p} = [ip,j]^v + [jp,i]^v$$

$$= ([km,l]^s + [lm,k]^s) \frac{\partial s^m}{\partial v^p} \frac{\partial s^k}{\partial v^i} \frac{\partial s^l}{\partial v^j} + 2 g_{kl}^s \frac{\partial^2 s^k}{\partial v^i \partial v^p} \frac{\partial s^l}{\partial v^j},$$

$$i,j,k,l,m,p = 1,\ldots,n.$$

Therefore, after computing by these formulas the following expression

$$-\frac{\partial g_{ij}^v}{\partial v^p} + \frac{\partial g_{pi}^v}{\partial v^j} + \frac{\partial g_{pj}^v}{\partial v^i}, \quad i,j,p = 1,\ldots,n,$$

we readily obtain using first (4.22) and then (4.24)

$$[ij,p]^v = [km,l]^s \frac{\partial s^l}{\partial v^p} \frac{\partial s^k}{\partial v^i} \frac{\partial s^m}{\partial v^j} + g_{kl}^s \frac{\partial^2 s^k}{\partial v^i \partial v^j} \frac{\partial s^l}{\partial v^p},$$

$$^v\Upsilon_{ij}^p = {}^s\Upsilon_{kl}^m \frac{\partial s^k}{\partial v^i} \frac{\partial s^l}{\partial v^j} \frac{\partial v^p}{\partial s^m} + \frac{\partial^2 s^k}{\partial v^i \partial v^j} \frac{\partial v^p}{\partial s^k}, \tag{4.29}$$

$$i,j,k,l,m,p = 1,\ldots,n.$$

Geometric Meanings of the Symbols. Let us consider the Christoffel symbols of a regular n-dimensional surface S^{rn} represented by (4.1) and whose covariant metric tensor is defined by (4.6). We know that the first partial derivatives of the parametrization $\boldsymbol{r}(\boldsymbol{s})$ are the tangent vectors forming the tangent plane to the surface and the elements of the covariant metric tensor (g_{ij}^{rs}). It appears that the second partial derivatives of $\boldsymbol{r}(\boldsymbol{s})$ are connected with the Christoffel symbols.

Let us designate by $\boldsymbol{P}$ the operator which projects the vectors from R^{n+l} on the tangent plane to the regular surface $S^{rn} \subset R^{n+l}$ at a point P. Now considering the vector-valued function

$$\boldsymbol{r}_{s^m s^p} = \frac{\partial^2 \boldsymbol{r}}{\partial s^m \partial s^p}, \quad m, p = 1, \ldots, n,$$

we can expand the vector $\boldsymbol{P}[\boldsymbol{r}_{s^m s^p}]$ (lying in the tangent n-dimensional plane) in both the base tangential $\boldsymbol{r}_{s^i}$, $i = 1, \ldots, n$, and normal $\boldsymbol{\nabla} s^i$, $i = 1, \ldots, n$, vectors. Applying the formula (2.6) in the case of the tangential vectors, i.e. assuming in (2.6) $\boldsymbol{a}_i = \boldsymbol{r}_{s^i}$, $i = 1, \ldots, n$, we find

$$\boldsymbol{P}[\boldsymbol{r}_{s^m s^p}] = a^{ij} (\boldsymbol{P}[\boldsymbol{r}_{s^m s^p}] \cdot \boldsymbol{r}_{s^j}) \boldsymbol{r}_{s^i}, \quad i, j, m, p = 1, \ldots, n, \quad (4.30)$$

where a^{ij} are the elements of the matrix which is inverse to the matrix (a_{ij})

$$a_{ij} = \boldsymbol{r}_{s^i} \cdot \boldsymbol{r}_{s^j}, \quad i, j = 1, \ldots, n.$$

Since $\boldsymbol{r}_{s^i} \cdot \boldsymbol{r}_{s^j} = g_{ij}^{rs}$, $i, j = 1, \ldots, n$, the functions a^{ij} in (4.30) are the elements of the contravariant metric tensor (g_{sr}^{ij}), i.e.

$$a^{ij} = g_{sr}^{ij}, \quad i, j = 1, \ldots, n.$$

Further, as the operator $\boldsymbol{P}$ projects the vector $\boldsymbol{r}_{s^m s^p}$ on the plane formed by the tangent vectors $\boldsymbol{r}_{s^i}$, $i = 1, \ldots, n$, we conclude that

$$\boldsymbol{P}[\boldsymbol{r}_{s^m s^p}] \cdot \boldsymbol{r}_{s^j} = \boldsymbol{r}_{s^m s^p} \cdot \boldsymbol{r}_{s^j}, \quad j, m, p = 1, \ldots, n.$$

Therefore equation (4.30) has the following form

$$\boldsymbol{P}[\boldsymbol{r}_{s^m s^p}] = g_{sr}^{ij} (\boldsymbol{r}_{s^m s^p} \cdot \boldsymbol{r}_{s^j}) \boldsymbol{r}_{s^i}, \quad i, j, m, p = 1, \ldots, n. \quad (4.31)$$

Now, applying (4.28) to this equation, we find

$$\boldsymbol{P}[\boldsymbol{r}_{s^m s^p}] = \Upsilon_{mp}^i \boldsymbol{r}_{s^i}, \quad i, m, p = 1, \ldots, n. \quad (4.32)$$

Thus the Christoffel symbol Υ_{mp}^i of the second kind is the ith component of the vector $\boldsymbol{P}[\boldsymbol{r}_{s^m s^p}]$ expanded in the base tangent vectors $\boldsymbol{r}_{s^i}$ (see Fig. 4.4 where the vector $\boldsymbol{r}_{s^m s^p}$ is identified with $\boldsymbol{r}_{mp}$).

Similarly we obtain

$$\boldsymbol{P}[\boldsymbol{r}_{s^m s^p}] = [mp, i] \boldsymbol{\nabla} s^i, \quad i, m, p = 1, \ldots, n. \quad (4.33)$$

This formula can also be inferred from (4.32). Indeed multiplying (4.10) by g_{ik}^{rs} gives

4. Multidimensional Geometry

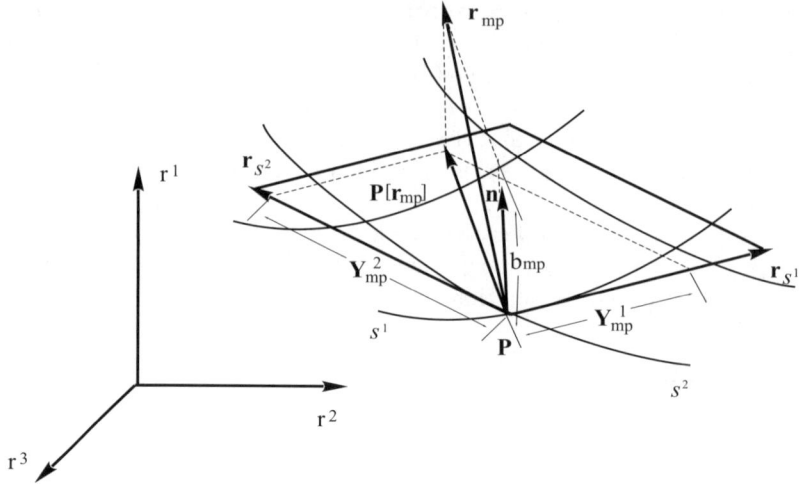

Fig. 4.4. Expension of the vector $r_{s^m s^p}$ in the base vectors

$$r_{s^k} = g_{ik}^{rs}\nabla s^i, \quad i, k = 1, \ldots, n,$$

and substituting this equation for r_{s^i} in (4.32), we readily come to (4.33).

So the Christoffel symbols of the first kind represent the components of the vector $P[r_{s^m s^p}]$ expanded in the base normal vectors to the coordinate hypersurfaces in S^{rn} (see Fig. 4.4 for $n = 2$).

4.4 Tensors

The theory of multidimensional geometry operates largely with the quantities called tensors. This section gives an introduction to such geometric objects.

4.4.1 Definition

Let M^n be some n-dimensional Riemannian manifold, in particular, a regular surface S^{rn}. A tensor of rank $k \geq 0$ at a point P of M^n is a set of values defined for each local coordinate system $s^1, \ldots, s^n$ around this point and indices $\boldsymbol{i} = (i_1, \ldots, i_k) \in R^k$, $i_j = 1, \ldots, n$, $j = 1, \ldots, k$, such that the values obey certain transformation laws when the coordinates are changed. The number k is called a tensor order or its rank, while the values with the indices are referred to as tensor components.

The tensors are also distinguished by their types. There are two basic tapes for the tensors of order $k > 0$: covariant and contravariant and the third mixed type having the features of the basic types.

Tensors of Order Zero. A quantity which has the same fixed value at the point P in an arbitrary coordinate system is called a scalar, or an invariant, or a tensor of order zero.

Covariant Tensors. In the pure covariant case the components of a tensor f of order k, whose designation is distinguished by indices being subscripts, for example, by $f^s_{i_1...i_k}$ for the indices $i_1,...,i_k$ and coordinates $s^1,...,s^n$, are subject to the following relations with respect to arbitrary coordinate systems $s^1,...,s^n$ and $v^1,...,v^n$:

$$f^s_{i_1...i_k} = f^v_{j_1...j_k} \frac{\partial v^{j_1}}{\partial s^{i_1}} \cdots \frac{\partial v^{j_k}}{\partial s^{i_k}}, \quad i_l, j_l = 1,...,n, \quad l = 1,...,k. \quad (4.34)$$

Thus it is sufficient to know the values of the tensor components for some one fixed coordinate system since its values for other systems can be computed by (4.34).

Contravariant Tensors. The components of the pure contravariant tensor f of rank $k > 0$, designated with the help of indices being superscripts: by $f^{i_1...i_k}_s$ in the coordinates $s^1,...,s^n$, obey the following law with respect to arbitrary coordinate systems $s^1,...,s^n$ and $v^1,...,v^n$:

$$f^{i_1...i_k}_s = f^{j_1...j_k}_v \frac{\partial s^{i_1}}{\partial v^{j_1}} \cdots \frac{\partial s^{i_k}}{\partial v^{j_k}}, \quad i_l, j_l = 1,...,n, \quad l = 1,...,k. \quad (4.35)$$

Mixed Tensors. The notion of the pure covariant and contravariant tensors gives rise to the concept of a mixed tensor if it is covariant in some indices and contravariant in the rest of them. Consequently there are used in designations superscripts for contravariant indices and subscripts for covariant indices. Namely, a mixed tensor f k times covariant and l times contravariant is a set of values $f^{j_1...j_l}_{i_1...i_k}(s)$ dependent on the coordinate system $s^1,...,s^n$ at the point of consideration such that for an arbitrary another coordinate system $v^1,...,v^n$

$$f^{j_1...j_l}_{i_1...i_k}(s) = f^{m_1...m_l}_{p_1...p_k}(v) \frac{\partial v^{p_1}}{\partial s^{i_1}} \cdots \frac{\partial v^{p_k}}{\partial s^{i_k}} \frac{\partial s^{j_1}}{\partial v^{m_1}} \cdots \frac{\partial s^{j_l}}{\partial v^{m_l}}, \quad (4.36)$$

$$i_a, p_a, j_b, m_b = 1,...,n, \quad a = 1,...,k, \quad b = 1,...,l.$$

4.4.2 Examples of Tensors

Covariant Tensors. A typical covariant tensor of the first order is the vector $\mathrm{grad}\varphi$ where φ is a tensor of order zero. The components of this vector designated as $(\mathrm{grad}\varphi)^s_i$ in the coordinates $s^1,...,s^n$ are computed as

$$(\mathrm{grad}\varphi)^s_i = \frac{\partial \varphi(s)}{\partial s^i} = \varphi_{s^i}, \quad i = 1,...,n. \quad (4.37)$$

It is obvious that the relations (4.34), for $k = 1$, are valid for these values.

74   4. Multidimensional Geometry

Note the second derivatives of φ, i.e. the set
$$\varphi^s_{ij} = \varphi_{s^i s^j}(s), \quad i,j = 1,\ldots,n,$$
do not form a tensor of the second rank since
$$\varphi_{s^i s^j}(s) = \varphi_{v^l v^m} \frac{\partial v^l}{\partial s^i} \frac{\partial v^m}{\partial s^j} + \varphi_{v^l} \frac{\partial^2 v^l}{\partial s^i \partial s^j}, \quad i,j,l,m = 1,\ldots,n, \quad (4.38)$$
and the second term in the right-hand part of these equations empedes satisfaction of (4.34) for $k = 2$. However using the Christoffel symbols of the second kind which are not tensors as well, since (4.29), produces the following covariant tensor $\nabla_{ij}(\varphi)$ of the second kind whose (ij)th component designated in the coordinates $s^1,\ldots,s^n$ as $\nabla^s_{ij}(\varphi)$ is computed as follows:
$$\nabla^s_{ij}(\varphi) = \varphi_{s^i s^j} - \varphi_{s^k} \Upsilon^k_{ij}, \quad i,j,k = 1,\ldots,n. \quad (4.39)$$
Formulas (4.29) and (4.38) readily yield that these components satisfy the condition (4.34) for $k = 2$. Indeed
$$\nabla^v_{ij}(\varphi) = \varphi_{v^i v^j} - \varphi_{v^k}{}^v\Upsilon^k_{ij}$$
$$= \varphi_{s^k s^m} \frac{\partial s^k}{\partial v^i} \frac{\partial s^m}{\partial v^j} + \varphi_{s^k} \frac{\partial^2 s^k}{\partial v^i \partial v^j}$$
$$- \varphi_{s^t} \left({}^s\Upsilon^m_{pl} \frac{\partial s^p}{\partial v^i} \frac{\partial s^l}{\partial v^j} \frac{\partial v^k}{\partial s^m} + \frac{\partial^2 s^p}{\partial v^i \partial v^j} \frac{\partial v^k}{\partial s^p}\right) \frac{\partial s^t}{\partial v^k} \quad (4.40)$$
$$= (\varphi_{s^k s^m} - \varphi_{s^p}{}^s\Upsilon^p_{km}) \frac{\partial s^k}{\partial v^i} \frac{\partial s^m}{\partial v^j} = \nabla^s_{km}(\varphi) \frac{\partial s^k}{\partial v^i} \frac{\partial s^m}{\partial v^j},$$
$$i,j,k,l,m,p,t = 1,\ldots,n,$$
i.e. the quantities $\nabla^s_{km}(\varphi)$ form a covariant tensor of the second kind. This tensor is called a tensor of mixed covariant derivatives of the invariant φ.

Analogous construction over an arbitrary covariant vector $\boldsymbol{f} = (f^s_i)$ defines a covariant tensor of the second kind called a covariant derivative of this vector. Its (ij)th component, designated by $(\nabla f)^s_{ij}$ in the coordinates $s^1,\ldots,s^n$, is computed by the following formula
$$(\nabla f)^s_{ij} = \frac{\partial}{\partial s^j} f^s_i - f^s_k \Upsilon^k_{ij}, \quad i,j,k = 1,\ldots,n. \quad (4.41)$$
Tensor relations (4.34), for $n = 2$, are verified for these components similarly as in (4.40).

It is obvious, comparing (4.39) and (4.41), that
$$\nabla^s_{ij}(\varphi) = (\nabla \mathrm{grad}\varphi)^s_{ij}, \quad i,j = 1,\ldots,n.$$

In the case of a regular surface $S^{rn} \subset R^{n+l}$ represented by (4.1) another example of a covariant vector is formed through an arbitrary fixed vector $\boldsymbol{P} \in R^{n+l}$ by the following formula for its components in the coordinates $s^1,\ldots,s^n$:

$$f_i^s = \boldsymbol{P} \cdot \boldsymbol{r}_{s^i}, \quad i = 1, \ldots, n, \tag{4.42}$$

where $\boldsymbol{r}_{s^i}$ is the ith basic tangent vector of S^{rn} in the coordinates $s^1, \ldots, s^n$. If $\boldsymbol{P} \in R^{n+l}$ is a vector orthogonal to S^{rn} at a point P then the quantities

$$b_{ij}^s = \boldsymbol{r}_{s^i s^j} \cdot \boldsymbol{P}, \quad i, j = 1, \ldots, n, \tag{4.43}$$

form a covariant tensor of the second order. Indeed

$$b_{ij}^v = \boldsymbol{r}_{v^i v^j} \cdot \boldsymbol{P} = \left(\frac{\partial s^k}{\partial v^i} \frac{\partial s^m}{\partial v^j} \boldsymbol{r}_{s^k s^m} + \frac{\partial^2 s^k}{\partial v^i \partial v^j} \boldsymbol{r}_{s^k} \right) \cdot \boldsymbol{P}$$

$$= \frac{\partial s^k}{\partial v^i} \frac{\partial s^m}{\partial v^j} \boldsymbol{r}_{s^k s^m} \cdot \boldsymbol{P} = b_{km}^s \frac{\partial s^k}{\partial v^i} \frac{\partial s^m}{\partial v^j}, \quad i, j, k, m = 1, \ldots, n \,.$$

Since (4.14) or (4.18), a typical example of a symmetric covariant tensor of the second order also gives the metric tensor of any regular m-dimensional surface S^{rn} defined by (4.1) or a manifold M^n.

There is an evident rule for forming a new covariant tensor from two original ones. Namely, let $\boldsymbol{f}$ and $\boldsymbol{v}$ be two covariant tensors of the rank k and l, respectively. Then the new tensor $\boldsymbol{f} \otimes \boldsymbol{v}$ is the covariant tensor of the rank $k + l$ whose components in the coordinates $s^1, \ldots, s^n$ are computed as

$$(\boldsymbol{f} \otimes \boldsymbol{v})_{i_1 \ldots i_{k+l}}^s = f_{i_1 \ldots i_k}^s v_{i_{k+1} \ldots i_{k+l}}^s, \quad i_j = 1, \ldots, n \,. \tag{4.44}$$

In particular, two smooth functions f and φ specified in the vicinity of a point $P \in S^{rn}$ produce a covariant tensor of the second rank, through the covariant vectors formulated by (4.37):

$$(f \otimes \varphi)_{ij}^s = f_{s^i} \varphi_{s^j}, \quad i, j = 1, \ldots, n \,. \tag{4.45}$$

Note, generally, this tensor is not symmetric.

Contravariant Tensors. Since (4.15) and (4.19), an example of the contravariant tensor of the second rank is represented by the contravariant metric tensor of S^{rn} and M^n, respectively.

By virtue of (4.15) we can readily conclude that for a fixed vector $\boldsymbol{P} \in R^{n+l}$ a set of values defined in the coordinates $s^1, \ldots, s^n$ as

$$\boldsymbol{P} \cdot \boldsymbol{\nabla} s^i, \quad i = 1, \ldots, n, \tag{4.46}$$

where $\boldsymbol{\nabla} s^i$ is the ith normal vector to the ith coordinate hypersurface in a regular surface S^{rn}, is a contravariant tensor of the first rank. Indeed, by (4.10) and (4.15)

$$\boldsymbol{P} \cdot \boldsymbol{\nabla} s^i = g_{sr}^{ij}(\boldsymbol{P} \cdot \boldsymbol{r}_{s^j}) = g_{vr}^{km} \frac{\partial s^i}{\partial v^k} \frac{\partial s^j}{\partial v^m} \left(\boldsymbol{P} \cdot \boldsymbol{r}_{v^l} \frac{\partial v^l}{\partial s^j} \right) = g_{vr}^{km}(\boldsymbol{P} \cdot \boldsymbol{r}_{v^m}) \frac{\partial v^i}{\partial s^k}$$

$$= \boldsymbol{P} \cdot \boldsymbol{\nabla} v^k \frac{\partial s^i}{\partial v^k}, \quad i, j, k, l, m = 1, \ldots, n \,.$$

Similarly to the case of the covariant tensors considered above two contravariant tensors $\boldsymbol{f}$ and $\boldsymbol{v}$ of rank k and l, respectively, form a contravariant tensor $\boldsymbol{f} \otimes \boldsymbol{v}$ of the order $k+l$, whose components in the coordinates $s^1, \ldots, s^n$ are computed by

$$(\boldsymbol{f} \otimes \boldsymbol{v})_{\boldsymbol{s}}^{i_1 \ldots i_{k+l}} = f_{\boldsymbol{s}}^{i_1 \ldots i_k} v_{\boldsymbol{s}}^{i_{k+1} \ldots i_{k+l}}, \quad i_j = 1, \ldots, n. \quad (4.47)$$

Mixed Tensors. Examples of the mixed tensors are readily constructed by the product of two tensors one of which is covariant and the order is contravariant. For instance, two tensors of the covariant and contravariant types formed by (4.42) and (4.46), respectively, produce, through a vector $\boldsymbol{P} \in R^{n+l}$, the following mixed tensor

$$P_j^i(\boldsymbol{s}) = P_{\boldsymbol{s}}^i P_j^{\boldsymbol{s}} = (\boldsymbol{P} \cdot \boldsymbol{\nabla} s^i)(\boldsymbol{P} \cdot \boldsymbol{r}_{s^j}) = g_{sr}^{il}(\boldsymbol{P} \cdot \boldsymbol{r}_{s^l})(\boldsymbol{P} \cdot \boldsymbol{r}_{s^j}) \quad (4.48)$$

$$i, j, l = 1, \ldots, n.$$

There also is an operation of covariant differentiation of a contravariant vector $\boldsymbol{f} = (f_s^i)$ which results in a mixed tensor by the following formula for its components designated by $(\nabla f)_j^i(\boldsymbol{s})$ in the coordinates $s^1, \ldots, s^n$:

$$(\nabla f)_j^i(\boldsymbol{s}) = \frac{\partial}{\partial s^j} f_{\boldsymbol{s}}^i + f_{\boldsymbol{s}}^k \Upsilon_{jk}^i, \quad i, j, k = 1, \ldots, n. \quad (4.49)$$

Using the relations (4.29), and (4.35) we easily find that the quantities $(\nabla f)_j^i(\boldsymbol{s})$ comprise a mixed tensor.

4.4.3 Tensor Operations

The operations over tensors defined at the same point of M^n are addition, multiplication and contraction.

Operation of Addition. The addition operation is carried out over tensors of the same order and type merely by adding the values of their components. A particular case of addition is the operation of subtraction.

Operation of Multiplication. The operation of multiplication is carried out over tensors of arbitrary order and type by multiplying each component of one tensor, say $\boldsymbol{f}_1$ by every component of another tensor, say $\boldsymbol{f}_2$, in particular, as in (4.44) and (4.47). As a result the order of the product equals the sum of the orders of the two original tensors $\boldsymbol{f}_1$ and $\boldsymbol{f}_2$. The same rule of summation is valid for the type of the product, namely, it is $k_1 + k_2$ times covariant and $l_1 + l_2$ times contravariant if the original tensor $\boldsymbol{f}_i$, $i = 1, 2$, is k_i times covariant and l_i times contravariant.

Operation of Contraction. The operation of contraction is carried out over mixed tensors only. Let us take a mixed tensor, say one time covariant and two times contravariant whose components in the coordinates $s^1, \ldots, s^n$ are designated, correspondingly, as $f_k^{ij}(\boldsymbol{s})$. Assume the indices j and k the same then summation over them gives quantities designated by $f_j^{ij}(\boldsymbol{s})$ which are dependent on one index i only. It is readily shown that these quantities form a contravariant tensor of order 1. Indeed

$$f_j^{ij}(\boldsymbol{s}) = f_k^{lm}(\boldsymbol{v})\frac{\partial v^k}{\partial s^j}\frac{\partial s^j}{\partial v^l}\frac{\partial s^i}{\partial v^m} = f_l^{lm}(\boldsymbol{v})\frac{\partial s^i}{\partial v^m}, \quad i,k,l,m=1,\ldots,n,$$

i.e. the system $b_s^i = f_j^{ij}(\boldsymbol{s})$ is a contravariant vector.

Analogously, the operation of contraction is defined for arbitrary mixed tensors by identifying some upper and lower indices in the components and producing summation over them. In particular, if $\boldsymbol{f}$ is a tensor k times covariant and k times contravariant then the operation of contraction over all indices produces an invariant (tensor of order zero).

For example the operation of contraction over the tensor (4.48) produces the following invariant

$$P_i^i(\boldsymbol{s}) = g_{sr}^{ij}(\boldsymbol{P}\cdot\boldsymbol{r}_{s^i})(\boldsymbol{P}\cdot\boldsymbol{r}_{s^j}), \quad i,j=1,\ldots,n. \tag{4.50}$$

With the operation of contraction one can define an invariant for arbitrary two tensors of the same order k provided one of them is covariant and another contravariant. This invariant is obtained by the composition of two operations: the first is multiplication of the tensors and the second is contraction of the obtained mixed tensor with respect to all indices. For example, the covariant and contravariant metric tensors yield the invariant

$$g_s^{ij}g_{ij}^s = n, \quad i,j=1,\ldots,n,$$

which equals n at all points of M^n.

4.5 Basic Invariants

This section reviews the most important invariants indispensable in the analysis of grid properties. They are formed by the operation of contraction over the contravariant metric tensor and some covariant tensors of order 2.

4.5.1 Beltrami's Differential Parameters

Mixed Differential Parameters. The successive operations of multiplication and contraction over the covariant tensor (4.45) and the contravariant metric tensor g_s^{ij} of a manifold M^n produce the invariant

$$\nabla(f,\varphi) = f_{s^i}\varphi_{s^j}g_s^{ij}, \quad i,j=1,\ldots,n, \tag{4.51}$$

called Beltrami's mixed differential parameter of f and φ. In accordance with this designation formula (4.12) is also read as

$$\boldsymbol{n} = \varphi_{s^j}g_{sr}^{ji}\boldsymbol{r}_{s^i} = \nabla(\varphi,\boldsymbol{r}), \quad i,j=1\ldots,n. \tag{4.52}$$

By putting in (4.51) f equal to φ the following invariant is formulated:

$$\nabla(f) = \nabla(f,f) = f_{s^i}f_{s^j}g_s^{ij}, \quad i,j=1,\ldots,n, \tag{4.53}$$

78   4. Multidimensional Geometry

which is referred to as Beltrami's first differential parameter of f. Thus formula (4.13) with this parameter is read as

$$\|\boldsymbol{n}\| = \sqrt{\nabla(\varphi)}, \tag{4.54}$$

i.e. the length of the normal $\boldsymbol{n}$ defined by (4.12) is an invariant.

Let $\varphi_1 = const$ and $\varphi_2 = const$ be two hypersurfaces in a regular surface S^{rn}. The angle θ between these hypersurfaces is defined as the angle between the corresponding normals $\boldsymbol{n}_1$ and $\boldsymbol{n}_2$ to them. In accordance with (4.52) and (4.54) the cosine of this angle is computed through the Beltrami's differential parameters:

$$\cos\theta = \frac{\nabla(\varphi_1,\boldsymbol{r})\cdot\nabla(\varphi_2,\boldsymbol{r})}{\sqrt{\nabla(\varphi_1)}\sqrt{\nabla(\varphi_2)}} = \frac{\nabla(\varphi_1,\varphi_2)}{\sqrt{\nabla(\varphi_1)}\sqrt{\nabla(\varphi_2)}}. \tag{4.55}$$

Second Differential Parameter. Another important Beltrami's differential parameter of a scalar φ is obtained by contracting the mixed tensor formed through the multiplication operation of the tensor of mixed derivatives (4.39) and the contravariant metric tensor. This invariant, designated by $\Delta_B[\varphi]$, is referred to as Beltrami's second differential parameter of φ. Namely

$$\Delta_B[\varphi] = g_s^{ij}\nabla_{ij}^s(\varphi), \quad i,j = 1,\ldots,n. \tag{4.56}$$

There exists one more important form of the invariant $\Delta_B[\varphi]$ helpful for its computing. To deduce it we note that, from (4.25) and (4.26),

$$\frac{1}{\sqrt{g^s}}\frac{\partial}{\partial s^j}(\sqrt{g^s}g_s^{ij}) = g_s^{ij}\Upsilon_{kj}^k - g_s^{im}\Upsilon_{jm}^j - g_s^{lj}\Upsilon_{lj}^i =$$
$$= -g_s^{lj}\Upsilon_{lj}^i, \quad i,j,k,l,m = 1,\ldots,n, \tag{4.57}$$

and consequently, taking advantage of these relations and (4.39) in (4.56), we obtain the following expression for $\Delta_B[\varphi]$

$$\Delta_B[\varphi] = g_s^{ij}\varphi_{s^is^j} - \varphi_{s^k}g_s^{ij}\Upsilon_{ij}^k$$
$$= g_s^{ij}\varphi_{s^is^j} + \frac{\varphi_{s^k}}{\sqrt{g^s}}\frac{\partial}{\partial s^j}(\sqrt{g^s}g_s^{jk}) \tag{4.58}$$
$$= \frac{1}{\sqrt{g^s}}\frac{\partial}{\partial s^j}(\sqrt{g^s}g_s^{jk}\varphi_{s^k}), \quad i,j,k = 1,\ldots,n.$$

4.5.2 Measure of Relative Spacing

In grid technology there is often a need for estimating grid spacing near some hypersurface in S^{rn}. Typically the hypersurface is specified by the equation $\varphi(\boldsymbol{s}) = 0$. This equation describes one more hypersurface in the parametric domain S^n as well. The parametric mapping $\boldsymbol{r}(\boldsymbol{s}) : S^n \to S^{rn}$ transforms a band of the thickness h around the hypersurface in S^n to a band in S^{rn} whose thickness l is computed by the following formula

4.5 Basic Invariants

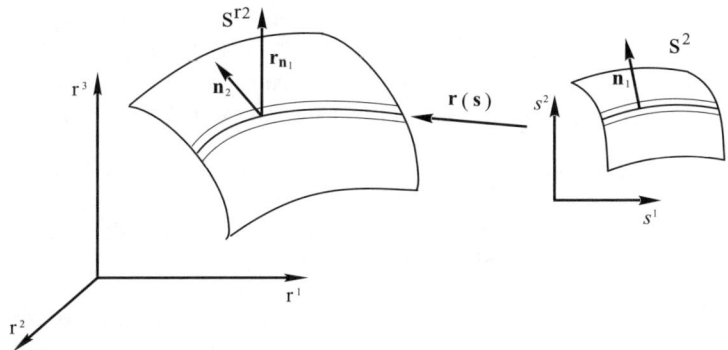

Fig. 4.5. Illustration for the measure of relative spacing

$$l = \left(\frac{\partial \boldsymbol{r}}{\partial \boldsymbol{n}_1} \cdot \boldsymbol{n}_2\right) h + O(h^2) \,, \tag{4.59}$$

where $\boldsymbol{n}_1$ is a unit normal to the hypersurface in S^n while $\boldsymbol{n}_2$ is a unit normal to the hypersurface in S^{rn} (Fig. 4.5 for $n = 2$). Since (4.52) and (4.54)

$$\begin{aligned}\boldsymbol{n}_1 &= (\varphi_{s^1}, \ldots, \varphi_{s^n})/\sqrt{\nabla^E(\varphi)} \,, \\ \boldsymbol{n}_2 &= \varphi_{s^k} g_{sr}^{ki} \boldsymbol{r}_{s^i}/\sqrt{\nabla(\varphi)} \,, \quad i, k = 1, \ldots, n \,,\end{aligned} \tag{4.60}$$

where

$$\nabla^E(\varphi) = \varphi_{s^i} \varphi_{s^j} \delta^i_j = (\varphi_{s^1})^2 + \cdots + (\varphi_{s^n})^2 \,, \quad i, j = 1, \ldots, n \,,$$

is Beltrami's first differential parameter of $\varphi(\boldsymbol{s})$ in the Eucledian metric δ^i_j of S^n. Substituting (4.60) and the relation

$$\frac{\partial \boldsymbol{r}}{\partial \boldsymbol{n}_1} = \frac{1}{\sqrt{\nabla^E(\varphi)}} \varphi_{s^j} \boldsymbol{r}_{s^j} \,, \quad j = 1, \ldots, n \,,$$

in (4.59) yields

$$\begin{aligned}l &= \frac{\varphi_{s^j} \boldsymbol{r}_{s^j} \cdot (\varphi_{s^k} g_{sr}^{ki} \boldsymbol{r}_{s^i})}{\sqrt{\nabla^E(\varphi)}\sqrt{\nabla(\varphi)}} h + O(h^2) \\ &= \sqrt{\frac{\nabla^E(\varphi)}{\nabla(\varphi)}} h + O(h^2) \,, \quad i, j, k = 1, \ldots, n \,.\end{aligned}$$

So the invariant

$$s(\varphi) = \sqrt{\nabla^E(\varphi)}/\sqrt{\nabla(\varphi)} \tag{4.61}$$

defined through Beltrami's first differential parameters of φ can be considered as a measure of relative spacing produced by the parametric transformation $\boldsymbol{r}(\boldsymbol{s})$ near the hypersurface $\varphi(\boldsymbol{s}) = 0$ in S^{rn}.

When a logical domain $\varXi^n$ introduced for generating grids is considered as one more parametric domain then the quantity (4.61) with the identification

$S^n = \Xi^n$ (i.e. $\nabla^E(\varphi) = (\varphi_{\xi^1})^2 + \cdots + (\varphi_{\xi^n})^2$) is referred to as a measure of relative grid spacing near the hypersurface $\varphi(\boldsymbol{s}) = 0$ in S^{rn}. In particular, let the hypersurface $\varphi(\boldsymbol{s}) = 0$ be the grid hypersurface $\xi^i(\boldsymbol{s}) - c = 0$. As $\nabla^E(\xi^i) = 1$, $\nabla(\xi^i) = g_{\xi r}^{ii}$ for i fixed, hence (4.61) has the following form

$$s(\xi^i) = 1/\sqrt{g_{\xi r}^{ii}}, \quad i \text{ fixed}. \tag{4.62}$$

4.5.3 Measure of Relative Clustering

The rate of change of relative spacing in the direction $\boldsymbol{n}_2$ normal to the hypersurface $\varphi(\boldsymbol{s}) = 0$ in S^{rn} is called a measure of relative clustering near this hypersurface. Designating this measure by v we find, since (4.60) and (4.61),

$$\begin{aligned} v(\varphi) &= \frac{\mathrm{d}}{\mathrm{d}n_2} s(\varphi) = \frac{1}{\sqrt{\nabla(\varphi)}} \varphi_{s^j} g_{sr}^{ji} \frac{\partial}{\partial s^i} s(\varphi) \\ &= \frac{1}{\sqrt{\nabla(\varphi)}} \nabla(\varphi, s(\varphi)), \quad i, j, k = 1, \ldots, n. \end{aligned} \tag{4.63}$$

i.e. it is defined through the Beltrami's first and mixed differential parameters.

In particular, if $\varphi(\boldsymbol{s}) \equiv \xi^i(\boldsymbol{s}) - c$, where ξ^i is the ith grid coordinate, using (4.62) and (4.63) yields

$$v(\xi^i) = \frac{1}{\sqrt{g_{\xi r}^{ii}}} \nabla\left(\xi^i, \frac{1}{\sqrt{g_{\xi r}^{ii}}}\right), \quad i \text{ fixed}. \tag{4.64}$$

It will be shown below that the measure (4.64) can also be expressed through Beltrami's second differential parameters and the so called mean curvature of the grid hypersurface $\xi^i = const$.

4.5.4 Mean Curvature

One more invariant of a regular surface S^{rn}, important in grid technology, is obtained from the covariant tensor (4.43) and contravariant metric tensor of S^{rn}

$$\sigma = g_{sr}^{ij} \boldsymbol{r}_{s^i s^j} \cdot \boldsymbol{P}, \quad i, j = 1, \ldots, n. \tag{4.65}$$

When the surface S^{rn} lies in a surface $S^{r(n+1)}$ and the vector $\boldsymbol{P}$ being orthogonal to S^{rn} belongs also to the tangent plane to $S^{r(n+1)}$ then the invariant σ from (4.65), scaled by the factor $1/(n\|\boldsymbol{P}\|)$ and designated as K_m, i.e.

$$K_m = \frac{1}{n\|\boldsymbol{P}\|} g_{sr}^{ij} \boldsymbol{r}_{s^i s^j} \cdot \boldsymbol{P}, \quad i, j = 1, \ldots, n, \tag{4.66}$$

is called the mean curvature of S^{rn} in $S^{r(n+1)}$ with the respect to the normal $\boldsymbol{P}$.

In particular for $n = 1$ (S^{r1} is a curve) the invariant K_m is referred to as the geodesic curvature of the curve S^{r1} in the surface S^{r2}.

In the following section a formula for this invariant will be established for an arbitrary hypersurface lying in some regular surface. If this hypersurface is found from the equation $\varphi(s) = 0$ then the mean curvature is defined by Beltrami's first and second differential parameters of φ.

4.6 Geometry of Hypersurfaces

We consider in this section the geometric characteristics which appear when a regular n-dimensional hypersurface S^{xn} represented by a set of local parametrizations

$$\boldsymbol{x}(\boldsymbol{s}) : S^n \to R^{n+l}, \quad S^n \subset R^n,$$

is a subset of an $(n+1)$-dimensional surface $S^{r(n+1)}$ specified by local parametrizations

$$\boldsymbol{r}(\boldsymbol{s}) : S^{n+1} \to R^{n+l}, \quad S^{n+1} \subset R^{n+1}.$$

This situation occurs in grid technology when a scalar-valued monitor function is considered or grid hypersurfaces are analyzed.

4.6.1 Normal Vector to a Hypersurface

A unit vector $\boldsymbol{n}$ at a point $\boldsymbol{x} \in S^{xn}$ which lies in the tangent plane to $S^{r(n+1)}$ at this point and orthogonal to the tangent plane to S^{xn} at the same point is called the unit normal vector to the surface S^{xn} in S^{rn}.

General Case. A normal vector is readily computed by the formula analogous to (2.25). Namely, let P be a point of the surface S^{xn} and let vectors $\boldsymbol{x}_1, \ldots, \boldsymbol{x}_n$ and $\boldsymbol{r}_1, \ldots, \boldsymbol{r}_{n+1}$ be the basic tangent vectors at this point of S^{xn} and S^{rn} respectively. Now we form, analogously to (2.25), the following $(n+1) \times (n+1)$ matrix

$$\boldsymbol{A} = \begin{pmatrix} \boldsymbol{r}_1 & \cdots & \boldsymbol{r}_{n+1} \\ \boldsymbol{x}_1 \cdot \boldsymbol{r}_1 & \cdots & \boldsymbol{x}_1 \cdot \boldsymbol{r}_{n+1} \\ \cdots & \cdots & \cdots \\ \boldsymbol{x}_n \cdot \boldsymbol{r}_1 & \cdots & \boldsymbol{x}_n \cdot \boldsymbol{r}_{n+1} \end{pmatrix}, \tag{4.67}$$

which derives the vector

$$\boldsymbol{b} = \det(\boldsymbol{A}). \tag{4.68}$$

The rank of the $n \times (n+1)$ matrix obtained from $\boldsymbol{A}$ by eliminating its top row equals n hence $\boldsymbol{b} \neq \boldsymbol{0}$. As

82   4. Multidimensional Geometry

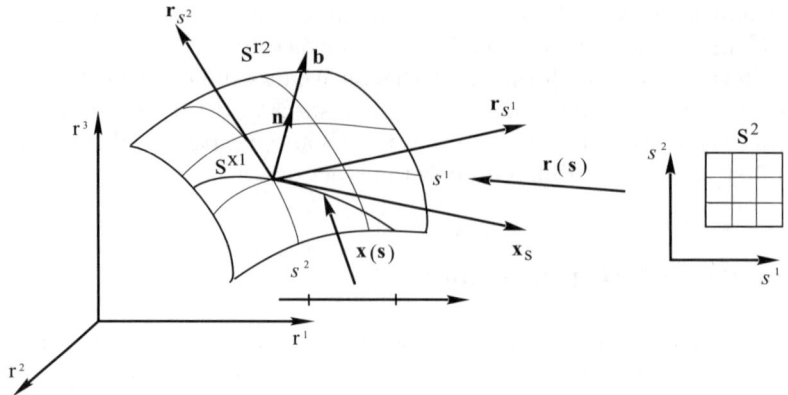

Fig. 4.6. Normal vector to the curve on the surface

$$b \cdot x_i = \begin{pmatrix} x_i \cdot r_1 & \cdots & x_i \cdot r_{n+1} \\ x_1 \cdot r_1 & \cdots & x_1 \cdot r_{n+1} \\ \cdots & \cdots & \cdots \\ x_n \cdot r_1 & \cdots & x_n \cdot r_{n+1} \end{pmatrix} = 0 \,,$$

The vector b is orthogonal to $x_1, \ldots, x_n$ and consequently to the surface S^{xn} at the point of consideration. Thus for the unit normal vector n to the surface S^{xn} in $S^{r(n+1)}$ we find

$$n = b/|b| \,. \tag{4.69}$$

In a one-dimensional case, i.e. when $n = 1$, we find from (4.67) and (4.68)

$$b = (x_1 \cdot r_2)r_1 - (x_1 \cdot r_1)r_2$$

(Fig. 4.6), while for $n = 2$ we obtain, similar to (2.26),

$$b = [(x_1 \cdot r_{k+1})(x_2 \cdot r_{k+2}) - (x_1 \cdot r_{k+2})(x_2 \cdot r_{k+1})]r_k \,, \quad k = 1, 2, 3 \,,$$

where any index, say l, is identified with $l \pm 3$. Thus the vector b obtained by the formula (4.68) can be considered as the tensor product of the vectors $x_1, \ldots, x_n$.

When the surface S^{xn} lies in R^{n+1} and is determined by the equation $\xi(x^1, \ldots, x^{n+1}) = c$ then there also is the well-known formula for the unit normal n:

$$n = \mathrm{grad}\xi/|\mathrm{grad}\xi| = (\xi_{x^1}, \ldots, \xi_{x^{n+1}})/\sqrt{(\xi_{x^1})^2 + \ldots + (\xi_{x^{n+1}})^2} \,. \tag{4.70}$$

Note if the surface S^{xn} is a coordinate hypersurface of $S^{r(n+1)}$ in some coordinates $s^1, \ldots, s^{n+1}$, for example, it is defined by the equation $s^1 = s_0^1$ then S^{xn} is represented in the coordinates $s^2, \ldots, s^{n+1}$ by

$$x(s^2, \ldots, s^{n+1}) = r(s_0^1, s^2, \ldots, s^{n+1}) \,.$$

Therefore

$$x_i = r_{s^i}(s_0^1, s^2, \ldots, s^{n+1}) , \quad i = 2, \ldots, n+1 ,$$

and consequently (4.67) and (4.68) result in

$$A = \begin{pmatrix} r_{s^1} & \cdots & r_{s^{n+1}} \\ g_{21}^{rs} & \cdots & g_{2n+1}^{rs} \\ \cdots & \cdots & \cdots \\ g_{n+11}^{rs} & \cdots & g_{n+1n+1}^{rs} \end{pmatrix} ,$$

$$b^1 = G^{1i} r_{s^i} , \quad i = 1, \ldots, n+1 ,$$

where G^{1i} is the $(1i)$th cofactor of the matrix (g_{ij}^{rs}), $i, j = 1, \ldots, n+1$. Since

$$G^{1i} = g^{rs} g_{sr}^{1i} , \quad i = 1, \ldots, n+1 ,$$

we obtain that

$$b^1 = g^{rs} g_{sr}^{1i} r_{s^i} , \quad i = 1, \ldots, n+1 ,$$

and comparing this expression with (4.10) gives, in this case,

$$b^1 = g^{rs} \nabla s^1 .$$

Analogously one readily shows that the vector b^i expressed as follows:

$$b^i = g^{rs} g_{sr}^{ij} r_{s^j} , \quad i, j = 1, \ldots, n+1 , \qquad (4.71)$$

or using (4.10)

$$b^i = g^{rs} \nabla s^i , \quad i = 1, \ldots, n+1 , \qquad (4.72)$$

is orthogonal to the coordinate hypersurface $s^i = s_0^i$.

Note by the vectors $x_1, \ldots, x_n$ and $r_1, \ldots, r_{n+1}$ in (4.67) for defining the unit normal vector n to S^{xn} in $S^{r(n+1)}$ by (4.69) one can use arbitrary sets of vectors if only they are independent and lie in the corresponding tangent planes.

Hypersurface in a Domain. Let $S^{r(n+1)} = D^{n+1} \subset R^{n+1}$ be an $(n+1)$-dimensional domain containing the n-dimensional surface S^{xn} represented by

$$x(s) : S^n \to R^{n+1} .$$

Since a unit normal vector n to S^{xn} in D^{n+1} is orthogonal to the tangential vectors x_{s^i}, $i = 1, \ldots, n$, the vectors

$$x_{s^1}, \ldots, x_{s^n}, n ,$$

constitute the basis of R^{n+1}. These vectors are called the base vectors of S^{xn} in R^{n+1} (Fig. 4.7 for $n = 2$). The vector $x_{s^m s^p}$, $m, p = 1, \ldots, n$, at a point $P \in S^{xn}$ is expanded in these vectors as

4. Multidimensional Geometry

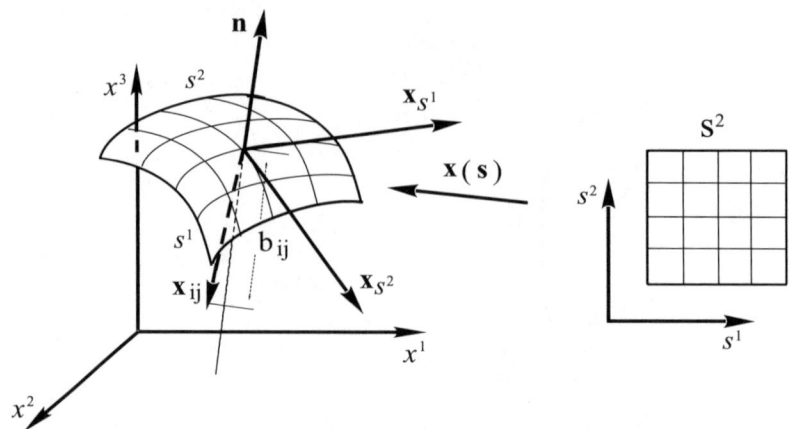

Fig. 4.7. Base vectors of the two-dimensional surface in R^3

$$\boldsymbol{x}_{s^m s^p} = \boldsymbol{P}[\boldsymbol{x}_{s^m s^p}] + (\boldsymbol{x}_{s^m s^p} \cdot \boldsymbol{n})\boldsymbol{n}, \quad m, p = 1, \ldots, n, \qquad (4.73)$$

Where $\boldsymbol{P}$ is the operator which projects vectors in R^{n+1} on the tangent plane to S^{xn}. Taking advantage of (4.32) yields

$$\boldsymbol{x}_{s^m s^p} = \Upsilon^i_{mp} \boldsymbol{x}_{s^i} + (\boldsymbol{x}_{s^m s^p} \cdot \boldsymbol{n})\boldsymbol{n}, \quad i, m, p = 1, \ldots, n, \qquad (4.74)$$

(see Fig. 4.4 for $n = 2$ with the identification $\boldsymbol{x} = \boldsymbol{r}$, $\boldsymbol{x}_{s^m s^p} = \boldsymbol{r}_{mp}$, $\boldsymbol{x}_{s^m s^p} \cdot \boldsymbol{n} = b_{mp}$). Similarly, using (4.33) in (4.73),

$$\boldsymbol{x}_{s^m s^p} = [mp, i]\boldsymbol{\nabla} s^i + (\boldsymbol{x}_{s^m s^p} \cdot \boldsymbol{n})\boldsymbol{n}, \quad i, m, p = 1, \ldots, n, \qquad (4.75)$$

where

$$\boldsymbol{\nabla} s^i = g^{ij}_{sx} \boldsymbol{x}_{s^j}, \quad i, j = 1, \ldots, n.$$

For example, let S^{xn} be a monitor surface over $S^n \subset R^n$ with a scalar-valued monitor function $f(\boldsymbol{s})$, i.e. S^{xn} is represented in the parametric coordinates $s^1, \ldots, s^n$ by

$$\boldsymbol{x}(\boldsymbol{s}) : S^n \to R^{n+1}, \quad \boldsymbol{x}(\boldsymbol{s}) = [\boldsymbol{s}, f(\boldsymbol{s})],$$

It is readily verified that for the elements of the covariant and contravariant metric tensors of S^{xn} in the coordinates $s^1, \ldots, s^n$ we have

$$\begin{aligned} g^{xs}_{ij} &= \boldsymbol{x}_{s^i} \cdot \boldsymbol{x}_{s^j} = \delta^i_j + f_{s_i} f_{s_j}, \quad i,j = 1, \ldots, n, \\ g^{ij}_{sx} &= \delta^i_j - \frac{1}{g^{xs}} \frac{\partial f}{\partial s_i} \frac{\partial f}{\partial s_j}, \quad i,j = 1, \ldots, n, \end{aligned} \qquad (4.76)$$

where

$$g^{xs} = \det(g^{xs}_{ij}) = 1 + (f_{s^1})^2 + \ldots + (f_{s^n})^2 = 1 + \nabla^E(f).$$

Since

$$\boldsymbol{x}_{s^i} = (\underbrace{0,\ldots,0}_{i-1}, 1, \underbrace{0,\ldots,0}_{n-i}, f_{s^i}), \quad i = 1,\ldots,n,$$

it is obvious that one unit normal vector $\boldsymbol{n}$ to S^{xn} in D^{n+1} can be computed as follows:

$$\boldsymbol{n} = \frac{1}{\sqrt{g^{xs}}}(f_{s^1},\ldots,f_{s^n},-1). \tag{4.77}$$

Therefore in this case the expansion (4.74) has the form

$$\boldsymbol{x}_{s^m s^p} = \varUpsilon^i_{mp}\boldsymbol{x}_{s^i} + \frac{1}{\sqrt{g^{xs}}} f_{s^m s^p}\boldsymbol{n}, \quad i,m,p = 1,\ldots,n. \tag{4.78}$$

Since (4.76)

$$g^{il}_{mp} f_{s^l} = \left(\delta^i_l - \frac{1}{g^{xs}} f_{s^i} f_{s^l}\right) f_{s^l} = \frac{1}{g^{xs}} f_{s^i}, \quad i,l,m,p = 1,\ldots,n,$$

therefore from (4.24) and (4.27) we find

$$\varUpsilon^i_{mp} = g^{li}_{sx}[mp,l] = g^{li}_{sx} f_{s^m s^p} f_{s^l} = \frac{1}{g^{xs}} f_{s^m s^p} f_{s^i}, \quad i,l,m,p = 1,\ldots,n.$$

So (4.78) results in

$$\boldsymbol{x}_{s^m s^p} = \frac{1}{g^{xs}} f_{s^m s^p}(f_{s^i}\boldsymbol{x}_{s^i} + \sqrt{g^{xs}}\boldsymbol{n}), \quad i,m,p = 1,\ldots,n.$$

4.6.2 Second Fundamental Form

Assuming in the coordinates $s^1,\ldots,s^n$, similarly to (4.43),

$$b_{ij} = \boldsymbol{x}_{s^i s^j} \cdot \boldsymbol{n}, \quad i,j = 1,\ldots,n, \tag{4.79}$$

where $\boldsymbol{n}$ is the unit normal to the surface S^{xn} in $S^{r(n+1)}$ (Figs. 4.6 and 4.7 for $n=2$) we define the so called second fundamental form of the surface S^{xn} in $S^{r(n+1)}$ by

$$b_{ij} ds^i ds^j, \quad i,j = 1,\ldots,n.$$

The covariant tensor (b_{ij}) reflects the local warping of the surface S^{xn} in $S^{r(n+1)}$.

4.6.3 Surface Curvatures

Multidimensional Case. The covariant tensor (b_{ij}) and the contravariant tensor (g^{ij}_{sx}) of the surface S^{xn} in $S^{r(n+1)}$ define the mixed tensor (K^i_j), where

$$K^i_j = g^{ik}_{sx} b_{kj}, \quad i,k,j = 1,\ldots,n. \tag{4.80}$$

An nth part of the trace of (K^i_j), namely, the quantity

86   4. Multidimensional Geometry

$$K_m = \frac{1}{n}\mathrm{tr}(K_j^i) = \frac{1}{n}g_{sx}^{ij}b_{ij}, \quad i,j = 1,\ldots,n, \quad (4.81)$$

is called the mean curvature of the surface S^{xn} in $S^{r(n+1)}$. From (4.79 – 4.81) we readily conclude that the quantity K_m is invariant of parametrizations of S^{xn} and $S^{r(n+1)}$.

Two-Dimensional Case. When $n = 2$ then the determinant of (K_j^i), i.e.

$$K_G = \det(K_j^i), \quad i,j = 1,2,$$

is called the Gaussian curvature of S^{x2} in S^{r3}. It is obvious that

$$K_G = \det(b_{ij})\det(g_{sx}^{ij}) = \frac{1}{g^{xs}}\det(b_{ij}), \quad (4.82)$$

where $g^{xs} = \det(g_{ij}^{xs}) = \det(\boldsymbol{x}_{s^i} \cdot \boldsymbol{x}_{s^j})$.

4.6.4 Formulas of the Mean Curvature

Formula for the Mean Curvature of a Monitor Surface Over a Domain. Let S^{xn} be the monitor surface over the domain S^n parametrized by

$$\boldsymbol{x}(\boldsymbol{s}) : S^n \to R^{n+1}, \quad \boldsymbol{x}(\boldsymbol{s}) = [\boldsymbol{s}, f(\boldsymbol{s})] \ .$$

Then, taking into account (4.77), we find

$$b_{ij} = \boldsymbol{x}_{s^i s^j} \cdot \boldsymbol{n} = -\frac{1}{\sqrt{g^s}} f_{s^i s^j}, \quad i,j = 1,\ldots,n \ .$$

where $g^s = \det(\boldsymbol{x}_{s^i} \cdot \boldsymbol{x}_{s^j})$. Using (4.76), (4.77), and (4.81) gives

$$K_m = \frac{1}{n\sqrt{g^s}}\left[\frac{1}{g^s}f_{s^i}f_{s^j}f_{s^i s^j} - \nabla^2[f]\right], \quad i,j = 1,\ldots,n \ . \quad (4.83)$$

Formula for the Mean Curvature of the Hypersurface Specified by the Equation $\varphi(\boldsymbol{s}) = 0$.

Multidimensional Case. Let the surface $S^{r(n+1)}$ be represented by a parametrization

$$\boldsymbol{r}(\boldsymbol{s}) : S^{n+1} \to R^{n+l}, \quad \boldsymbol{s} = (s^1, \ldots, s^{n+1}), \quad l \geq 1,$$

while the n-dimensional surface S^{xn} in $S^{r(n+1)}$ is the image through $\boldsymbol{r}(\boldsymbol{s})$ of the n-dimensional surface in S^{n+1} defined from the equation $\varphi(\boldsymbol{s}) = 0$, i.e. the hypersurface S^{xn} consists of the points $\boldsymbol{r}(\boldsymbol{s})$, $\boldsymbol{s} \in S^{n+1}$, such that $\varphi(\boldsymbol{s}) = 0$. In this paragraph we will find a formula for the mean curvature of this hypersurface S^{xn} in $S^{r(n+1)}$.

Let the equation $\varphi(\boldsymbol{s}) = 0$ be locally resolved with respect to s^{n+1}, i.e. there is a function $s^{n+1}(s^1, \ldots, s^n)$ such that

$$\varphi[s^1, \ldots, s^n, s^{n+1}(s^1, \ldots, s^n)] \equiv 0, \quad (s^1, \ldots, s^n) \in S^n,$$

4.6 Geometry of Hypersurfaces

for some n-dimensional domain S^n. This is possible if $\varphi_{s^{n+1}} \neq 0$ at some point $s \in S^{n+1}$. To make things definite we assume that $\varphi_{s^{n+1}} < 0$. Then the hypersurface S^{xn} is represented locally by the coordinates $s^1, \ldots, s^n$ from the domain S^n as

$$x(s^1, \ldots, s^n) : S^n \to R^{n+l}, \tag{4.84}$$

where

$$x(s^1, \ldots, s^n) = r(s^1, \ldots, s^n, s^{n+1}(s^1, \ldots, s^n)) .$$

Since

$$\frac{\partial s^{n+1}}{\partial s^i} = -\frac{\varphi_{s^i}}{\varphi_{s^{n+1}}}, \quad i = 1, \ldots, n ,$$

the basic tangent vectors to S^{xn} in the coordinates $s^1, \ldots, s^n$ are subject to the equations

$$x_{s^i} = r_{s^i} + \frac{\partial s^{n+1}}{\partial s^i} r_{s^{n+1}} = \frac{1}{\varphi_{s^{n+1}}} (\varphi_{s^{n+1}} r_{s^i} - \varphi_{s^i} r_{s^{n+1}}) , \tag{4.85}$$
$$i = 1, \ldots, n ,$$

therefore the values for the elements g^{xs}_{ij}, $i, j = 1, \ldots, n$, of the covariant metric tensor of S^{xn} in the coordinates $s^1, \ldots, s^n$ can be computed by the following formula

$$g^{xs}_{ij} = x_{s^i} \cdot x_{s^j}$$
$$= \frac{1}{(\varphi_{s^{n+1}})^2} [(\varphi_{s^{n+1}})^2 g^{rs}_{ij} - \varphi_{s^{n+1}}(\varphi_{s^j} g^{rs}_{n+1i} + \varphi_{s^i} g^{rs}_{n+1j}) \tag{4.86}$$
$$+ \varphi_{s^i} \varphi_{s^j} g^{rs}_{n+1n+1}], \quad i, j = 1, \ldots, n ,$$

where the quantities g^{rs}_{kl}, $k, l = 1, \ldots, n+1$, are the elements of the covariant metric tensor of $S^{r(n+1)}$ in the coordinates $s^1, \ldots, s^{n+1}$ computed by the formula

$$g^{rs}_{kl} = r_{s^k} \cdot r_{s^l} , \quad k, l = 1, \ldots, n+1 .$$

For the unit normal vector n to S^{xn} in $S^{r(n+1)}$ we have from (4.12)

$$n = b/|b| , \tag{4.87}$$

where

$$b = \varphi_{s^l} g^{lk}_{sr} r_{s^k} , \quad l, k = 1, \ldots, n+1 ,$$
$$|b| = \sqrt{\varphi_{s^l} \varphi_{s^j} g^{lj}_{sr}} = \sqrt{\nabla(\varphi)} , \quad j, l = 1, \ldots, n+1 . \tag{4.88}$$

Since our assumption that $\varphi_{s^{n+1}} < 0$, equations (4.87) and (4.88) result in

$$n \cdot r_{s^{n+1}} = \frac{1}{\sqrt{\nabla \varphi}} \varphi_{s^l} g^{lk}_{sr} r_{s^k} \cdot r_{s^{n+1}} = \frac{1}{\sqrt{\nabla \varphi}} \varphi_{s^{n+1}} < 0 ,$$

88   4. Multidimensional Geometry

i.e. the vectors $\boldsymbol{n}$ and $\boldsymbol{r}_{s^{n+1}}$ have opposite directions with respect to the tangent plane to S^{xn} in $S^{r(n+1)}$.

Now we compute the elements g_{sx}^{ij} of the contravariant metric tensor of the hypersurface S^{xn} in the coordinates $s^1, \ldots, s^n$.

First note that the following $n+1$ vectors

$$\boldsymbol{x}_1, \boldsymbol{x}_2, \ldots, \boldsymbol{x}_n, \boldsymbol{n},$$

where $\boldsymbol{n}$ is defined by (4.87), comprise a basis of R^{n+1}. As $\boldsymbol{n}$ is orthogonal to $\boldsymbol{x}_i$, $i = 1, \ldots, n$, so, in accordance with (2.6), any vector $\boldsymbol{v} \in R^{n+1}$ is represented through the vectors of this basis by

$$\boldsymbol{v} = g_{sx}^{ij}(\boldsymbol{v} \cdot \boldsymbol{x}_j)\boldsymbol{x}_i + (\boldsymbol{v} \cdot \boldsymbol{n})\boldsymbol{n}, \quad i, j = 1, \ldots, n. \tag{4.89}$$

Note every basic normal vector $\boldsymbol{\nabla} s^i$, $i = 1, \ldots, n$, to the coordinate surface $s^i = c$ in $S^{r(n+1)}$ defined, in accordance with (4.10), by

$$\boldsymbol{\nabla} s^i = g_{sr}^{il}\boldsymbol{r}_{s^l}, \quad l = 1, \ldots, n+1, \tag{4.90}$$

is orthogonal to $\boldsymbol{x}_k$, $k = 1, \ldots, n$, and $k \neq i$. Indeed, from (4.85) and (4.89),

$$\boldsymbol{x}_k \cdot \boldsymbol{\nabla} s^i = \frac{1}{\varphi_{s^{n+1}}}(\varphi_{s^{n+1}}\boldsymbol{r}_{s^k} - \varphi_{s^k}\boldsymbol{r}_{s^{n+1}}) \cdot g_s^{il}\boldsymbol{r}_{s^l}$$

$$= \delta_i^k - \frac{\varphi_{s^k}}{\varphi_{s^{n+1}}}\delta_i^{n+1} = 0, \tag{4.91}$$

$$l = 1, \ldots, n+1, \quad i, k = 1, \ldots, n, \quad i \neq k.$$

Therefore, from (4.87 – 4.91), we obtain

$$\boldsymbol{\nabla} s^i = g_{sx}^{ij}(\boldsymbol{\nabla} s^i \cdot \boldsymbol{x}_{s^i})\boldsymbol{x}_{s^j} + (\boldsymbol{\nabla} s^i \cdot \boldsymbol{n})\boldsymbol{n}$$

$$= g_{sx}^{ij}\boldsymbol{x}_{s^j} + \frac{1}{\sqrt{\nabla(\varphi)}}\varphi_{s^l}g_{sr}^{il}\boldsymbol{n}, \tag{4.92}$$

$$l = 1, \ldots, n+1, \quad i, j = 1, \ldots, n, \quad i \text{ fixed}.$$

Thus by expanding the basic tangent vectors $\boldsymbol{r}_{s^l}$, $l = 1, \ldots, n$, through $\boldsymbol{x}_1, \ldots, \boldsymbol{x}_n$, and $\boldsymbol{n}$ and availing us of this expansion in (4.90) we can compute the contravariant elements g_{sx}^{ij}, $i, j = 1, \ldots, n$, of S^{xn} in the coordinates $s^1, \ldots, s^n$.

In order to find this expansion we first note that, in accordance with (4.85),

$$\boldsymbol{r}_{s^k} = \frac{1}{\varphi_{s^{n+1}}}(\varphi_{s^{n+1}}\boldsymbol{x}_{s^k} + \varphi_{s^k}\boldsymbol{r}_{s^{n+1}}), \quad k = 1, \ldots, n. \tag{4.93}$$

Multiplying this equation by

$$\varphi_{s^l}g_{sr}^{lk}/\sqrt{\nabla(\varphi)}, \quad l = 1, \ldots, n+1, \quad k = 1, \ldots, n,$$

and using (4.87) and (4.88) we find

4.6 Geometry of Hypersurfaces

$$\boldsymbol{n} - \frac{\varphi_{s^l} g_{sr}^{ln+1}}{\sqrt{\nabla(\varphi)}} \boldsymbol{r}_{s^{n+1}} = \frac{\varphi_{s^l} g_{sr}^{lk}}{\sqrt{\nabla(\varphi)}} \boldsymbol{x}_{s^k} + \frac{\nabla(\varphi) - \varphi_{s^l} \varphi_{s^{n+1}} g_{sr}^{ln+1}}{\sqrt{\nabla(\varphi)} \varphi_{s^{n+1}}} \boldsymbol{r}_{s^{n+1}},$$
$$l = 1, \ldots, n+1, \quad k = 1, \ldots, n.$$

Therefore

$$\boldsymbol{r}_{s^{n+1}} = \frac{\varphi_{s^{n+1}}}{\sqrt{\nabla(\varphi)}} \left(\boldsymbol{n} - \frac{\varphi_{s^l} g_{sr}^{lk}}{\sqrt{\nabla(\varphi)}} \boldsymbol{x}_{s^k} \right), \tag{4.94}$$
$$l = 1, \ldots, n+1, \quad k = 1, \ldots, n,$$

and using this equation in (4.93) gives

$$\boldsymbol{r}_{s^j} = \boldsymbol{x}_{s^j} + \frac{1}{\sqrt{\nabla(\varphi)}} \left(\varphi_{s^j} \boldsymbol{n} - \frac{\varphi_{s^j} \varphi_{s^l} g_{sr}^{lk}}{\sqrt{\nabla(\varphi)}} \boldsymbol{x}_{s^k} \right), \tag{4.95}$$
$$l = 1, \ldots, n+1, \quad j, k = 1, \ldots, n.$$

Substituting (4.94) and (4.95) in (4.90) we obtain

$$\nabla s^i = g_{sr}^{ij} \boldsymbol{r}_{s^j} + g_{sr}^{in+1} \boldsymbol{r}_{s^{n+1}} = g_{sr}^{ij} \boldsymbol{x}_{s^j} + \frac{1}{\sqrt{\nabla(\varphi)}} \varphi_{s^j} g_{sr}^{ji} \boldsymbol{n}$$
$$- \frac{1}{\nabla(\varphi)} \varphi_{s^j} \varphi_{s^l} g_{sr}^{ij} g_{sr}^{lk} \boldsymbol{x}_{s^k} + \frac{1}{\sqrt{\nabla(\varphi)}} \varphi_{s^{n+1}} g_{sr}^{in+1} \boldsymbol{n}$$
$$- \frac{1}{\nabla(\varphi)} \varphi_{s^{n+1}} \varphi_{s^l} g_{sr}^{lk} g_{sr}^{in+1} \boldsymbol{x}_{s^k}$$
$$= \left(g_{sr}^{ij} - \frac{1}{\nabla(\varphi)} \varphi_{s^p} \varphi_{s^l} g_{sr}^{ip} g_{sr}^{lj} \right) \boldsymbol{x}_{s^j} + \frac{1}{\sqrt{\nabla(\varphi)}} \varphi_{s^l} g_{sr}^{il} \boldsymbol{n},$$
$$i, j, k = 1, \ldots, n, \quad l, p = 1, \ldots, n+1.$$

Comparing this expansion with (4.92) we find

$$g_{sx}^{ij} = g_{sr}^{ij} - \frac{1}{\nabla(\varphi)} \varphi_{s^p} \varphi_{s^l} g_{sr}^{ip} g_{sr}^{jl}, \tag{4.96}$$
$$i, j = 1, \ldots, n, \quad l, p = 1, \ldots, n+1.$$

This very matrix is the contravariant metric tensor of the subsurface S^{xn} in the coordinates $s^1, \ldots, s^n$. In order to check this statement we note first that for arbitrary symmetric matrices (a_{kl}) and (b^{kl}), $k, l = 1, \ldots, n+1$, the following relations are valid:

$$b^{ij} a_{jk} = b^{il} a_{lk} - b^{in+1} a_{n+1k},$$
$$b^{ij} \varphi_{s^j} a_{kn+1} = b^{il} \varphi_{s^l} a_{kn+1} - b^{in+1} \varphi_{s^{n+1}} a_{kn+1}, \tag{4.97}$$
$$b^{ij} \varphi_{s^k} a_{n+1j} = b^{il} \varphi_{s^k} a_{n+1l} - b^{in+1} \varphi_{s^k} a_{n+1k+1},$$
$$i, j, k = 1, \ldots, n, \quad l = 1, \ldots, n+1.$$

Availing ourselves of these relations we obtain

90   4. Multidimensional Geometry

$$g^{ij}_{sx}g^{xs}_{jk} = \left(g^{ij}_{sr} - \frac{1}{\nabla(\varphi)}\varphi_{s^p}\varphi_{s^l}g^{ip}_{sr}g^{jl}_{sr}\right)$$

$$\times \left[g^{rs}_{jk} - \frac{1}{\varphi_{s^{n+1}}}(\varphi_{s^j}g^{rs}_{n+1k} + \varphi_{s^k}g^{rs}_{n+1j}) + \frac{\varphi_{s^j}\varphi_{s^k}}{\varphi_{s^{n+1}}}g^{rs}_{n+1n+1}\right]$$

$$= g^{il}_{sr}g^{rs}_{lk} - g^{in+1}_{sr}g^{rs}_{n+1k} - \frac{1}{\varphi_{s^{n+1}}}g^{rs}_{n+1k}(g^{il}_{sr}\varphi_{s^l} - g^{in+1}_{sr}\varphi_{s^{n+1}})$$

$$-\frac{\varphi_{s^k}}{\varphi_{s^{n+1}}}(g^{il}_{sr}g^{rs}_{ln+1} - g^{in+1}_{sr}g^{rs}_{n+1n+1})$$

$$+\frac{\varphi_{s^k}}{(\varphi_{s^{n+1}})^2}g^{rs}_{n+1n+1}(g^{il}_{sr}\varphi_{s^l} - g^{in+1}_{sr}\varphi_{s^{n+1}})$$

$$-\frac{\varphi_{s^p}\varphi_{s^m}}{\nabla(\varphi)}g^{ip}_{sr}\left[g^{ml}_{sr}g^{rs}_{lk} - g^{mn+1}_{sr}g^{rs}_{n+1k}\right.$$

$$-\frac{1}{\varphi_{s^{n+1}}}g^{rs}_{kn+1}(g^{ml}_{sr}\varphi_{s^l} - g^{mn+1}_{sr}\varphi_{s^{n+1}})$$

$$-\frac{\varphi_{s^k}}{\varphi_{s^{n+1}}}(g^{ml}_{sr}g^{rs}_{ln+1} - g^{mn+1}_{sr}g^{rs}_{n+1n+1})$$

$$\left.-\frac{\varphi_{s^k}}{(\varphi_{s^{n+1}})^2}g^{rs}_{n+1n+1})(g^{ml}_{sr}\varphi_{s^l} - g^{mn+1}_{sr}\varphi_{s^{n+1}})\right]$$

$$= \delta^i_k - \frac{\varphi_{s^l}}{\varphi_{s^{n+1}}}g^{il}_{sr}g^{rs}_{kn+1} + \frac{\varphi_{s^k}\varphi_{s^l}}{(\varphi_{s^{n+1}})^2}g^{il}_{sr}g^{rs}_{n+1n+1}$$

$$+\frac{\varphi_{s^p}}{\varphi_{s^{n+1}}}g^{ip}_{sr}g^{rs}_{kn+1} - \frac{\varphi_{s^p}\varphi_{s^k}}{(\varphi_{s^{n+1}})^2}g^{ip}_{sr}g^{rs}_{n+1n+1} = \delta^i_k,$$

$$i,j,k = 1,\ldots,n, \quad m,l,p = 1,\ldots,n+1,$$

i.e. the matrix (g^{ij}_{sx}), $i,j = 1,\ldots,n$, whose elements are defined by (4.96) is the inverse to the covariant metric tensor (g^{xs}_{ij}), $i,j = 1,\ldots,n$, of S^{xn}, expressed by (4.86). Consequently the elements g^{ij}_{sx} specified by (4.96) comprise the contravariant metric tensor of S^{xn} in the coordinates $s^1,\ldots,s^n$.

Now we proceed to the computation of the quantity $\boldsymbol{x}_{s^is^j} \cdot \boldsymbol{n}$. From (4.85) we find

$$\boldsymbol{x}_{s^is^j} = \frac{1}{(\varphi_{s^{n+1}})^2}L_{ij}[\boldsymbol{r}] - \frac{1}{(\varphi_{s^{n+1}})^3}L_{ij}[\varphi]\boldsymbol{r}_{s^{n+1}}, \quad i,j = 1,\ldots,n, \quad (4.98)$$

where L_{ij} is the operator defined at a function $v(s^1,\ldots,s^{n+1})$ as

$$L_{ij}[v] = (\varphi_{s^{n+1}})^2 v_{s^is^j} - \varphi_{s^{n+1}}(\varphi_{s^j}v_{s^is^{n+1}} + \varphi_{s^i}v_{s^js^{n+1}})$$
$$+\varphi_{s^i}\varphi_{s^j}v_{s^{n+1}s^{n+1}}, \quad i,j = 1,\ldots,n. \quad (4.99)$$

Therefore, using (4.87) and (4.88),

4.6 Geometry of Hypersurfaces

$$L_{ij}[r] \cdot n = \frac{1}{\sqrt{\nabla(\varphi)}}[(\varphi_{s^{n+1}})^2 \varphi_{s^l} g^{lk}_{sr} r_{s^i s^j} \cdot r_{s^k}$$
$$-\varphi_{s^{n+1}} \varphi_{s^l} g^{lk}_{sr} (\varphi_{s^j} r_{s^i s^{n+1}} \cdot r_{s^k} + \varphi_{s^i} r_{s^j s^{n+1}} \cdot r_{s^k})$$
$$+\varphi_{s^i} \varphi_{s^j} \varphi_{s^l} g^{lr}_{sr} r_{s^{n+1} s^{n+1}} \cdot r_{s^k}]$$
$$= \frac{\varphi_{s^l}}{\sqrt{\nabla(\varphi)}}[(\varphi_{s^{n+1}})^2 \Upsilon^l_{ij} - \varphi_{s^{n+1}}(\varphi_{s^j} \Upsilon^l_{in+1} + \varphi_{s^i} \Upsilon^l_{jn+1})$$
$$+\varphi_{s^i} \varphi_{s^j} \Upsilon^l_{n+1 n+1}],$$
$$i, j = 1, \ldots, n, \quad l, k = 1, \ldots, n+1.$$
(4.100)

where Υ^l_{ij}, $k, l, p = 1, \ldots, n+1$, are the Christoffel symbols of the second kind of $S^{r(n+1)}$ in the coordinates $s^1, \ldots, s^{n+1}$. Analogously

$$L_{ij}[\varphi] r_{s^{n+1}} \cdot n = \frac{\varphi_{s^{n+1}}}{\sqrt{\nabla(\varphi)}} L_{ij}[\varphi], \quad i, j = 1, \ldots, n. \qquad (4.101)$$

Thus for the mean curvature of the hypersurface S^{xn} in $S^{r(n+1)}$ we have, from (4.81) and (4.96),

$$K_m = \frac{1}{n} g^{ij}_{sx} x_{s^i s^j} \cdot n$$
$$= \frac{1}{2(\varphi_{s^{n+1}})^2 \sqrt{\nabla(\varphi)}} \left(g^{ij}_{sr} - \frac{1}{\nabla(\varphi)} \varphi_{s^p} \varphi_{s^l} g^{ip}_{sr} g^{jl}_{sr} \right)$$
$$\times \{(\varphi_{s^{n+1}})^2 (\varphi_{s^l} \Upsilon^l_{ij} - \varphi_{s^i s^j}) - \varphi_{s^{n+1}} \varphi_{s^j} (\varphi_{s^l} \Upsilon^l_{in+1} - \varphi_{s^i s^{n+1}})$$
$$- \varphi_{s^{n+1}} \varphi_{s^i} (\varphi_{s^l} \Upsilon^l_{jn+1} - \varphi_{s^j s^{n+1}}) \qquad (4.102)$$
$$+ \varphi_{s^i} \varphi_{s^j} (\varphi_{s^l} \Upsilon^l_{n+1 n+1} - \varphi_{s^{n+1} s^{n+1}})\}$$
$$= -\frac{1}{n(\varphi_{s^{n+1}})^2 \sqrt{\nabla(\varphi)}} \left(g^{ij}_{sr} - \frac{1}{\nabla(\varphi)} \varphi_{s^p} \varphi_{s^l} g^{ip}_{sr} g^{jl}_{sr} \right) d_{ij},$$
$$i, j = 1, \ldots, n, \quad l, p = 1, \ldots, n+1,$$

where

$$d_{kl} = (\varphi_{s^{n+1}})^2 \nabla_{kl}(\varphi) - \varphi_{s^{n+1}} \varphi_{s^k} \nabla_{n+1 l}(\varphi) - \varphi_{s^{n+1}} \varphi_{s^l} \nabla_{n+1 k}(\varphi)$$
$$+ \varphi_{s^k} \varphi_{s^l} \nabla_{n+1 n+1}(\varphi), \quad k, l, = 1, n+1,$$

while

$$\nabla_{kl}(\varphi) = \varphi_{s^k s^l} - \varphi_{s^p} \Upsilon^p_{kl}, \quad k, l, = 1, n+1,$$

is the mixed covariant derivative of φ with respect to s^k and s^l in the metric of $S^{r(n+1)}$.

In order to compute (4.102) we use an analog of the formulas (4.97) which states that the following combinations of the same matrices (a_{kl}) and (b^{kl}), as in (4.97), are subject to the relations

4. Multidimensional Geometry

$$b^{ij}a_{ij} = b^{kl}a_{kl} - 2b^{kn+1}a_{kn+1} + b^{n+1n+1}a_{n+1n+1},$$

$$b^{ij}\varphi_{s^i}a_{jn+1} = b^{ij}\varphi_{s^j}a_{in+1} = a_{ln+1}(b^{kl}\varphi_{s^k} - b^{n+1l}\varphi_{s^{n+1}})$$
$$-a_{n+1n+1}(b^{ln+1}\varphi_{s^l} - b^{n+1n+1}\varphi_{s^{n+1}}),$$
$$i,j = 1,\ldots,n, \quad k,l = 1,\ldots,n+1.$$
(4.103)

Therefore

$$b^{ij}\left(a_{ij} - \frac{1}{\varphi_{s^{n+1}}}(\varphi_{s^i}a_{n+1j} + \varphi_{s^j}a_{n+1i}) - \frac{\varphi_{s^i}\varphi_{s^j}}{(\varphi_{s^n})^2}a_{n+1n+1}\right)$$
$$= b^{kl}a_{kl} - 2b^{kn+1}a_{kn+1} + b^{n+1n+1}a_{n+1n+1}$$

$$-\frac{2}{\varphi_{s^{n+1}}}[a_{ln+1}(b^{kl}\varphi_{s^k} - b^{n+1l}\varphi_{s^{n+1}})$$
$$-a_{n+1n+1}(b^{ln+1}\varphi_{s^l} - b^{n+1n+1}\varphi_{s^{n+1}})]$$
(4.104)

$$+\frac{a_{n+1n+1}}{(\varphi_{s^{n+1}})^2}(b^{kl}\varphi_{s^k}\varphi_{s^l} - 2b^{kn+1}\varphi_{s^k}\varphi_{s^{n+1}} + b^{n+1n+1}\varphi_{s^{n+1}}\varphi_{s^{n+1}})$$

$$= b^{kl}a_{kl} - \frac{2}{\varphi_{s^{n+1}}}b^{kl}\varphi_{s^k}a_{ln+1} + \frac{a_{n+1n+1}}{(\varphi_{s^{n+1}})^2}b^{kl}\varphi_{s^k}\varphi_{s^l},$$
$$i,j = 1,\ldots,n, \quad k,l = 1,\ldots,n+1.$$

Assuming now

$$b^{kl} = g^{kl}_{sr} - \frac{1}{\nabla(\varphi)}\varphi_{s^p}\varphi_{s^t}g^{pk}_{sr}g^{tl},$$
$$a_{kl} = \nabla_{kl}(\varphi), \quad k,l,p,t = 1,\ldots,n+1,$$

we obtain from (4.102) and (4.104)

$$K_m = \frac{1}{n}g^{ij}_{sx}x_{s^is^j}\cdot n$$

$$= -\frac{1}{2\sqrt{\nabla(\varphi)}}[(g^{kl}_{sr} - \frac{1}{\nabla(\varphi)}\varphi_{s^p}\varphi_{s^t}g^{pk}_{sr}g^{tl})\nabla_{kl}(\varphi)$$

$$+\frac{2}{\varphi_{s^{n+1}}}\left(g^{kl}_{sr}\varphi_{s^k} - \frac{1}{\nabla(\varphi)}\varphi_{s^p}\varphi_{s^k}g^{pk}_{sr}\varphi_{s^t}g^{lt}\right)\nabla_{ln+1}(\varphi)$$
(4.105)

$$-\frac{1}{(\varphi_{s^{n+1}})^2}\left(g^{kl}_{sr}\varphi_{s^k}\varphi_{s^l} - \frac{1}{\nabla(\varphi)}\varphi_{s^p}\varphi_{s^k}g^{pk}_{sr}\varphi_{s^l}\varphi_{s^t}g^{lt}\right)\nabla_{n+1n+1}(\varphi)$$

$$= \frac{1}{n(\nabla(\varphi))^{3/2}}[\varphi_{s^p}\varphi_{s^t}g^{pk}_{sr}g^{tl}_{sr} - \nabla(\varphi)g^{kl}_{sr}]\nabla_{kl}(\varphi),$$
$$i,j = 1,\ldots,n, \quad k,l,p,t = 1,\ldots,n+1.$$

It is obvious that the same formula for K_m is obtained if $\varphi_{s^i}(s) \neq 0$ for some i, $1 \leq i \leq n$, and $s \in S^{n+1}$.

Hypersurface in a Domain. In the case $S^{r(n+1)}$ is an $(n+1)$-dimensional domain with the Eucledian metric $g^{rs}_{kl} = \delta^k_l$ the equation (4.105) results in

$$K_m = \frac{1}{n|\text{grad }\varphi|^3}[\varphi_{s^k}\varphi_{s^l}\varphi_{s^k s^l} - |\text{grad }\varphi|^2 \varphi_{s^p s^p}], \qquad (4.106)$$

$$k, l, p = 1, \ldots, n+1.$$

In particular, let S^{xn} be an n-dimensional sphere of a radius ρ, i.e.

$$\varphi(s) \equiv \rho^2 - \sum_{i=1}^{n+1}(s^i)^2.$$

Then, at the points of the sphere,

$$|\text{grad}\varphi|^2 = 4\rho^2,$$

$$\varphi_{s^k s^k} = -2(n+1), \quad k = 1, \ldots, n+1,$$

$$\varphi_{s^k}\varphi_{s^l}\varphi_{s^k s^l} = -8\rho^2, k, l = 1, \ldots, n+1.$$

Thus, from (4.106),

$$K_m = \frac{1}{8n\rho^3}[-8\rho^2 + 8(n+1)\rho^2] = \frac{1}{\rho}.$$

If S^{xn} is a monitor surface over a domain S^n defined by the values of a scalar-valued function $f(s)$, then this surface in R^{n+1} is also specified by the equation

$$\varphi(s^1, \ldots, s^{n+1}) \equiv f(s^1, \ldots, s^n) - s^{n+1} = 0.$$

Consequently

$$|\text{grad}\varphi| = \sqrt{g^s} = \sqrt{1 + f_{s^i}f_{s^i}}, \quad i = 1, \ldots, n,$$

$$\varphi_{s^k s^l} = \begin{cases} 0, & k = n+1 \text{ or } l = n+1, \\ f_{s^k s^l}, & k \neq n+1 \text{ and } l \neq n+1, \end{cases}$$

therefore, from (4.106), we have the following formula for K_m

$$K_m = \frac{1}{n(g^s)^{3/2}}(f_{s^i}f_{s^j}f_{s^i s^j} - g^s f_{s^i s^i}), \quad i = 1, \ldots, n,$$

which coincides with (4.83).

Expression Through Beltrami's Differential Parameters. Note that similarly to the formula (4.25)

$$\frac{\partial}{\partial s^k}g^{ij}_{sr} = -g^{im}_{sr}\Upsilon^j_{mk} - g^{lj}_{sr}\Upsilon^i_{lk}, \quad i,j,l,m = 1, \ldots, n+1.$$

Taking advantage of these relations yields

94   4. Multidimensional Geometry

$$\frac{\partial}{\partial s^i}\nabla(\varphi) = \frac{\partial}{\partial s^i}(\varphi_{sk}\varphi_{sl}g^{kl}_{sr}) = 2g^{kl}_{sr}\varphi_{sl}(\varphi_{sks^i} - \varphi_{sm}\Gamma^m_{ki})$$

$$= 2g^{kl}_{sr}\varphi_{sl}\nabla_{ki}(\varphi), \quad i,k,l,m = 1,\ldots,n+1,$$
(4.107)

and consequently

$$\varphi_{sp}\varphi_{st}g^{pk}_{sr}g^{tl}_{sr}\nabla_{kl}(\varphi) = \frac{1}{2}\varphi_{sp}g^{pk}_{sr}\frac{\partial}{\partial s^k}\nabla(\varphi)$$

$$= \frac{1}{2}\nabla(\varphi,\nabla(\varphi)), \quad k,l,p,t = 1,\ldots,n+1.$$
(4.108)

Further, in accordance with (4.56),

$$g^{kl}_{sr}\nabla_{kl}(\varphi) = \Delta_B[\varphi], \quad k,l = 1,\ldots,n+1.$$
(4.109)

Therefore (4.105) is transformed, with the help of (4.108) and (4.109), to

$$K_m = -\frac{1}{n}\left[\frac{\Delta_B[\varphi]}{\sqrt{\nabla(\varphi)}} + \nabla\left(\varphi,\frac{1}{\sqrt{\nabla(\varphi)}}\right)\right].$$
(4.110)

This formula does not require the knowledge of a normal to the hypersurface $\varphi(s) = 0$ in S^{rn} so it is used for determining the mean curvature of such a hypersurface in an arbitrary Riemannian manifold.

Another Form. One more formula for the mean curvature is found from the following relation

$$\Delta_B[\varphi] + \sqrt{\nabla(\varphi)}\nabla\left(\varphi,\frac{1}{\sqrt{\nabla(\varphi)}}\right) = \frac{1}{\sqrt{g^{rs}}}\frac{\partial}{\partial s^j}(\sqrt{g^{rs}}g^{ij}_{sr}\varphi_{si})$$

$$+\sqrt{\nabla(\varphi)}g^{ij}_{sr}\varphi_{si}\frac{\partial}{\partial s^j}\frac{1}{\sqrt{\nabla(\varphi)}} = \sqrt{\frac{\nabla(\varphi)}{g^{rs}}}\frac{\partial}{\partial s^j}\left(\sqrt{\frac{g^{rs}}{\nabla(\varphi)}}g^{ij}_{sr}\varphi_{si}\right),$$

$i,j = 1,\ldots,n+1.$

Hence equation (4.110) also becomes

$$K_m = -\frac{1}{n\sqrt{g^{rs}}}\frac{\partial}{\partial s^j}\left(\sqrt{\frac{g^{rs}}{\nabla(\varphi)}}g^{ij}_{sr}\varphi_{si}\right), \quad i,j = 1,\ldots,n+1.$$
(4.111)

In particular, for the n-dimensional coordinate hypersurface $\varphi \equiv s^i - c_0 = 0$, we have

$$\nabla(\varphi) = g^{ii}_{sr}, \quad i \text{ fixed},$$

therefore, in accordance with (4.111), the mean curvature of the coordinate hypersurface $s^i = c_0$ is expressed as follows:

$$K_m = -\frac{1}{n\sqrt{g^{rs}}}\frac{\partial}{\partial s^j}\left(\sqrt{\frac{g^{rs}}{g^{ii}_{sr}}}g^{ij}_{sr}\right), \quad i,j = 1,\ldots,n+1, \quad i \text{ fixed}.$$
(4.112)

4.6 Geometry of Hypersurfaces

One-Dimensional Case. When $n = 1$, i.e. S^{xn} is a curve, while $S^{r(n+1)}$ is a two-dimensional surface then (g_{ij}^{rs}) is a 2×2 matrix and its elements satisfy the relations

$$g_{ij}^{rs} = (-1)^{i+j} g^{rs} g_{sr}^{3-i3-j}, \quad i,j = 1,2,$$

where $g^{rs} = \det(g_{ij}^{rs})$. Substituting these relations in (4.86) for $n = 1$ we find the following expression for the metric element g_{11}^{xs} of the curve S^{x1} in S^{r2} represented by the equation $\varphi(s) = 0$, $s = (s^1, s^2)$,

$$g_{11}^{xs} = \frac{1}{(\varphi_{s^2})^2} g^{rs} \varphi_{s^l} \varphi_{s^k} g_{sr}^{lk} = \frac{1}{(\varphi_{s^2})^2} g^{rs} \nabla(\varphi), \quad k,l = 1,2.$$

Therefore the contravariant metric element of S^{x1} is expressed as

$$g_{sx}^{11} = (\varphi_{s^2})^2 / [g^{rs} \nabla(\varphi)].$$

Further note that the formula (4.99) with $n = 1$ has the following form

$$L_{11}[v] = |\text{grad}\varphi|^2 v_{s^l s^l} - \varphi_{s^k} \varphi_{s^l} v_{s^k s^l}, \quad k,l = 1,2,$$

hence equation (4.105) for $n = 1$ is transformed to

$$\begin{aligned}\sigma &= \frac{1}{g^{rs}(\nabla(\varphi))^{3/2}} [|\text{grad}\varphi|^2 (\varphi_{s^l} \Upsilon_{kk}^l - \varphi_{s^k s^k}) \\ &\quad - \varphi_{s^k} \varphi_{s^p} (\varphi_{s^l} \Upsilon_{kp}^l - \varphi_{s^k s^p})] \\ &= \frac{1}{g^{rs}(\nabla(\varphi))^{3/2}} [\varphi_{s^k} \varphi_{s^l} \nabla_{kl}(\varphi) - |\text{grad}\varphi|^2 \nabla_{kk}(\varphi)] \\ &\quad - (-1)^{k+l} \frac{1}{g^{rs}(\nabla(\varphi))^{3/2}} \varphi_{s^{3-k}} \varphi_{s^{3-l}} \nabla_{kl}(\varphi), \quad k,l,p = 1,2,\end{aligned} \quad (4.113)$$

where $\sigma = K_m$ is the geodesic curvature of S^{x1} in S^{r2}.

The geodesic curvature of the coordinate line $s^i = c_0$ is also computed from (4.112) by the formula

$$\sigma_i = -(-1)^{i+j} \frac{1}{\sqrt{g^{rs}}} \frac{\partial}{\partial s^j} \left[\frac{1}{\sqrt{g_{3-i3-i}^{rs}}} g_{3-i3-j}^{rs} \right], \quad (4.114)$$

$i,j = 1,2, \quad i$ fixed,

i.e.

$$\sigma_1 = -\frac{1}{\sqrt{g^{rs}}} \left[\frac{\partial}{\partial s^1} \left(\frac{1}{\sqrt{g_{22}^{rs}}} g_{22}^{rs} \right) - \frac{\partial}{\partial s^2} \left(\frac{1}{\sqrt{g_{22}^{rs}}} g_{12}^{rs} \right) \right],$$

$$\sigma_2 = -\frac{1}{\sqrt{g^{rs}}} \left[\frac{\partial}{\partial s^2} \left(\frac{1}{\sqrt{g_{11}^{rs}}} g_{11}^{rs} \right) - \frac{\partial}{\partial s^1} \left(\frac{1}{\sqrt{g_{11}^{rs}}} g_{12}^{rs} \right) \right],$$

where σ_i, $i = 1,2$, is the geodesic curvature of the coordinate curve $s^i = c_0$.

Computation of the Mean Curvature in the Case of a Parametric Representation.

Multidimensional Case. Let the coordinates $s^1, \ldots, s^{n+1}$ of the points of the n-dimensional surface S^{xn} in $S^{r(n+1)}$ be represented locally by the following transformation

$$s(t) : T^n \to S^{n+1}, \quad t = (t^1, \ldots, t^n), \qquad (4.115)$$

where $T^n \subset R^n$ is some n-dimensional domain of R^n. Then S^{xn} is parametrized in the coordinates $t^1, \ldots, t^n$ by

$$x(t) = r[s(t)] : T^n \to R^{n+l}, \qquad (4.116)$$

where

$$r(s) : S^{n+1} \to R^{n+l}, \quad l \geq 1,$$

is the parametrization of $S^{r(n+1)}$. So the basic tangent vectors x_i, $i = 1, \ldots, n$, to the hypersurface S^{xn} in $S^{r(n+1)}$ are expressed as follows:

$$x_i = r_{s^k} \frac{ds^k}{dt^i}, \quad i = 1, \ldots, n, \quad k = 1, \ldots, n+1. \qquad (4.117)$$

For finding a normal to S^{xn} in $S^{r(n+1)}$ we use formulas (4.68) and (4.69) which require the computation of the determinant of the matrix (4.67) which has, in accordance with (4.117), the following form

$$A = \begin{pmatrix} r_{s^1} & \cdots & r_{s^{n+1}} \\ g^{rs}_{1l} \dfrac{ds^l}{dt^1} & \cdots & g^{rs}_{n+1\,l} \dfrac{ds^l}{dt^1} \\ \cdot & \cdots & \cdot \\ g^{rs}_{1l} \dfrac{ds^l}{dt^n} & \cdots & g^{rs}_{n+1\,l} \dfrac{ds^l}{dt^n} \end{pmatrix}, \quad l = 1, \ldots, n+1. \qquad (4.118)$$

We readily see that the matrix A is composed as the product of the covariant metric tensor and the following matrix B

$$B = \begin{pmatrix} g^{j1}_{sr} r_{s^j} & \cdots & g^{jn+1}_{sr} r_{s^j} \\ \dfrac{ds^1}{dt^1} & \cdots & \dfrac{ds^{n+1}}{dt^1} \\ \cdot & \cdots & \cdot \\ \dfrac{ds^1}{dt^n} & \cdots & \dfrac{ds^{n+1}}{dt^n} \end{pmatrix}, \quad j = 1, \ldots, n+1, \qquad (4.119)$$

namely,

$$A = B(g^{rs}_{ij}).$$

4.6 Geometry of Hypersurfaces

Therefore, analogously to formula (4.68), a normal vector $\boldsymbol{b}$ to S^{xn} in $S^{r(n+1)}$ is also computed by

$$\boldsymbol{b} = \det \boldsymbol{B} .$$

In accordance with the rule of the computation of the determinant of a matrix we find from (4.119)

$$\boldsymbol{b} = -(-1)^i g^{ji}_{sr}(\det \boldsymbol{D}_i)\boldsymbol{r}_{s^j} , \quad i,j = 1,\ldots,n+1 , \qquad (4.120)$$

where $\boldsymbol{D}_i$ is the $n \times n$ matrix obtained by deleting the ith column of the $n \times (n+1)$ matrix (ds^i/dt^j), $i = 1,\ldots,n+1$, $j = 1,\ldots,n$, i.e.

$$\boldsymbol{D}_i = \begin{pmatrix} \dfrac{ds^1}{dt^1} & \cdots & \dfrac{ds^{i-1}}{dt^1} & \dfrac{ds^{i+1}}{dt^1} & \cdots & \dfrac{ds^{n+1}}{dt^1} \\ \cdot & \cdots & \cdot & \cdot & \cdots & \cdot \\ \dfrac{ds^1}{dt^n} & \cdots & \dfrac{ds^{i-1}}{dt^n} & \dfrac{ds^{i+1}}{dt^n} & \cdots & \dfrac{ds^{n+1}}{dt^n} \end{pmatrix} .$$

Hence

$$|\boldsymbol{b}|^2 = \boldsymbol{b} \cdot \boldsymbol{b} = (-1)^{i+k} g^{ki}_{sr} \det \boldsymbol{D}_i \det \boldsymbol{D}_k , \quad i,k = 1,\ldots,n+1 . \qquad (4.121)$$

As for

$$\boldsymbol{x}_{t^i t^j} = \frac{\partial^2 \boldsymbol{x}(t)}{\partial t^i \partial t^j} , \quad i,j = 1,\ldots,n ,$$

we have from (4.116)

$$\boldsymbol{x}_{t^i t^j} = \boldsymbol{r}_{s^m s^l} \frac{ds^m}{dt^i} \frac{ds^l}{dt^j} + \boldsymbol{r}_{s^m} \frac{d^2 s^m}{dt^i dt^j} , \quad i,j = 1,\ldots,n , \; l,m = 1,\ldots,n+1 ,$$

therefore

$$b_{ij} = \boldsymbol{x}_{t^i t^j} \cdot \boldsymbol{n} = \boldsymbol{x}_{t^i t^j} \cdot \frac{\boldsymbol{b}}{|\boldsymbol{b}|}$$

$$= -(-1)^p \frac{1}{|\boldsymbol{b}|} g^{pk}_{sr} \det \boldsymbol{D}_p \left(\frac{ds^m}{dt^i} \frac{ds^l}{dt^j} \boldsymbol{r}_{s^m s^l} \cdot \boldsymbol{r}_{s^k} + \frac{d^2 s^m}{dt^i dt^j} \boldsymbol{r}_{s^m} \cdot \boldsymbol{r}_{s^k} \right)$$

$$= -(-1)^p \frac{1}{|\boldsymbol{b}|} \det \boldsymbol{D}_p \left(\Upsilon^p_{ml} \frac{ds^m}{dt^i} \frac{ds^l}{dt^j} + \frac{d^2 s^p}{dt^i dt^j} \right) ,$$

$$i,j = 1,\ldots,n , \quad k,l,m,p = 1,\ldots,n+1 ,$$

and consequently

$$K_m = \frac{1}{2} g^{ij}_{tx} b_{ij} = \frac{1}{2|\boldsymbol{b}|} \det \boldsymbol{C} , \quad i,j = 1,\ldots,n , \qquad (4.122)$$

where g^{ij}_{tx} is the (ij)th element of the contravariant metric tensor of S^{xn} in the coordinates $t^1,\ldots,t^n$, while

98   4. Multidimensional Geometry

$$
C = \begin{pmatrix}
g_{tx}^{ij}\left(\Upsilon^1_{ml}\dfrac{\partial s^m}{\partial t^i}\dfrac{\partial s^l}{\partial t^j} + \dfrac{\partial^2 s^1}{\partial t^i \partial t^j}\right) & \cdots & g_{tx}^{ij}\left(\Upsilon^{n+1}_{ml}\dfrac{\partial s^m}{\partial t^i}\dfrac{\partial s^l}{\partial t^j} + \dfrac{\partial^2 s^{n+1}}{\partial t^i \partial t^j}\right) \\
\dfrac{\partial s^1}{\partial t^1} & \cdots & \dfrac{\partial s^{n+1}}{\partial t^1} \\
\cdot & \cdots & \cdot \\
\dfrac{\partial s^1}{\partial t^n} & \cdots & \dfrac{\partial s^{n+1}}{\partial t^n}
\end{pmatrix}.
$$

Now we establish a relation between the elements of the contravariant metric tensor (g_{tx}^{ij}), $i,j = 1, \ldots, n$, of S^{xn} in the coordinates $t^1, \ldots, t^n$ and the metric elements of $S^{r(n+1)}$ in the coordinates $s^1, \ldots, s^{n+1}$. For this purpose we notice that the vectors

$$\boldsymbol{x}_1, \ldots, \boldsymbol{x}_n, \boldsymbol{n},$$

where $\boldsymbol{n} = \boldsymbol{b}/|\boldsymbol{b}|$, constitute a basis for the tangent plane to $S^{r(n+1)}$. The vectors $\boldsymbol{r}_{s^k}$, $k = 1, \ldots, n+1$, are expanded in these basis by the formula (2.6), namely,

$$\boldsymbol{r}_{s^k} = g_{tx}^{ij}(\boldsymbol{r}_{s^k} \cdot \boldsymbol{x}_i) \cdot \boldsymbol{x}_j + (\boldsymbol{r}_{s^k} \cdot \boldsymbol{n})\boldsymbol{n}, \quad i,j = 1, \ldots, n, \quad k = 1, \ldots, n+1,$$

therefore, using (4.117) and (4.120),

$$
\begin{aligned}
g_{kl}^{rs} &= \boldsymbol{r}_{s^k} \cdot \boldsymbol{r}_{s^l} \\
&= [g_{tx}^{ij}(\boldsymbol{r}_{s^k} \cdot \boldsymbol{x}_i)\boldsymbol{x}_j + (\boldsymbol{r}_{s^k} \cdot \boldsymbol{n})\boldsymbol{n}] \cdot [g_{tx}^{kp}(\boldsymbol{r}_{s^l} \cdot \boldsymbol{x}_p)\boldsymbol{x}_k + (\boldsymbol{r}_{s^l} \cdot \boldsymbol{n})\boldsymbol{n}] \\
&= g_{tx}^{ij} g_{tx}^{kp} g_{kj}^{xt}(\boldsymbol{r}_{s^k} \cdot \boldsymbol{x}_i)(\boldsymbol{r}_{s^l} \cdot \boldsymbol{x}_p) + (\boldsymbol{r}_{s^k} \cdot \boldsymbol{n})(\boldsymbol{r}_{s^l} \cdot \boldsymbol{n}) \\
&= g_{tx}^{ij}(\boldsymbol{r}_{s^k} \cdot \boldsymbol{x}_i)(\boldsymbol{r}_{s^l} \cdot \boldsymbol{x}_j) + \dfrac{1}{|\boldsymbol{b}|^2}(\boldsymbol{r}_{s^k} \cdot \boldsymbol{b})(\boldsymbol{r}_{s^l} \cdot \boldsymbol{b}) \\
&= g_{tx}^{ij} g_{km}^{rs} \dfrac{\partial s^m}{\partial t^i} g_{lp}^{rs} \dfrac{\partial s^p}{\partial t^j} + (-1)^{k+l}\dfrac{(g^{rs})^2}{|\boldsymbol{b}|^2}\det \boldsymbol{D}_k \det \boldsymbol{D}_l,
\end{aligned}
$$

$i,j = 1, \ldots, n$, $k,l = 1, \ldots, n+1$, k,l fixed.

Multiplying these relations by g_{sr}^{lh} yields

$$\delta_h^k = g_{tx}^{ij} g_{km}^{rs} \dfrac{\partial s^m}{\partial t^i}\dfrac{\partial s^h}{\partial t^j} + (-1)^{k+l}\dfrac{(g^{rs})^2}{|\boldsymbol{b}|^2}\det \boldsymbol{D}_k \det \boldsymbol{D}_l g_{sr}^{lh}, \qquad (4.123)$$

$i,j = 1, \ldots, n$, $k,l,h = 1, \ldots, n+1$, k, fixed.

Further multiplication of (4.123) by g_{sr}^{kp} gives

$$g_{sr}^{hp} = g_{tx}^{ij}\dfrac{\partial s^p}{\partial t^i}\dfrac{\partial s^h}{\partial t^j} + (-1)^{m+l}\dfrac{(g^{rs})^2}{|\boldsymbol{b}|^2} g_{sr}^{hl} g_{sr}^{mp} \det \boldsymbol{D}_m \det \boldsymbol{D}_l, \qquad (4.124)$$

$i,j = 1, \ldots, n$, $h,l,m,p = 1, \ldots, n+1$.

Therefore

4.6 Geometry of Hypersurfaces   99

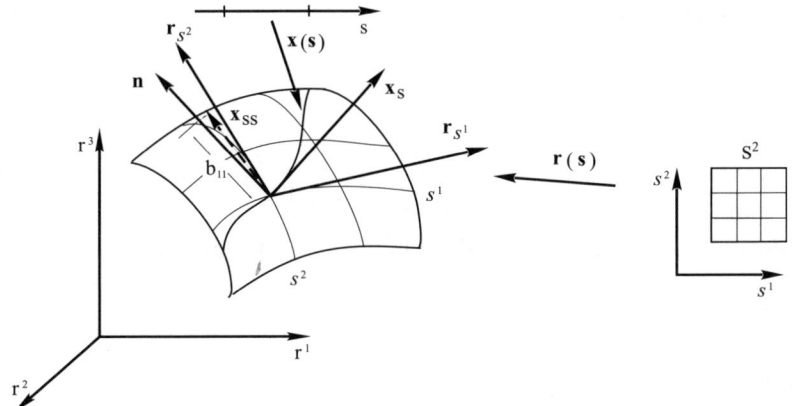

Fig. 4.8. Illustration for a curve on the surface

$$g_{tx}^{ij}\Upsilon_{ml}^{p}\frac{\partial s^m}{\partial t^i}\frac{\partial s^l}{\partial t^j} = g_{sr}^{ml}\Upsilon_{ml}^{p} - d^m d^l \Upsilon_{ml}^{p},$$

$$i,j = 1,\ldots,n, \quad l,m,p = 1,\ldots,n+1,$$

where

$$d^k = (-1)^m \frac{g^{rs}}{|\boldsymbol{b}|} g_{sr}^{mk} \det \boldsymbol{D}_m, \quad k,m = 1,\ldots,n+1.$$

One-Dimensional Subsurface. In the case $n = 1$ (Fig. 4.8), when the surface S^{x1} is a curvilinear line in S^{r2}, the twice mean curvature is called the geodesic curvature of the curve S^{x1} in S^{r2}. Typically the geodesic curvature is designated by σ. It is obvious that in this case $K_G = \sigma$. Besides this it is evident that the geodesic curvature of S^{x1} in S^{r2} is its curvature determined by (3.8) if S^{r2} is a two-dimensional domain.

The geodesic curvature of S^{x1} in S^{r2} can be computed through the elements of the first groundforms of S^{x1} and S^{r2}. In order to prove this we assume that the coordinates s^1 and s^2 of S^{x1} are specified by the function

$$\boldsymbol{s}(t) : [a,b] \to S^2,$$

i.e. S^{x1} is parametrized as

$$\boldsymbol{r}[\boldsymbol{s}(t)] : [a,b] \to R^{l+2},$$

where

$$\boldsymbol{r}(\boldsymbol{s}) : S^2 \to R^{2+l}, \quad l \geq 0$$

is the parametrization of S^{r2}. The geodesic curvature of S^{x1} in S^{r2} is defined through the scalar product of

$$\boldsymbol{r}_{tt} = \frac{d^2 \boldsymbol{r}[\boldsymbol{s}(t)]}{dt^2} = \boldsymbol{r}_{s^i s^j}\frac{ds^i}{dt}\frac{ds^j}{dt} + \boldsymbol{r}_{s^i}\frac{d^2 s^i}{dt^2}, \quad i,j = 1,2,$$

100   4. Multidimensional Geometry

and a normal to this curve in S^{r2}. Namely,

$$\sigma = \frac{1}{g^{rt}} \mathbf{r}_{tt} \cdot \mathbf{n},$$

where

$$g^{rt} = \frac{d\mathbf{r}[s(t)]}{dt} \cdot \frac{d\mathbf{r}[s(t)]}{dt} = g^{rs}_{ij} \frac{ds^i}{dt} \frac{ds^j}{dt}, \quad i,j = 1,2,$$

while the normal $\mathbf{n}$ may be computed by (4.69). This formula yields

$$\mathbf{n} = \mathbf{b}/|\mathbf{b}|$$

where

$$\mathbf{b} = \left(\frac{ds^i}{dt}\mathbf{r}_{s^i} \cdot \mathbf{r}_{s^2}\right)\mathbf{r}_{s^1} - \left(\frac{ds^i}{dt}\mathbf{r}_{s^i} \cdot \mathbf{r}_{s^1}\right)\mathbf{r}_{s^2}$$

$$= (g^{rs}_{2i}\mathbf{r}_{s^1} - g^{rs}_{1i}\mathbf{r}_{s^2})\frac{ds^i}{dt} = -(-1)^i g^{rs} g^{ik}_{sr} \frac{ds^{3-i}}{dt}\mathbf{r}_{s^k}, \quad i,k = 1,2.$$

From this equation we readily find

$$|\mathbf{b}|^2 = \mathbf{b} \cdot \mathbf{b} = (-1)^{i+j}(g^{rs})^2 g^{ik}_{sr} g^{rs}_{kl} g^{jl}_{sr} \frac{ds^{3-i}}{dt}\frac{ds^{3-j}}{dt}$$

$$= (-1)^{i+j}(g^{rs})^2 g^{ij}_{sr}\frac{ds^{3-i}}{dt}\frac{ds^{3-j}}{dt} = g^{rs}g^{rt}, \quad i,j,k,l = 1,2,$$

and consequently

$$\mathbf{n} = -(-1)^i \sqrt{\frac{g^{rs}}{g^{rt}}} g^{ik}_{sr} \frac{ds^{3-i}}{dt}\mathbf{r}_{s^k}, \quad i,k = 1,2.$$

Thus using the above formulas gives the following expression for the geodesic curvature of S^{x1} in S^{r2}

$$\sigma = -(-1)^k \frac{\sqrt{g^{rs}}}{(g^{rt})^{3/2}}\left(\mathbf{r}_{s^i s^j}\frac{ds^i}{dt}\frac{ds^j}{dt} + \mathbf{r}_{s^i}\frac{d^2 s^i}{dt^2}\right) \cdot \left(g^{lk}_{sr}\frac{ds^{3-k}}{dt}\mathbf{r}_{s^l}\right)$$

$$= -(-1)^k \frac{\sqrt{g^{rs}}}{(g^{rt})^{3/2}}\frac{ds^{3-k}}{dt}\left(\frac{d^2 s^k}{dt^2} + \Upsilon^k_{ij}\frac{ds^i}{dt}\frac{ds^j}{dt}\right), \quad i,j,k,l = 1,2,$$

or using the matrix notation

$$\sigma = \frac{\sqrt{g^{rs}}}{(g^{rt})^{3/2}} \det(A) \qquad (4.125)$$

where

$$A = \begin{pmatrix} \frac{d^2 s^1}{dt^2} + \Upsilon^1_{ij}\frac{ds^i}{dt}\frac{ds^j}{dt} & \frac{d^2 s^2}{dt^2} + \Upsilon^2_{ij}\frac{ds^i}{dt}\frac{ds^j}{dt} \\ \frac{ds^1}{dt} & \frac{ds^2}{dt} \end{pmatrix}.$$

4.7 Relations to the Principal Curvatures of Two-Dimensional Surfaces

Note formula (4.125) is determined through the elements of the first groundform only, therefore it is used to formulated the geodesic curvature of curves in arbitrary two-dimensional Riemannian manifolds too.

4.7 Relations to the Principal Curvatures of Two-Dimensional Surfaces

4.7.1 Second Fundamental Form

The coefficients of the second fundamental form

$$b_{ij} \mathrm{d} s^i \mathrm{d} s^j \,, \quad i,j = 1,2 \,,$$

of the surface S^{x2} in R^3 represented by

$$\boldsymbol{x}(\boldsymbol{s}) : S^2 \to R^3 \,, \quad \boldsymbol{x} = (x^1, x^2, x^3) \,, \quad \boldsymbol{s} = (s^1, s^2) \,, \tag{4.126}$$

are defined by the dot products of the second derivatives of the vector function $\boldsymbol{x}(\boldsymbol{s})$ and the unit normal vector $\boldsymbol{n}$ to the surface at the point $\boldsymbol{s}$ under consideration:

$$b_{ij} = \boldsymbol{x}_{s^i s^j} \cdot \boldsymbol{n} \,, \quad i,j = 1,2 \,.$$

Thus, from (2.26) we obtain for b_{ij}, $i,j = 1,2$,

$$b_{ij} = \frac{1}{\sqrt{g^{rs}}} \left[\frac{\partial^2 x^l}{\partial s^i \partial s^j} \left(\frac{\partial x^{l+1}}{\partial s^1} \frac{\partial x^{l+2}}{\partial s^2} - \frac{\partial x^{l+2}}{\partial s^1} \frac{\partial x^{l+1}}{\partial s^2} \right) \right] \,, \quad l = 1,2,3 \,, \tag{4.127}$$

with the identification convention for the superscripts that k is equivalent to $k \pm 3$. Correspondingly, for the monitor surface with the scalar-valued monitor function $f(\boldsymbol{s})$, we obtain

$$b_{ij} = \frac{1}{\sqrt{1 + (f_{s^1})^2 + (f_{s^2})^2}} f_{s^i s^j} \,, \quad i,j = 1,2 \,. \tag{4.128}$$

The tensor (b_{ij}) reflects the local warping of the surface, namely its deviation from the tangent plane at the point under consideration. In particular, if $(b_{ij}) \equiv 0$ at all points of S^2 then the surface is a plane.

4.7.2 Principal Curvatures

Let a curve on the surface $S^{x2} \subset R^3$ be defined by the intersection of a plane containing the normal $\boldsymbol{n}$ with the surface. Taking into account (3.8), we obtain for the curvature of this curve

$$k = \frac{b_{ij} \mathrm{d} s^i \mathrm{d} s^j}{g^{rs}_{ij} \mathrm{d} s^i \mathrm{d} s^j} \,, \quad i,j = 1,2 \,. \tag{4.129}$$

Here (ds^1, ds^2) is the direction of the curve, i.e. $ds^i = c(ds^i/d\varphi)$, where $s(\varphi)$ is a curve parametrization. The two extreme quantities K_I and K_{II} of the values of k are called the principal curvatures of the surface at the point under consideration. In order to compute the principal curvatures, we consider the following relation for the value of the curvature:

$$(b_{ij} - kg_{ij}^{rs})ds^i ds^j = 0, \quad i,j = 1,2, \tag{4.130}$$

which follows from (4.129). In order to find the maximum and minimum values of k, the usual method of equating to zero the derivative with respect to ds^i is applied. Thus the components of the (ds^1, ds^2) direction giving an extreme value of k are subject to the restriction

$$(b_{ij} - kg_{ij}^{rs})ds^j = 0, \quad i,j = 1,2,$$

which, in fact, is the eigenvalue problem for curvature. One finds the eigenvalues k by setting the determinant of this equation equal to zero, obtaining thereby the secular equation for k:

$$\det(b_{ij} - kg_{ij}^{rs}) = 0, \quad i,j = 1,2.$$

This equation, written out in full, is a quadratic equation

$$k^2 - g_{sr}^{ij}b_{ij}k + [b_{11}b_{22} - (b_{12})^2]/g^{rs} = 0,$$

with two roots, which are the maximum and minimum values K_I and K_{II} of the curvature k:

$$K_{I,II} = \frac{1}{2}g_{sr}^{ij}b_{ij} \pm \sqrt{\frac{1}{4}(g_{sr}^{ij}b_{ij})^2 - \frac{1}{g^{rs}}[b_{11}b_{22} - (b_{12})^2]}. \tag{4.131}$$

Mean Curvature. One half of the sum of the principal curvatures is, in fact, the mean surface curvature:

$$K_m = \frac{1}{2}g_{sr}^{ij}b_{ij} = \frac{1}{2}(K_I + K_{II}), \quad i,j = 1,2. \tag{4.132}$$

In the case of the monitor surface represented by the scalar-valued function $f(s^1, s^2)$, we obtain using (4.128)

$$K_m = \frac{f_{s^1s^1}[1+(f_{s^2})^2] + f_{s^2s^2}[1+(f_{s^1})^2] - 2f_{s^1}f_{s^2}f_{s^1s^2}}{2[1+(f_{s^1})^2+(f_{s^2})^2]^{3/2}}.$$

A surface whose mean curvature is zero, i.e. $K_I = -K_{II}$, possesses the following unique property. Namely, if a surface bounded by a specified contour has a minimum area then its mean curvature is zero. Conversely, of all the surfaces bounded by a curve whose length is sufficiently small, the minimum area is possessed by the surface whose mean curvature is zero. So the surface whose mean curvature is equal zero at all its points is referred to as a minimal surface.

4.7 Relations to the Principal Curvatures of Two-Dimensional Surfaces

Gaussian Curvature. Taking into account (4.82), we readily see that the Gaussian curvature is the product of the two principal curvatures K_{I} and K_{II}, i.e.

$$K_{\text{G}} = K_{\text{I}} K_{\text{II}} = \frac{1}{g^{rs}}[b_{11}b_{22} - (b_{12})^2] . \qquad (4.133)$$

In terms of the derivatives of a scalar-valued function $f(s)$ representing the monitor surface S^{r2} we have, from (4.128) and (4.133),

$$K_{\text{G}} = \frac{f_{s^1 s^1} f_{s^2 s^2} - (f_{s^1 s^2})^2}{[1 + (f_{s^1})^2 + (f_{s^2})^2]^2} . \qquad (4.134)$$

Formula Dependent on Christoffel Symbols. In the case of a general two-dimensional surface represented by (4.126) an expression for the quantity $b_{11}b_{22} - (b_{12})^2$ can be obtained through derivatives of the elements of the metric tensor and the coefficients of the second fundamental form of the surface. This is accomplished by using the expansion (4.74) which yields the following relation:

$$\boldsymbol{x}_{s^1 s^1} \cdot \boldsymbol{x}_{s^2 s^2} - \boldsymbol{x}_{s^1 s^2} \cdot \boldsymbol{x}_{s^1 s^2} = g_{ij}^{rs}(\Upsilon_{11}^i \Upsilon_{22}^j - \Upsilon_{12}^i \Upsilon_{12}^j) + b_{11}b_{22} - (b_{12})^2 ,$$

$$i, j = 1, 2 . \qquad (4.135)$$

The left-hand part of (4.135) equals

$$\boldsymbol{x}_{s^1 s^1} \cdot \boldsymbol{x}_{s^2 s^2} - \boldsymbol{x}_{s^1 s^2} \cdot \boldsymbol{x}_{s^1 s^2} = \frac{\partial}{\partial s^1}(\boldsymbol{x}_{s^2 s^2} \cdot \boldsymbol{x}_{s^1}) - \frac{\partial}{\partial s^2}(\boldsymbol{x}_{s^1 s^2} \cdot \boldsymbol{x}_{s^1}) . \qquad (4.136)$$

Since

$$\boldsymbol{x}_{s^2 s^2} \cdot \boldsymbol{x}_{s^1} = [22, 1] = \frac{1}{2}\left(2\frac{\partial g_{12}^{rs}}{\partial s^2} - \frac{\partial g_{22}^{rs}}{\partial s^1}\right) ,$$

$$\boldsymbol{x}_{s^1 s^2} \cdot \boldsymbol{x}_{s^1} = [12, 1] = \frac{1}{2}\frac{\partial g_{11}^{rs}}{\partial s^2} ,$$

we obtain from (4.136)

$$\boldsymbol{x}_{s^1 s^1} \cdot \boldsymbol{x}_{s^2 s^2} - \boldsymbol{x}_{s^1 s^2} \cdot \boldsymbol{x}_{s^1 s^2} = \frac{1}{2}\left(2\frac{\partial^2 g_{12}^{rs}}{\partial s^1 \partial s^2} - \frac{\partial^2 g_{11}^{rs}}{\partial s^2 \partial s^2} - \frac{\partial^2 g_{22}^{rs}}{\partial s^1 \partial s^1}\right) .$$

Therefore (4.133) results in

$$K_{\text{G}} = \frac{1}{g^{rs}}\left[\frac{1}{2}\left(2\frac{\partial^2 g_{12}^{rs}}{\partial s^1 \partial s^2} - \frac{\partial^2 g_{11}^{rs}}{\partial s^2 \partial s^2} - \frac{\partial^2 g_{22}^{rs}}{\partial s^1 \partial s^1}\right) - g_{ij}^{rs}(\Upsilon_{11}^i \Upsilon_{22}^j - \Upsilon_{12}^i \Upsilon_{12}^j)\right] ,$$

$$i, j = 1, 2 .$$

Applying (4.8) transforms this equation to

$$K_G = -\frac{1}{g^{rs}} \left[\frac{1}{2} \frac{\partial^2}{\partial s^i \partial s^j} (g^{rs} g^{ij}_{sr}) + g^{rs}_{ij} (\Upsilon^i_{11} \Upsilon^j_{22} - \Upsilon^i_{12} \Upsilon^j_{12}) \right],$$
$$i, j = 1, 2.$$
(4.137)

This equation depends on the elements of the first surface groundform only. Therefore it can be applied to compute the Gauss curvature of two-dimensional Riemannian manifolds with arbitrary metric tensors.

Since (4.24)

$$g^{rs}_{ij}(\Upsilon^i_{kl} \Upsilon^j_{mp}) = g^{ij}_{sr}[kl, i][mp, j], \quad i, j, k, l, m, p = 1, 2,$$

therefore (4.137) has also the following form

$$K_G = -\frac{1}{g^{rs}} \left\{ \frac{1}{2} \frac{\partial^2}{\partial s^i \partial s^j} (g^{rs} g^{ij}_{sr}) + g^{rs}_{ij}([11, i][22, j] \right.$$
$$\left. - [12, i][12, j]) \right\}, \quad i, j = 1, 2.$$
(4.138)

In particular, in the case of the spherical metric

$$g^s_{ij} = v(\mathbf{s}) \delta^i_j, \quad i, j = 1, 2,$$

we readily find from (4.138)

$$K_G = -\frac{1}{[v(\mathbf{s})]^2} \left\{ \frac{1}{2} \nabla^2 [v] + \frac{1}{v(\mathbf{s})} ([11, i][22, i] - [12, i][12, i]) \right\}$$
$$= -\frac{1}{2[v(\mathbf{s})]^2} \left\{ \nabla^2 [v] - \frac{1}{v(\mathbf{s})} [(v_{s^1})^2 + (v_{s^2})^2] \right\}$$
(4.139)
$$= -\frac{1}{2v(\mathbf{s})} \nabla^2 [\ln v], \quad i = 1, 2.$$

Formula Dependent on Metric Elements. Now we establish one more important expression for K_G. For this purpose we compute the quantity

$$d = \det A$$

where

$$A = \begin{pmatrix} g^{rs}_{11} & g^{rs}_{12} & g^{rs}_{22} \\ \dfrac{\partial g^{rs}_{11}}{\partial s^1} & \dfrac{\partial g^{rs}_{12}}{\partial s^1} & \dfrac{\partial g^{rs}_{22}}{\partial s^1} \\ \dfrac{\partial g^{rs}_{11}}{\partial s^2} & \dfrac{\partial g^{rs}_{12}}{\partial s^2} & \dfrac{\partial g^{rs}_{22}}{\partial s^2} \end{pmatrix}.$$

Using (4.22 – 4.24) gives

4.7 Relations to the Principal Curvatures of Two-Dimensional Surfaces

$$d = g_{11}^{rs}\left(\frac{\partial g_{12}^{rs}}{\partial s^1}\frac{\partial g_{22}^{rs}}{\partial s^2} - \frac{\partial g_{12}^{rs}}{\partial s^2}\frac{\partial g_{22}^{rs}}{\partial s^1}\right) - g_{12}^{rs}\left(\frac{\partial g_{11}^{rs}}{\partial s^1}\frac{\partial g_{22}^{rs}}{\partial s^2} - \frac{\partial g_{11}^{rs}}{\partial s^2}\frac{\partial g_{22}^{rs}}{\partial s^1}\right)$$

$$+ g_{22}^{rs}\left(\frac{\partial g_{11}^{rs}}{\partial s^1}\frac{\partial g_{12}^{rs}}{\partial s^2} - \frac{\partial g_{11}^{rs}}{\partial s^2}\frac{\partial g_{12}^{rs}}{\partial s^1}\right)$$

$$= 2g_{11}^{rs}\{([11,2]+[21,1])[22,2] - ([12,2]+[22,1])[21,2]\}$$

$$- 4g_{12}([11,1][22,2] - [12,1][21,2])$$

$$+ 2g_{22}^{rs}\{[11,1]([12,2]+[22,1]) - [12,1]([11,2]+[21,1])\}$$

$$= 2g_{11}^{rs}[(\Upsilon_{11}^k g_{k2}^{rs} + \Upsilon_{21}^k g_{k1}^{rs})\Upsilon_{22}^l g_{l2}^{rs} - (\Upsilon_{12}^k g_{k2}^{rs} + \Upsilon_{22}^k g_{k1}^{rs})\Upsilon_{21}^l g_{l2}^{rs}]$$

$$- 4g_{12}^{rs}(\Upsilon_{11}^k g_{k1}^{rs}\Upsilon_{22}^l g_{l2}^{rs} - \Upsilon_{12}^k g_{k1}^{rs}\Upsilon_{21}^l g_{l2}^{rs})$$

$$+ 2g_{22}^{rs}[\Upsilon_{11}^k g_{k1}^{rs}(\Upsilon_{12}^l g_{l2}^{rs} + \Upsilon_{22}^l g_{l1}^{rs}) - \Upsilon_{12}^k g_{k1}^{rs}(\Upsilon_{11}^l g_{l2}^{rs} + \Upsilon_{21}^l g_{l1}^{rs})] \quad (4.140)$$

$$= 2(\Upsilon_{11}^k \Upsilon_{22}^l - \Upsilon_{12}^k \Upsilon_{12}^l)(g_{11}^{rs} g_{k2}^{rs} g_{l2}^{rs} - 2g_{12}^{rs} g_{k1}^{rs} g_{l2}^{rs} + g_{22}^{rs} g_{k1}^{rs} g_{l1}^{rs})$$

$$+ 2(g_{11}^{rs}\Upsilon_{21}^k \Upsilon_{22}^l + g_{22}^{rs}\Upsilon_{11}^k \Upsilon_{12}^l)(g_{k1}^{rs} g_{l2}^{rs} - g_{l1}^{rs} g_{k2}^{rs})$$

$$= 2g^{rs}[g_{11}^{rs}(\Upsilon_{11}^1 \Upsilon_{22}^1 - \Upsilon_{12}^1 \Upsilon_{12}^1) + g_{22}^{rs}(\Upsilon_{11}^2 \Upsilon_{22}^2 - \Upsilon_{12}^2 \Upsilon_{12}^2)$$

$$+ 2g_{12}^{rs}(\Upsilon_{11}^2 \Upsilon_{22}^1 - \Upsilon_{12}^2 \Upsilon_{12}^1) + g_{11}^{rs}(\Upsilon_{21}^1 \Upsilon_{22}^2 - \Upsilon_{21}^2 \Upsilon_{22}^1)$$

$$+ g_{22}^{rs}(\Upsilon_{11}^1 \Upsilon_{12}^2 - \Upsilon_{11}^2 \Upsilon_{12}^1)]$$

$$= 2g^{rs}[g_{ij}^{rs}(\Upsilon_{11}^i \Upsilon_{22}^j - \Upsilon_{12}^i \Upsilon_{12}^j) + g_{12}^{rs}(\Upsilon_{11}^2 \Upsilon_{22}^1 - \Upsilon_{11}^1 \Upsilon_{22}^2)$$

$$+ g_{11}^{rs}(\Upsilon_{21}^1 \Upsilon_{22}^2 - \Upsilon_{21}^2 \Upsilon_{22}^1) + g_{22}^{rs}(\Upsilon_{11}^1 \Upsilon_{12}^2 - \Upsilon_{11}^2 \Upsilon_{12}^1)] , \quad k,l = 1,2 .$$

On the other hand

$$\Upsilon_{ki}^k \frac{\partial}{\partial s^j}(g^{rs} g_{sr}^{ij}) = (-1)^{i+1}\Upsilon_{ki}^k\left(\frac{\partial}{\partial s^1}g_{23-i}^{rs} - \frac{\partial}{\partial s^2}g_{13-i}^{rs}\right)$$

$$= \Upsilon_{k1}^k([21,2]-[22,1]) + \Upsilon_{k2}^k([12,1]-[11,2])$$

$$= \Upsilon_{k1}^k(\Upsilon_{21}^l g_{l2}^{rs} - \Upsilon_{22}^l g_{l1}^{rs}) + \Upsilon_{k2}^k(\Upsilon_{12}^l g_{l1}^{rs} - \Upsilon_{11}^l g_{l2}^{rs})$$

$$= g_{l2}^{rs}(\Upsilon_{k1}^k \Upsilon_{21}^l - \Upsilon_{k2}^k \Upsilon_{11}^l) + g_{l1}^{rs}(\Upsilon_{k2}^k \Upsilon_{12}^l - \Upsilon_{k1}^k \Upsilon_{22}^l)$$

$$= g_{12}^{rs}[\Upsilon_{21}^1(\Upsilon_{11}^1 + \Upsilon_{21}^2) - \Upsilon_{11}^1(\Upsilon_{12}^1 + \Upsilon_{22}^2)] \quad (4.141)$$

$$+ g_{21}^{rs}[\Upsilon_{12}^2(\Upsilon_{12}^1 + \Upsilon_{22}^2) - \Upsilon_{22}^2(\Upsilon_{11}^1 + \Upsilon_{21}^2)]$$

$$= 2g_{12}^{rs}(\Upsilon_{21}^1 \Upsilon_{21}^2 - \Upsilon_{11}^1 \Upsilon_{22}^2) + g_{22}^{rs}[\Upsilon_{21}^2(\Upsilon_{11}^1 + \Upsilon_{21}^2) - \Upsilon_{11}^2(\Upsilon_{12}^1 + \Upsilon_{22}^2)]$$

$$+ g_{11}^{rs}[\Upsilon_{12}^1(\Upsilon_{12}^1 + \Upsilon_{22}^2) - \Upsilon_{22}^1(\Upsilon_{11}^1 + \Upsilon_{21}^2)]$$

$$= g_{ij}^{rs}(\Upsilon_{12}^i \Upsilon_{12}^j - \Upsilon_{11}^i \Upsilon_{22}^j) + g_{11}^{rs}(\Upsilon_{12}^1 \Upsilon_{22}^2 - \Upsilon_{22}^1 \Upsilon_{21}^2)$$

$$+ g_{22}^{rs}(\Upsilon_{21}^2 \Upsilon_{11}^1 - \Upsilon_{11}^2 \Upsilon_{12}^1) + g_{12}^{rs}(\Upsilon_{11}^2 \Upsilon_{22}^1 - \Upsilon_{11}^1 \Upsilon_{22}^2) .$$

Availing us of (4.140) and (4.141) we find

$$d - 2g^{rs}\Upsilon_{ki}^{k}\frac{\partial}{\partial s^j}(g^{rs}g_{sr}^{ij}) = 4g^{rs}g_{ij}^{rs}(\Upsilon_{11}^{i}\Upsilon_{22}^{j} - \Upsilon_{12}^{i}\Upsilon_{12}^{j}), \quad i,j = 1,2. \quad (4.142)$$

Since (4.25)

$$\Upsilon_{ki}^{k} = -\sqrt{g^{rs}}\frac{\partial}{\partial s^i}\left(\frac{1}{\sqrt{g^{rs}}}\right), \quad i,k = 1,2,$$

therefore, using (4.137), the equation (4.142) becomes

$$K_G = -\frac{1}{g^{rs}}\left[\frac{1}{2}\frac{\partial^2}{\partial s^i \partial s^j}(g^{rs}g_{sr}^{ij}) + \frac{d}{4g^{rs}}\right.$$
$$\left. + \frac{\sqrt{g^{rs}}}{2}\frac{\partial}{\partial s^i}\left(\frac{1}{\sqrt{g^{rs}}}\right)\frac{\partial}{\partial s^j}(g^{rs}g_{sr}^{ij})\right] \quad (4.143)$$
$$= -\frac{d}{4(g^{rs})^2} - \frac{1}{2\sqrt{g^{rs}}}\frac{\partial}{\partial s^i}\left[\frac{1}{\sqrt{g^{rs}}}\frac{\partial}{\partial s^j}(g^{rs}g_{sr}^{ij})\right], \quad i,j = 1,2.$$

In particular, when the coordinate system s^1, s^2 is orthogonal, then $d = 0$, $g_{sr}^{12} = 0$ and (4.143) yields in this case

$$K_G = -\frac{1}{2\sqrt{g^{rs}}}\left[\frac{\partial}{\partial s^1}\left(\frac{1}{\sqrt{g^{rs}}}\frac{\partial}{\partial s^1}g_{22}^{rs}\right) + \frac{\partial}{\partial s^2}\left(\frac{1}{\sqrt{g^{rs}}}\frac{\partial}{\partial s^2}g_{11}^{rs}\right)\right]. \quad (4.144)$$

Classification of Points of Two-Dimensional Surfaces. A surface point is called elliptic if $K_G > 0$, i.e. both K_I and K_{II} are both negative or both positive at the point of consideration. A saddle or hyperbolic point has principal curvatures of opposite sign, and therefore has negative Gaussian curvature. A parabolic point has one principal curvature vanishing and, consequently, a vanishing Gaussian curvature. This classification of points is prompted by the form of the curve which is obtained by the intersection of the surface with a slightly offset tangent plane. For an elliptic point the curve is an ellipse; for a saddle point it is a hyperbola. It is a pair of lines (degenerate conic) at a parabolic point, and it vanishes at a planar point, where both principal curvatures are zero.

Part II

Application to Advanced Grid Technology

The current part of the monograph including Chaps. 5–9 gives an account of the geometrization of the popular comprehensive grid methods and presents an important extension of the methods, related to the application of the theory of Riemannian manifolds to the formulation of grid equations and grid functionals by developing some procedures for the construction of monitor metric tensors.

As a comprehensive tool for generating grids there is chosen an operator of Beltrami in an arbitrary metric. This operator gives rise to the formulation of generalized Laplace equations. In Chap. 5 it is established that each one-to-one, smooth multidimensional transformation deriving a numerical grid in a domain or on a surface by the mapping approach, is realized by a solution of a system of the generalized Laplace equations in a suitable monitor metric specified in the physical geometry. The system can be interpreted as the multidimensional equidistribution principle in which the monitor metric tensor is an extension of a scalar-valued weight function. With such an interpretation for a comprehensive mathematical model for generating grids in domains or on surfaces there are chosen the generalized Laplace equations. The required grid properties are realized through the specification of suitable metric tensors that can be formulated as a combination of metric components with weights each of which is responsible for providing one individual grid property.

Contrary to the classical geometric studies which analyze geometric features and characteristics of specified Riemannian manifolds the problem of finding appropriate monitor metrics producing the numerical grids with required properties is somewhat an inverse problem of creation of Riemannian manifolds with desirable geometric characteristics. In accordance with the concept of the inverse problem Chaps. 5 and 6 discuss some new techniques aimed at the construction and analysis of special monitor metrics in physical geometries. The techniques are designed by generalizing the projection approach in which the monitor metric in an n-dimensional physical geometry is borrowed from a natural metric of the n-dimensional regular surface derived by a height monitor function over the geometry. By this technology the required metric is defined through the original metric of the physical geometry and several monitor functions with weights. The chapters also establish and review some of the relations of the Riemannian geometry for the purpose of obtaining new equations with implemented metric characteristics aimed at facilitating control for the generation of structured grids with the required quality properties.

Taking advantage of the relations established, the generalized Laplace equations are converted in Chap. 7 into a compact form convenient for numerical treatment by the available algorithms. Some relations of the mean curvature of the monitor surfaces to the equations for grid generation are also exhibited in this chapter.

In Chap. 8 the technique of the multidimensional differential geometry is applied to analysis of the qualitative behavior of the grids generated by the mapping approach. For this purpose a new characteristic of grid clustering is formulated. A relation between this grid measure and some geometric characteristics of grid hypersurfaces and the monitor functions forming the monitor metric is established. The well-known results for grids generated by Laplace equations about node clustering near concave boundary segments of domains and node rarefaction near convex ones are extended, using the relation, to arbitrary boundary segments and to more general elliptic equations formulated for generating grids. On the basis of this formula monitor functions are readily estimated in the popular Poisson, diffusion, and generalized Laplace elliptic models of grid equations to provide grid clustering or, if it is reasonable, grid rarefaction near arbitrary segments of physical geometries.

The final chapter gives a description of a numerical code for generating grids with the numerical solution in the logical domain of the elliptic equations obtained in Chap. 7 by changing mutually dependent and independent variables in the generalized Laplace equations.

5. Comprehensive Grid Models

This chapter formulates differential and variational models for generating grids on surfaces and in domains of arbitrary dimensions. The formulations are based on both the operator of Beltrami and the energy (smoothness) functional with respect to monitor metrics.

For the purpose of unification of formulation of grid models for generating meshes in arbitrary physical geometries a physical domain, surface, or curve is considered in a unified manner as a single geometric object referred to as an n-dimensional regular surface locally represented by a parametrization

$$\boldsymbol{x}(\boldsymbol{s}): S^n \to \mathbb{R}^{n+k}, \ \boldsymbol{x} = (x^1, \ldots, x^{n+k}), \ \boldsymbol{s} = (s^1, \ldots, s^n), \ n \geq 1, \quad (5.1)$$

where S^n is an n-dimensional parametric domain an interval if $n = 1$), while $\boldsymbol{x}(\boldsymbol{s})$ is a smooth vector-valued function of rank n at all points $\boldsymbol{s} \in S^n$. We shall designate by S^{xn} the regular surface parametrized by (5.1). Note, when $k = 0$ then S^{xn} is a domain $X^n \subset \mathbb{R}^n$ which itself can naturally be considered as a parametric domain S^n for X^n with $\boldsymbol{x}(\boldsymbol{s})$ being the identical parametrization, i.e., $\boldsymbol{x}(\boldsymbol{s}) \equiv \boldsymbol{s}$.

The process of numerical grid generation on S^{xn} is turned to finding an intermediate smooth nondegenerate transformation

$$\boldsymbol{s}(\boldsymbol{\xi}): \Xi^n \to S^n, \quad \boldsymbol{\xi} = (\xi^1, \ldots, \xi^n) \quad (5.2)$$

between S^n and a suitable computational (logical) domain Ξ^n and consequently the mesh points on S^{xn} are generated as images through

$$\boldsymbol{x}[\boldsymbol{s}(\boldsymbol{\xi})]: \Xi^n \to \mathbb{R}^{n+k}$$

of the nodes of a reference grid in Ξ^n (see Fig. 5.1).

The computational domain Ξ^n and the cells of the reference mesh can be either rectangular (Fig. 5.1) or of a different shape (for example triangular as in Fig. 1.9) matching gualitatively the physical geometry.

The coordinates $\xi^1, \ldots, \xi^n$ of the surface S^{xn} specified by composition of the parametrizations (5.1) and (5.2) are called logical or grid coordinates, while the coordinates $s^1, \ldots, s^n$ in (5.1) are referred to as parametric coordinates. We assume further that the parametric coordinates $s^1, \ldots, s^n$ in S^n are the Cartesian coordinates.

For the purpose of providing efficient control of grid generation in the geometry S^{xn} we introduce a notion of a monitor manifold M^n over S^{xn}. The

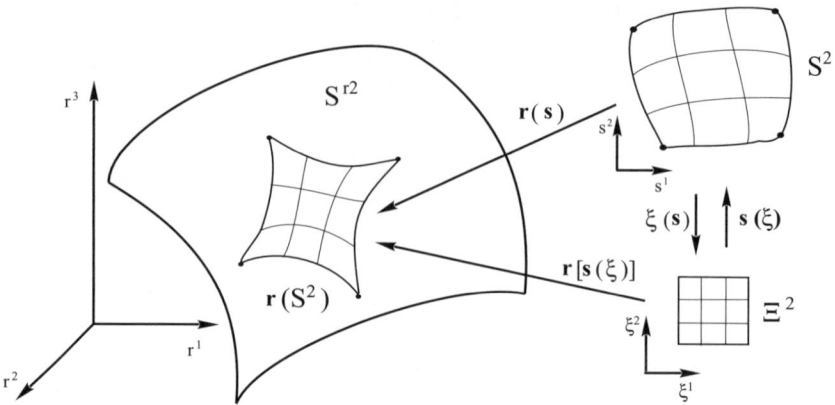

Fig. 5.1. Illustration for grid generation

points of M^n are the points of S^{xn} while the metric tensor in the coordinates $s^1, \ldots, s^n$ is defined in the form (4.21).

5.1 Formulation of a Differential Grid Generator

This section reviews a mathematical model for robust grid generators based on the differential operator of Beltrami.

5.1.1 Beltramian Operator

The operator is formulated on a set of twice differentiable functions $\boldsymbol{f}(\boldsymbol{x})$ defined on an arbitrary Riemannian manifold M^n with a covariant metric tensor (g^x_{ij}) in local coordinates x^i, $i = 1, \ldots, n$, as the following elliptic operator

$$\Delta_B[\boldsymbol{f}] = \frac{1}{\sqrt{g^x}} \frac{\partial}{\partial x^j} \left(\sqrt{g^x} g^{jk}_x \frac{\partial \boldsymbol{f}(\boldsymbol{x})}{\partial x^k} \right), \quad j,k = 1, \ldots, n, \quad (5.3)$$

where $\boldsymbol{x} = (x^1, \ldots, x^n)$, (g^{jk}_x), $j,k = 1, \ldots, n$, is the contravariant metric tensor of the manifold in the coordinates x^i, $i = 1, \ldots, n$, $g^x = \det(g^x_{ij})$.

Remind we hold a convention that a summation is carried out over repeated indices unless otherwise noted thus (5.3) is, in fact,

$$\Delta_B[\boldsymbol{f}] = \frac{1}{\sqrt{g^x}} \sum_{j=1}^{n} \frac{\partial}{\partial x^j} \left(\sqrt{g^x} \sum_{k=1}^{n} g^{jk}_x \frac{\partial \boldsymbol{f}(\boldsymbol{x})}{\partial x^k} \right).$$

Notice the matrix (g^{ij}_x) in (5.3) is the inverse of the matrix (g^x_{ij}) and vice versa, hence the elements of the covariant and contravariant metric tensors are connected by the relations

5.1 Formulation of a Differential Grid Generator

$$g_x^{ij} g_{jk}^x = g_{ij}^x g_x^{jk} = \delta_i^k, \quad i, j, k = 1, \ldots, n,$$

where $\delta_i^k = 0, 1$ if $k \neq i$, $k = i$.

It was shown in Sect. 4.5.1 (see equations (4.56) and (4.58)) that the formula (5.3) is invariant of parametrizations of the manifold M^n. We present here one more inference of this fact by virtue of the identity (2.48). Let s^i, $i = 1, \ldots, n$, be another local coordinate system of M^n and (g_{ij}^s) and (g_s^{ij}), $i, j = 1, \ldots, n$, be the covariant and contravariant metric tensors, respectively, of M^n in these coordinates. Using the metric relations

$$g_x^{ij} = g_s^{kl} \frac{\partial x^i}{\partial s^k} \frac{\partial x^j}{\partial s^l}, \quad i, j, k, l = 1, \ldots, n,$$

$$g^x J^{-2} = g^s, \quad g^x = g^s (J)^2,$$

where

$$J = \det\left(\frac{\partial s^i}{\partial x^j}\right), \quad g^s = \det(g_{ij}^s), \quad i, j = 1, \ldots, n,$$

and the fundamental identity

$$\frac{\partial}{\partial x^j}\left(J \frac{\partial x^j}{\partial s^i}\right) = 0, \quad i, j = 1, \ldots, n,$$

(see 2.48) gives

$$\Delta_B[\boldsymbol{f}](\boldsymbol{x}(\boldsymbol{s})) = \frac{1}{\sqrt{g^x}} \frac{\partial}{\partial x^j}\left(\sqrt{g^x} J^{-1} J g_s^{mk} \frac{\partial x^j}{\partial s^m} \frac{\partial x^i}{\partial s^k} \frac{\partial \boldsymbol{f}[\boldsymbol{x}(\boldsymbol{s})]}{\partial s^p} \frac{\partial s^p}{\partial x^i}\right)$$

$$= \frac{1}{\sqrt{g^x}} \frac{\partial}{\partial x^j}\left(\sqrt{g^s} J \frac{\partial x^j}{\partial s^m} g_s^{mk} \frac{\partial \boldsymbol{f}[\boldsymbol{x}(\boldsymbol{s})]}{\partial s^k}\right)$$

$$= \frac{J}{\sqrt{g^x}} \frac{\partial}{\partial x^j}\left(\sqrt{g^s} g_s^{mk} \frac{\partial \boldsymbol{f}[\boldsymbol{x}(\boldsymbol{s})]}{\partial s^k}\right) \frac{\partial x^j}{\partial s^m}$$

$$= \frac{1}{\sqrt{g^s}} \frac{\partial}{\partial s^m}\left(\sqrt{g^s} g_s^{mp} \frac{\partial \boldsymbol{f}[\boldsymbol{x}(\boldsymbol{s})]}{\partial s^p}\right) = \Delta_B[\boldsymbol{f}(\boldsymbol{x})](\boldsymbol{s}),$$

$$i, j, k, m, p = 1, \ldots, n.$$

So the value of the Beltramian operator is an invariant of the choice of a parametrization of the manifold M^n. This invariant is called Beltrami's second differential parameter of $\boldsymbol{f}$.

5.1.2 Boundary Value Problem for Grid Equations

The Beltramian operator allows one to formulate a mathematical model for generating grids on arbitrary Riemannian manifolds with smooth local metric tensors. Though for the practical purpose it is sufficient to consider the manifolds as the monitor manifolds over a physical geometry S^{xn}. Let M^n be such n-dimensional manifold with the metric tensor (g_{ij}^s) in the local coordinates

s^i, $i = 1, \ldots, n$, whose values lie in some simply connected parametrization domain $S^n \subset R^n$. Thus there is a local map $\boldsymbol{x}(\boldsymbol{s}) : S^n \to M^n$. Analogously to the case of the physical geometry S^{xn}, a local grid in M^n is found by mapping a reference grid in a standard logical domain Ξ^n into M^n by the composition of $\boldsymbol{x}(\boldsymbol{s})$ and some one-to-one intermediate smooth transformation

$$\boldsymbol{s}(\boldsymbol{\xi}) : \Xi^n \to S^n\ , \quad \boldsymbol{\xi} = (\xi^1, \ldots, \xi^n)\ , \quad \boldsymbol{s} = (s^1, \ldots, s^n)\ ,$$

i.e. by $\boldsymbol{x}(\boldsymbol{s}(\boldsymbol{\xi})) : \Xi^n \to M^n$ (see Fig. 5.1).

Note the parametrization $\boldsymbol{x}(\boldsymbol{s})$ also generates a grid in M^n by mapping some grid in S^n. However, this grid may be unsatisfactory and as a rule it is not independent of parametrizations. Besides this, if the geometry of S^n is complex, the grid generation in S^n may require serious efforts. The role of the intermediate transformation $\boldsymbol{s}(\boldsymbol{\xi})$ is to make the grid on M^n satisfy the necessary properties, in particular, the property of independence of the choice of a parametrization. While the role of the logical domain Ξ^n is to replace the parametrization domain S^n with a standard parametric domain (n-dimensional cube, simplex, prism, etc) having a simpler shape.

The logical domain Ξ^n, its reference grid, and the parametrization $\boldsymbol{x}(\boldsymbol{s})$ are chosen by the user. Therefore the local grid in M^n with the required properties is defined when a suitable intermediate transformation $\boldsymbol{s}(\boldsymbol{\xi})$ is found. One of the ways to find this transformation is to use the operator of Beltrami in the metric of M^n. Namely, $\boldsymbol{s}(\boldsymbol{\xi})$ is determined as the inverse of the transformation

$$\boldsymbol{\xi}(\boldsymbol{s}) : S^n \to \Xi^n\ , \quad \boldsymbol{\xi}(\boldsymbol{s}) = [\xi^1(\boldsymbol{s}), \ldots, \xi^n(\boldsymbol{s})]$$

which is a solution of the following Dirichlet boundary value problem

$$\Delta_B[\boldsymbol{\xi}] \equiv \frac{1}{\sqrt{g^s}} \frac{\partial}{\partial s^j}\left(\sqrt{g^s} g_s^{jk} \frac{\partial \boldsymbol{\xi}}{\partial s^k}\right) = 0\ , \quad j, k = 1, \ldots, n\ ,$$

$$\Gamma[\boldsymbol{\xi}] \equiv \boldsymbol{\xi}|_{\partial S^n} = \boldsymbol{\varphi}(\boldsymbol{s}) : \partial S^n \to \partial \Xi^n\ , \quad \boldsymbol{\varphi}(\boldsymbol{s}) = [\varphi^1(\boldsymbol{s}), \ldots, \varphi^n(\boldsymbol{s})]\ ,$$

or in a component form

$$\Delta_B[\xi^i] \equiv \frac{1}{\sqrt{g^s}} \frac{\partial}{\partial s^j}\left(\sqrt{g^s} g_s^{jk} \frac{\partial \xi^i}{\partial s^k}\right) = 0\ , \quad i, j, k = 1, \ldots, n\ ,$$

$$\Gamma[\xi^i] \equiv \xi^i|_{\partial S^n} = \varphi^i(\boldsymbol{s})\ , \quad i = 1, \ldots, n\ , \tag{5.4}$$

where g_s^{jk} is the (jk)th element of the contravariant metric tensor of M^n in the parametric coordinates $s^1, \ldots, s^n$, $g^s = \det(g_{ij}^s)$, ∂S^n and $\partial \Xi^n$ is the boundary of S^n and Ξ^n, respectively, while $\boldsymbol{\varphi}(\boldsymbol{s})$ is a one-to-one continuous transformation between the boundaries of S^n and Ξ^n. So we see that the transformation $\boldsymbol{\xi}(\boldsymbol{s})$ is the function whose Beltrami's second differential parameter is equal to the zero vector $\boldsymbol{0}$. We shall call the equations in (5.4) as grid equations. In the theory of Riemannian manifolds these equations are called generalized Laplace equations.

5.1 Formulation of a Differential Grid Generator

The functions $\xi^1, \ldots, \xi^n$ satisfying (5.4) form a curvilinear coordinate system in S^n and M^n. These curvilinear coordinates are further referred to as grid coordinates.

Since $\Delta_B[\boldsymbol{\xi}]$ is independent of parametrizations of M^n we obtain by the solution of (5.4) the same grid in M^n regardless of the original coordinate system in M^n provided the boundary conditions in (5.4) for different coordinates are consistent. Assuming $s^i = \xi^i$, $i = 1, \ldots, n$, in (5.4) we also find that this system is equivalent to the following system

$$\Delta_B[\xi^i] \equiv \frac{1}{\sqrt{g^{\boldsymbol{\xi}}}} \frac{\partial}{\partial \xi^j} (\sqrt{g^{\boldsymbol{\xi}}} g_{\boldsymbol{\xi}}^{ji}) = 0, \quad i, j = 1, \ldots, n,$$

$$\Gamma[\xi^i] \equiv \xi^i|_{\partial S^n} = \varphi^i(\boldsymbol{s}), \quad i = 1, \ldots, n,$$

(5.5)

where $(g_{\boldsymbol{\xi}}^{ji})$, $i, j = 1, \ldots, n$, is the contravariant metric tensor of M^n in the grid coordinates $\xi^1, \ldots, \xi^n$,

$$g^{\boldsymbol{\xi}} = J^2 g^s, \quad J = \det\left(\frac{\partial s^i}{\partial \xi^j}\right) = 1/\det\left(\frac{\partial \xi^i}{\partial s^j}\right).$$

Note, in order to find grid nodes through the intermediate transformation $\boldsymbol{s}(\boldsymbol{\xi})$ the inverse of which is a solution of the system (5.4) or (5.5) there is no necessity to compute the transformation $\boldsymbol{s}(\boldsymbol{\xi})$ at all points $\boldsymbol{\xi} \in \Xi^n$. It is sufficient to solve numericaly the boundary value problem obtained from (5.4) by interchanging mutually dependent and independent variables. The transformed problem with respect to the components $s^i(\boldsymbol{\xi})$, $i = 1, \ldots, n$, of the intermediate transformation $\boldsymbol{s}(\boldsymbol{\xi})$ should be solved on the reference grid in Ξ^n. The values of this very numerical solution

$$\boldsymbol{s}(\boldsymbol{\xi}) = [s^1(\boldsymbol{\xi}), \ldots, s^n(\boldsymbol{\xi})]$$

determine intermediate grid nodes in S^n and consequently on M^n by mapping them through $\boldsymbol{x}(\boldsymbol{s})$.

The elliptic system in (5.4) is of divergence form hence its solution is subject to the maximum principle. Therefore the grid nodes produced through (5.4) will be inside of S^{xn} if the computational domain Ξ^n is convex. Moreover, for $n = 2$ there is valid a general theorem of Rado from which a particular corollary follows that the transformation $\boldsymbol{\xi}(\boldsymbol{s})$ obtained by the solution of the Dirichlet boundary value problem for the system in (5.4) with an arbitrary metric is nondegenerate if Ξ^2 is convex and the boundary mapping determined by the Dirichlet condition is a one-to-one map between the boundaries of S^2 and Ξ^2. This property is one of the major reasons why the formulation of the grid system in (5.4) is made with respect to the function $\boldsymbol{\xi}(\boldsymbol{s})$, though its formulation with respect to the intermediate transformation $\boldsymbol{s}(\boldsymbol{\xi})$ seamed to be more natural as such system would be linear and of divergence form and therefore more tractable for finding this transformation and consequently for implementations of the grid equations into numerical codes. However such divergent model, owing to the maximum principle, does

5.1.3 Interpretation as a Multidimensional Equidistribution Principle

In a one-dimensional case the equation in (5.4) is as follows:

$$\Delta_B[\xi] \equiv \frac{1}{\sqrt{g^s}} \frac{d}{ds}\left(\sqrt{g^s} g_s^{11} \frac{d\xi}{ds}\right) \equiv \frac{1}{\sqrt{g^s}} \frac{d}{ds}\left(\sqrt{g_s^{11}} \frac{d\xi}{ds}\right) = 0,$$

since $g^s = g_{11}^s = 1/g_s^{11}$. This equation presents the well known differential formulation of the equidistribution principle with the weight function $w(s) = \sqrt{g_s^{11}}$ defined through the monitor metric g_{11}^s. The multidimensional equation in (5.4) may also be interpreted as a formulation of the multidimensional equidistribution principle with the matrix-valued weight function

$$\boldsymbol{w}(\boldsymbol{s}) = (\sqrt{g^s} g_s^{jk}), \quad j, k = 1, \ldots, n,$$

in the coordinates s^i, $i = 1, \ldots, n$, defined through the metric g_{ij}^s, $i, j = 1, \cdots, n$. By this interpretation the problem of grid generation with the required grid properties is naturally solved by the choice of the matrix-valued weight function or, what is the same, with a specification of the metric, called a monitor metric, in M^n.

5.1.4 Examples of Familiar Grid Equations

Laplace System. If M^n is a domain S^n in R^n with the Euclidean metric, i.e. in the Cartesian coordinates $s^1, \ldots, s^n$,

$$g_{ij}^s = \delta_j^i, \quad i, j = 1, \ldots, n,$$

then

$$g^s = 1, \quad g_s^{ij} = \delta_j^i, \quad i, j = 1, \ldots, n,$$

and consequently the grid equations in (5.4) for this metric are the Laplace equations

$$\nabla^2[\xi^i] \equiv \frac{\partial}{\partial s^j} \frac{\partial}{\partial s^j} \xi^i(\boldsymbol{s}) = 0, \quad i, j = 1, \ldots, n. \tag{5.6}$$

So the equations in (5.4) being a generalization of the Laplace equations are called generalized Laplace equations.

The ordinary Laplace equations (5.6) were proposed for $n = 2$ by Crowley (1962) and Winslow (1967) for generating fixed grids in two-dimensional domains. Therefore the method for generating grids on surfaces S^{xn} by solving the boundary value problem (5.4) can also be considered as an extension of the Crowley–Winslow approach.

Diffusive System. A generalization of the Eucledian metric in S^n to the spherical metric specified in the original parametric Cartesian coordinates $s^1, \ldots, s^n$ by

$$g^s_{ij} = v(\boldsymbol{s})\delta^i_j, \quad i,j = 1,\ldots,n, \quad v(\boldsymbol{s}) > 0, \tag{5.7}$$

produces the grid equations for (5.4) in the following form

$$\frac{1}{[v(\boldsymbol{s})]^{n/2}} \frac{\partial}{\partial s^j}\left[(v(\boldsymbol{s}))^{(n-2)/2}\frac{\partial \xi^i}{\partial s^j}\right] = 0, \quad i,j = 1,\ldots,n, \tag{5.8}$$

since

$$g^s = [v(\boldsymbol{s})]^n, \quad g^{ij}_s = \frac{1}{v(\boldsymbol{s})}\delta^i_j, \quad i,j = 1,\ldots,n.$$

Thus the popular diffusive form

$$\frac{\partial}{\partial s^j}\left(w(\boldsymbol{s})\frac{\partial \xi^i}{\partial s^j}\right) = 0, \quad i,j = 1,\ldots,n, \tag{5.9}$$

of the elliptic grid generator for constructing adaptive grids in a domain S^n is, in fact, equivalent to the grid system (5.8) for $n \neq 2$ in the metric (5.7) with

$$v(\boldsymbol{s}) = w(\boldsymbol{s})^{2/(n-2)}$$

imposed in the domain S^n.

Mind equations (5.8) for $n = 2$ are equivalent to Laplace equations (5.6) regardless of the form of the function $v(\boldsymbol{s})$ in (5.7). Therefore equations (5.9), with $w(\boldsymbol{s}) \neq const$, are not equivalent to the grid equations in (5.4) for $n = 2$.

Note the diffusive equations (5.9) for $n = 2$ were introduced for generating adaptive grids by Danaev, Liseikin, and Yanenko (1980) and Winslow (1981). An extension of the diffusive form (5.9) by introducing an individual weight function $w^i(\boldsymbol{s})$ for each direction $\xi^i(\boldsymbol{s})$ was realized by Eiseman (1987) and Reed, Hsu, and Shiau (1988).

Surface Grid System. Let $S^{x2} \subset R^3$ be a surface represented locally by a parametrization

$$\boldsymbol{x}(\boldsymbol{s}): S^2 \to R^3, \quad \boldsymbol{s} = (s^1, s^2), \quad \boldsymbol{x} = (x^1, x^2, x^3).$$

The elements of the covariant metric tensor (g_{ij}) of S^{x2} in the coordinates s^1, s^2 are defined by

$$g_{ij} = \frac{\partial \boldsymbol{x}}{\partial s^i} \cdot \frac{\partial \boldsymbol{x}}{\partial s^j}, \quad i,j = 1,2. \tag{5.10}$$

The equations

$$\frac{\partial}{\partial s^j}\left(\sqrt{g}g^{kj}\frac{\partial \xi^i}{\partial s^k}\right) = 0, \quad i,j,k = 1,2, \tag{5.11}$$

where $g = \det(g_{ij})$, (g^{kj}) is the contravariant metric tensor of S^{x2}, are equivalent to the grid equations in (5.4) for $n = 2$ with respect to the metric

(5.10). These equations popular for generating fixed grids on boundaries of domains were proposed by Warsi [1981]. The equations (5.11) were also used by Danaev (1986) for generating isometric grids on surfaces. The property of invariancy guarantees that the distribution of the grid nodes found through (5.11) does not depend on a parametrization of S^{x2}.

Relation to Beltramian Equations. The Beltramian system for generating quasiconformal mappings $\boldsymbol{\xi}(s) : S^n \to \varXi^n$ between the parametric domain S^n of a Riemannian manifold M^n with the metric g_{ij}^s and $\varXi^n$ with the Eucledian metric has the following form

$$\sqrt{g^{\xi s}}\frac{\partial \boldsymbol{s}}{\partial \xi^i} = \sqrt{g^s}\boldsymbol{\nabla}^i, \quad i = 1, \ldots, n, \tag{5.12}$$

where

$$\boldsymbol{\nabla}^i = (g_s^{1m}\frac{\partial \xi^i}{\partial s^m}, \ldots, g_s^{nm}\frac{\partial \xi^i}{\partial s^m}), \quad i, m = 1, \ldots, n,$$

$$g^{\xi s} = \det\left(\frac{\partial \boldsymbol{\xi}}{\partial s^i} \cdot \frac{\partial \boldsymbol{\xi}}{\partial s^j}\right).$$

Note if $n = 2$, $g_{ij}^s = \delta_j^i$, then (5.12) is the Couchy–Riemann system

$$\frac{\partial \xi^1}{\partial s^1} = \frac{\partial \xi^2}{\partial s^2}, \quad \frac{\partial \xi^1}{\partial s^2} = -\frac{\partial \xi^2}{\partial s^1},$$

i.e. $\boldsymbol{\xi}(s)$, in this case, is a conformal mapping. So the system (5.12) is generally considered as a condition of conformality of the mapping $\boldsymbol{\xi}(s) : S^n \to \varXi^n$ with respect to the Riemannian metric g_{ij}^s.

From the identity (2.48) we have

$$\mathrm{div}(\sqrt{g^{\xi s}}\frac{\partial \boldsymbol{s}}{\partial \xi^i}) = \frac{\partial}{\partial s^j}\left(\sqrt{g^{\xi s}}\frac{\partial s^j}{\partial \xi^i}\right) \equiv 0, \quad i, j = 1, \ldots, n,$$

therefore the system (5.12) results in the system of the generalized Laplace equations presented in (5.4).

Note that the contrary assertion is not valid, i. e. the system in (5.4) does not lead, in general, to the system (5.12).

5.1.5 Realization of Specified Grids

Grids in Domains. Let M^n be an n-dimensional domain $X^n \subset R^n$ which has a local grid obtained with the aid of a nondegenerate smooth transformation

$$\boldsymbol{x}(\boldsymbol{\xi}) : \varXi^n \to X^n, \quad \boldsymbol{x} = (x^1, \ldots, x^n), \quad \boldsymbol{\xi} = (\xi^1, \ldots, \xi^n), \tag{5.13}$$

i.e. the local grid in X^n is the image of a reference grid in $\varXi^n$ into X^n through the coordinate transformation $\boldsymbol{x}(\boldsymbol{\xi})$. Note the reference grid in $\varXi^n$ can be either structured or unstructured.

Let S^n be the image of the domain Ξ^n in X^n through $\boldsymbol{x}(\boldsymbol{\xi})$. Then S^n can be formally considered as a local parametric domain for X^n with the coordinates $s^i = x^i$, $i = 1, \ldots, n$. Let

$$\boldsymbol{r}(\boldsymbol{s}) : S^n \to R^n, \quad \boldsymbol{r} = (r^1, \ldots, r^n), \quad \boldsymbol{s} = (s^1, \ldots, s^n)$$

be the mapping defined by

$$\boldsymbol{r}(\boldsymbol{s}) \equiv \boldsymbol{\xi}(\boldsymbol{s}), \quad \boldsymbol{s} \in S^n,$$

where $\boldsymbol{\xi}(\boldsymbol{s})$ is the inverse of $\boldsymbol{x}(\boldsymbol{\xi}) : \Xi^n \to S^n$. Now imposing in X^n a local metric in the coordinates s^i, $i = 1, \ldots, n$, by

$$g_{ij}^s = \boldsymbol{r}_{s^i} \cdot \boldsymbol{r}_{s^j} = \frac{\partial \xi^l}{\partial s^i} \frac{\partial \xi^l}{\partial s^j}, \quad i, j, l = 1, \ldots, n, \tag{5.14}$$

we readily establish the following formulas

$$g_s^{ij} = \frac{\partial s^i}{\partial \xi^l} \frac{\partial s^j}{\partial \xi^l}, \quad i, j, l = 1, \ldots, n,$$

$$g^s = J^2, \quad J = \det\left(\frac{\partial \xi^i}{\partial s^j}\right), \quad i, j = 1, \ldots, n. \tag{5.15}$$

Applying the equations in (5.4) to the corresponding components of the function $\boldsymbol{\xi}(\boldsymbol{s})$, which is the inverse of (5.13) with $\boldsymbol{s} = \boldsymbol{x}$, and taking into account (5.15) and (2.48), we obtain

$$\begin{aligned}
\Delta_B[\xi^i] &= \frac{1}{\sqrt{g^s}} \frac{\partial}{\partial s^j}\left(\sqrt{g^s} g_s^{jk} \frac{\partial \xi^i}{\partial s^k}\right) \\
&= \frac{1}{\sqrt{g^s}} \frac{\partial}{\partial s^j}\left(\sqrt{g^s} \frac{\partial s^j}{\partial \xi^l} \frac{\partial s^k}{\partial \xi^l} \frac{\partial \xi^i}{\partial s^k}\right) \\
&= \frac{1}{\sqrt{g^s}} \frac{\partial}{\partial s^j}\left(J \frac{\partial s^j}{\partial \xi^i}\right) = 0, \quad i, j, k, l = 1, \ldots, n.
\end{aligned}$$

As a solution of the Dirichlet problem (5.4) with the boundary conditions specified by the given mapping $\boldsymbol{\xi}(\boldsymbol{s})$ is unique, we obtain that this solution is the transformation that is the inverse of the original intermediate transformation $\boldsymbol{x}(\boldsymbol{\xi})$. So we see that an arbitrary nondegenerate grid in X^n found by the mapping approach is realized, with a proper choice of the metric in X^n, by the comprehensive grid generator. This statement appears to be useful for two purposes. First, one can use it to verify the reliability of any numerical algorithm for the solution of the system in (5.4). Second, this fact prompts one on how to find a coordinate system with necessary local coordinates by specifying the mapping $\boldsymbol{\xi}(\boldsymbol{s})$ in selected zones. For example, it can help one build an orthogonal system near boundary segments or in the vicinity of interfaces of blocks.

Grids on Surfaces. Let now S^{x2} be a surface lying in R^3 and locally represented by a parametrization

$$\boldsymbol{x}(\boldsymbol{s}) : S^2 \to R^3 \, , \quad \boldsymbol{x} = (x^1, x^2, x^3) \, , \quad \boldsymbol{s} = (s^1, s^2) \, ,$$

where S^2 is a two-dimensional parametric domain with the Cartesian coordinates s^1, s^2, while $\boldsymbol{x}(\boldsymbol{s})$ is a smooth nondegenerate vector-valued function. Note the transformation $\boldsymbol{x}(\boldsymbol{s})$ engenders in S^{x2} a natural metric tensor in the coordinates s^1, s^2:

$$g_{ij}^{xs} = \boldsymbol{x}_{s^i} \cdot \boldsymbol{x}_{s^j} = \frac{\partial x^k}{\partial s^i} \frac{\partial x^k}{\partial s^j} \, , \quad i,j = 1,2 \, , \quad k = 1,2,3 \, .$$

If a grid on the surface S^{x2} is constructed with the aid of a transformation

$$\boldsymbol{x}_1(\boldsymbol{\xi}) : \Xi^2 \to R^3 \, ,$$

then there is a diffeomorphism

$$\boldsymbol{\xi}(\boldsymbol{s}) : S^2 \to \Xi^2$$

such that

$$\boldsymbol{x}(\boldsymbol{s}) = \boldsymbol{x}_1[\boldsymbol{\xi}(\boldsymbol{s})] : S^2 \to R^3 \, .$$

It is easily shown, likewise to the case of the domain discussed above, that the components of the transformation $\boldsymbol{\xi}(\boldsymbol{s})$ are the solutions of the grid equations in (5.4) with respect to the covariant metric tensor specified in the coordinates s^1, s^2 as

$$g_{ij}^s = \frac{\partial \boldsymbol{\xi}}{\partial s^i} \cdot \frac{\partial \boldsymbol{\xi}}{\partial s^j} \, , \quad i,j = 1,2 \, .$$

Examples of Metrics Deriving Classical Grid Coordinates. Here we find, using (5.14), the expressions for the metric elements by which polar and spherical coordinate systems are realized through the solution of the problem (5.4). For the two-dimensional polar system of coordinates

$$x = \rho \cos \varphi \, ,$$
$$y = \rho \sin \varphi \, ,$$
(5.16)

we find, assuming $x = s^1$, $y = s^2$, $\rho = \xi^1$, $\varphi = \xi^2$,

$$\frac{\partial \boldsymbol{s}}{\partial \xi^1} = \left(\frac{s^1}{\rho}, \frac{s^2}{\rho} \right) \, , \quad \frac{\partial \boldsymbol{s}}{\partial \xi^2} = (-s^2, s^1) \, ,$$

$$\frac{\partial \boldsymbol{\xi}}{\partial s^1} = \left(\frac{s^1}{\rho}, -\frac{s^2}{\rho^2} \right) \, , \quad \frac{\partial \boldsymbol{\xi}}{\partial s^2} = \left(\frac{s^2}{\rho}, \frac{s^1}{\rho^2} \right) \, ,$$

where $\rho = \sqrt{(s^1)^2 + (s^2)^2}$. So the elements of the corresponding metric covariant and contravariant tensors (5.14) and (5.15) in the coordinates s^1, s^2,

5.1 Formulation of a Differential Grid Generator

are as follows:

$$g_{ij}^s = \frac{\partial \boldsymbol{\xi}}{\partial s^i} \cdot \frac{\partial \boldsymbol{\xi}}{\partial s^j} = g^s[\delta_j^i + s^i s^j (1 - g^s)] , \quad i,j = 1,2 ,$$

$$g_s^{ij} = \frac{\partial s^i}{\partial \xi^k} \frac{\partial s^j}{\partial \xi^k} = \frac{1}{g^s}\delta_j^i - s^i s^j (1 - g^s) , \quad i,j,k = 1,2 ,$$

(5.17)

where $g^s = \det(g_{ij}^s) = 1/((s^1)^2 + (s^2)^2) = 1/\rho^2$. We readily see that this metric is considerably different from the Eucledian metric. The equations in (5.4) with respect to the metric (5.17) realizing the polar coordinate system have the form

$$\frac{1}{\sqrt{g^s}} \frac{\partial}{\partial s^i} \left\{ \sqrt{g^s} \left[\frac{1}{g^s} \delta_j^k - (1 - g^s) s^k s^j \right] \frac{\partial \xi^i}{\partial s^k} \right\} = 0 , \quad i,j,k = 1,2 . \quad (5.18)$$

Analogously, for the three-dimensional spherical coordinate system

$$x = \rho \cos \varphi \cos \phi ,$$
$$y = \rho \sin \varphi \cos \phi , \quad (5.19)$$
$$z = \rho \sin \phi ,$$

we find, assuming $x = s^1, y = s^2, z = s^3, \rho = \xi^1, \varphi = \xi^2, \phi = \xi^3$,

$$\frac{\partial \boldsymbol{s}}{\partial \xi^1} = \left(\frac{s^1}{\rho}, \frac{s^2}{\rho}, \frac{s^3}{\rho} \right) ,$$

$$\frac{\partial \boldsymbol{s}}{\partial \xi^2} = (-s^2, s^1, 0) ,$$

$$\frac{\partial \boldsymbol{s}}{\partial \xi^3} = \left(-\frac{s^1 s^3}{\sqrt{(s^1)^2 + (s^2)^2}}, -\frac{s^2 s^3}{\sqrt{(s^1)^2 + (s^2)^2}}, \sqrt{(s^1)^2 + (s^2)^2} \right) ,$$

$$\frac{\partial \boldsymbol{\xi}}{\partial s^1} = \left(\frac{s^1}{\rho}, -\frac{s^2}{(s^1)^2 + (s^2)^2}, -\frac{s^1 s^3}{\rho^2 \sqrt{(s^1)^2 + (s^2)^2}} \right) ,$$

$$\frac{\partial \boldsymbol{\xi}}{\partial s^2} = \left(\frac{s^2}{\rho}, \frac{s^1}{(s^1)^2 + (s^2)^2}, -\frac{s^2 s^3}{\rho^2 \sqrt{(s^1)^2 + (s^2)^2}} \right) ,$$

$$\frac{\partial \boldsymbol{\xi}}{\partial s^3} = \left(\frac{s^3}{\rho}, 0, \frac{\sqrt{(s^1)^2 + (s^2)^2}}{\rho^2} \right) ,$$

where $\rho = \sqrt{(s^1)^2 + (s^2)^2 + (s^3)^2}$. The elements of the corresponding metric tensors (g_{ij}^s), $i,j = 1,2,3$, and (g_s^{ij}), $i,j = 1,2,3$, in the coordinates s^1, s^2, and s^3 realizing the spherical coordinates through the solution of (5.4) have the following form

$$g_{11}^s = \frac{1}{\rho^2} + (s^1)^2\left(\frac{1}{\rho^2} - \frac{1}{\rho^4}\right) + \frac{(s^2)^2}{[(s^1)^2 + (s^2)^2]^2} - \frac{(s^2)^2}{\rho^2[(s^1)^2 + (s^2)^2]},$$

$$g_{12}^s = s^1 s^2\left(\frac{1}{\rho^2} - \frac{1}{\rho^4}\right) + \frac{s^1 s^2}{\rho^2[(s^1)^2 + (s^2)^2]} - \frac{s^1 s^2}{[(s^1)^2 + (s^2)^2]^2},$$

$$g_{13}^s = s^1 s^3\left(\frac{1}{\rho^2} - \frac{1}{\rho^4}\right),$$

$$g_{22}^s = \frac{1}{\rho^2} + (s^2)^2\left(\frac{1}{\rho^2} - \frac{1}{\rho^4}\right) + \frac{(s^1)^2}{[(s^1)^2 + (s^2)^2]^2} - \frac{(s^1)^2}{\rho^2[(s^1)^2 + (s^2)^2]}, \quad (5.20)$$

$$g_{23}^s = s^2 s^3\left(\frac{1}{\rho^2} - \frac{1}{\rho^4}\right),$$

$$g_{33}^s = \frac{1}{\rho^2} + (s^3)^2\left(\frac{1}{\rho^2} - \frac{1}{\rho^4}\right).$$

5.1.6 Formulation of Monitor Metrics

The expressions (5.14) and (5.17) for the metrics realizing both arbitrary and particular polar coordinate systems, respectively, through the solution of the grid equations in (5.4) prompt us one way on how to design a suitable formula for specifying necessary metrics in general geometries in order to obtain suitable intermediate transformations for generating grids. Namely let g_{ij}, $i,j = 1,\ldots,n$, and g^{ij}, $i,j = 1,\ldots,n$, be any covariant and contravariant metric tensor, respectively, of a physical geometry (in general of a Riemannian manifold) M in the coordinates $s^1,\ldots,s^n$. Then the metric g_{ij} and three smooth functions $z(s)$, $v(s)$, and $f(s)$ define the following new covariant and contravariant monitor metric tensors g_{ij}^s and g_s^{ij}, respectively, in the coordinates $s^1,\ldots,s^n$:

$$g_{ij}^s = z g_{ij} + v\frac{\partial f}{\partial s^i}\frac{\partial f}{\partial s^j}, \quad i,j = 1,\ldots,n,$$

$$g_s^{ij} = \frac{1}{z}g^{ij} - d g^{im}\frac{\partial f}{\partial s^m}g^{jp}\frac{\partial f}{\partial s^p}, \quad i,j,m,p = 1,\ldots,n, \quad (5.21)$$

where

$$d = \frac{v}{z[z + v\nabla(f)]},$$

$$\nabla(f) = g^{ij}\frac{\partial f}{\partial s^i}\frac{\partial f}{\partial s^j}, \quad i,j = 1,\ldots,n.$$

The points of the geometry M with this metric comprise a Rimennian manifold M^n called a monitor manifold over M.

The formula in (5.21) for the contravariant metric tensor of M^n will be proved below in Sect. 6.2.1. Note the functions z, v, and f in (5.21) must be such that $z \geq 0$ and $\det(g_{ij}^s) \neq 0$.

5.1 Formulation of a Differential Grid Generator

The covariant and contravariant metric tensors g_{ij}^s and g_s^{ij}, respectively, in (5.17) in the coordinates s^1, s^2, considered above for generating the polar coordinate system (5.16), are realized by the metric tensors (5.21) with

$$g_{ij} = g^{ij} = \delta_j^i, \quad i,j = 1,2,$$

$$z(s) = \frac{1}{\rho^2}, \quad v(s) = \frac{1}{\rho^2}\left(1 - \frac{1}{\rho^2}\right), \quad f(s) = \frac{1}{2}\rho^2, \tag{5.22}$$

where $\rho^2 = (s^1)^2 + (s^2)^2 = |s|^2$.

A coordinate transformation $s(\xi)$ representing a more general polar coordinate system

$$s^1 = b(\xi^1)\cos\xi^2, \quad s^2 = b(\xi^1)\sin\xi^2,$$

when $b'(\xi^1) > 0$ is realized as the inverse of the solution of (5.4) in the metric (5.21) if

$$g_{ij} = g^{ij} = \delta_j^i, \quad i,j = 1,2,$$

$$z(s) = \frac{1}{\rho^2}, \quad f(s) = \frac{1}{2}\rho^2,$$

$$v(s) = z(s)\left[\frac{1}{[b'(b^{-1}(\rho))]^2} - z(s)\right],$$

where $\rho^2 = (s^1)^2 + (s^2)^2$.

The functions $z(s)$ and $v(s)$ in (5.21) can be interpreted as weight functions which control the influence of the original metric g_{ij} and the monitor function $f(s)$ on the grid behavior. A generalization of the metric (5.21) is naturally carried out by the following formula

$$g_{ij}^s = z(s)g_{ij} + v^k(s)f_{s^i}^k(s)f_{s^j}^k(s), \quad i,j = 1,\ldots,n, \quad k = 1,\ldots,l, \tag{5.23}$$

with weight functions $z(s) \geq 0$ and $v^k(s) \geq 0$, $k = 1,\ldots,l$, and monitor functions $f^k(s)$, $k = 1,\ldots,l$. This metric is referred to as a monitor metric. Remind it is assumed that a summation is carried out in (5.23) over the index k.

Note when

$$g_{ij} = \mathbf{x}_{s^i} \cdot \mathbf{x}_{s^j}, \quad i,j = 1,\ldots,n,$$

where $\mathbf{x}(s)$ is the parametrization (5.1) of S^{xn} then the metric (5.23) for $z(s) = v^k(s) = 1$, $k = 1,\ldots,l$, is the metric of an n-dimensional regular surface represented in the form (5.1) as

$$\mathbf{r}(s): S^n \to \mathbb{R}^{n+k+l}, \quad \mathbf{r}(s) = [\mathbf{x}(s), f^1(s), \ldots, f^l(s)]. \tag{5.24}$$

This surface designated by S^{rn} is referred to as a monitor surface over S^{xn}.

The approach for formulating monitor metrics in the form (5.23) is somewhat similar to the variational approach in which the grid functional is formulated as a combination of several functionals with weights (see (1.24)) each of which is responsible for providing some individual grid property. Note the boundary value problem (5.4) with respect to an arbitrary metric is well-posed. Moreover it is well-posed in the monitor metric (5.23) with arbitrary weight and monitor functions contrary to the problem of the minimization of the combined functional that is well-posed only for special weight functions. Besides this the grid obtained from the minimization of the combined functional may be dependent on a parametrization of the physical geometry.

The general monitor metrics in the forms (5.23) and (5.24) were introduced by Liseikin (2002a, 2003) and Liseikin (1991), respectively.

The monitor metric (5.23) can be extended to a more general metric through a set of covariant tensors of the first rank

$$\boldsymbol{F}^k(\boldsymbol{s}) = [F_1^k(\boldsymbol{s}), \ldots, F_n^k(\boldsymbol{s})] \, , \quad k = 1, \ldots, l \, ,$$

by the following formula in the parametric coordinates $s^1, \ldots, s^n$:

$$g_{ij}^s = z(\boldsymbol{s})g_{ij} + F_i^k(\boldsymbol{s})F_j^k(\boldsymbol{s}) \, , \quad i,j = 1, \ldots, n \, , \quad k = 1, \ldots, l \, . \tag{5.25}$$

The monitor metric (5.23) is realized by this metric if

$$F_i^k(\boldsymbol{s}) = \sqrt{v^k(\boldsymbol{s})} \frac{\partial f^k}{\partial s^i} \, , \quad i = 1, \ldots, n \, , \quad k = 1, \ldots, l \, , \quad k \text{ fixed} \, .$$

Formulation of the monitor metric in the form (5.25) may be useful for generating a family of grid coordinates orthogonal to a boundary segment. For example, let a boundary curve of a two-dimensional domain S^2 be defined from the equation $\varphi(s^1, s^2) = 0$. If we assume the variable $\varphi(\boldsymbol{s})$ as a logical grid coordinate then the second grid coordinate $\psi(\boldsymbol{s})$ which is orthogonal to the curve $\varphi(\boldsymbol{s}) = 0$ is subject to the relation

$$\mathrm{grad}\varphi \cdot \mathrm{grad}\psi = 0 \, ,$$

i.e.

$$(\psi_{s^1}, \psi_{s^2}) = w(\boldsymbol{s})(\varphi_{s^2}, -\varphi_{s^1}) \, , \tag{5.26}$$

at the points of the curve. As the coordinate system φ, ψ is realized by the solution of the Beltrami's equations in (5.4) with respect to the metric defined by (5.14) for $n = 2$, $\xi^1 = \varphi$, $\xi^2 = \psi$, we find, availing us of (5.26) that this metric should be as follows:

$$g_{ij}^s = \varphi_{s^i}\varphi_{s^j} + \psi_{s^i}\psi_{s^j} = \varphi_{s^i}\varphi_{s^j} + (-1)^{i+j}w^2\varphi_{s^{3-i}}\varphi_{s^{3-j}} \, , \tag{5.27}$$

$$i,j = 1, 2 \, , \quad i,j \text{ fixed} \, .$$

Since

$$(-1)^{i+j}\varphi_{s^{3-i}}\varphi_{s^{3-j}} = [(\varphi_{s^1})^2 + (\varphi_{s^2})^2]\delta_j^i - \varphi_{s^i}\varphi_{s^j} \, , \quad i,j = 1, 2 \, , \quad i,j \text{ fixed},$$

so the expression (5.27) has the following equivalent form

$$g_{ij}^s = [(\varphi_{s^1})^2 + (\varphi_{s^2})^2]w^2(\boldsymbol{s})\delta_j^i + [1 - w^2(\boldsymbol{s})]\varphi_{s^i}\varphi_{s^j} \, , \quad i,j = 1, 2 \, , \tag{5.28}$$

which is, in fact, of the form (5.21).

5.2 Variational Formulation

Here we discuss a relation of the grid equations in (5.4) to a variational grid generation approach.

5.2.1 Functional of Grid Smoothness

The system of the generalized Laplace equations in (5.4) is equivalent to the Euler–Lagrange equations for the following functional

$$I[\boldsymbol{\xi}] = \int_{S^n} \sqrt{g^s} g_s^{jk} \frac{\partial \xi^i}{\partial s^k} \frac{\partial \xi^i}{\partial s^j} d\boldsymbol{s} \ . \tag{5.29}$$

Indeed the Euler–Lagrange equations for the functional in a general form

$$I[\boldsymbol{\xi}] = \int_{S^n} G\left[\boldsymbol{s}, \boldsymbol{\xi}(\boldsymbol{s}), \frac{\partial \boldsymbol{\xi}}{\partial s^1}(\boldsymbol{s}), \ldots, \frac{\partial \boldsymbol{\xi}}{\partial s^n}(\boldsymbol{s})\right] d\boldsymbol{s} \tag{5.30}$$

are as follows:

$$G_{\xi^i} - \frac{\partial}{\partial s^j}\left\{G_{\partial \xi^i/\partial s^j}\left[\boldsymbol{s}, \boldsymbol{\xi}(\boldsymbol{s}), \frac{\partial \boldsymbol{\xi}}{\partial s^1}(\boldsymbol{s}), \ldots, \frac{\partial \boldsymbol{\xi}}{\partial s^n}(\boldsymbol{s})\right]\right\} = 0 \ ,$$
$$i, j = 1, \ldots, n \ . \tag{5.31}$$

The quantities g^s and g_s^{jk} in (5.29) are specified in the parametric coordinates $s^1, \ldots, s^n$ therefore they remain unchanged when the functions $\xi^i(\boldsymbol{s})$ are varied. So the system of the Euler–Lagrange equations (5.31) derived from the functional (5.29) is readily obtained and has the following form

$$2\frac{\partial}{\partial s^j}\left(\sqrt{g^s} g_s^{jk} \frac{\partial \xi^i}{\partial s^k}\right) = 0 \ , \quad i, j, k = 1, \ldots, n \ ,$$

which is equivalent to the system of the grid equations in (5.4), since $\sqrt{g^s} \neq 0$. Thus the technique based on the minimization of the functional (5.29) produces the very grids obtained by the differential grid generator proposed above.

In the particular case when

$$\sqrt{g^s} g_s^{ij} = w(\boldsymbol{s}) \delta_j^i \ , \quad i, j = 1, \ldots, n \ ,$$

the expression (5.29) represents the diffusion-adaptation functional

$$I[\boldsymbol{\xi}] = \int_{S^n} w(\boldsymbol{s}) \sum_{i=1}^n g_s^{ii} d\boldsymbol{s} \ ,$$

with the weight function $w(\boldsymbol{s})$. This functional was originally proposed by Danaev, Liseikin, and Yanenko (1980) and Winslow (1981) for generating adaptive grids in domains.

126   5. Comprehensive Grid Models

Using the relations (4.19) the functional (5.29) is also expressed as

$$I[\boldsymbol{\xi}] = \int_{S^n} \sqrt{g^s}\mathrm{tr}(g_\xi^{ij})\mathrm{d}s ,\qquad(5.32)$$

where

$$\mathrm{tr}(g_\xi^{ij}) = \sum_{i=1}^n g_\xi^{ii} .$$

Taking into account that

$$\mathrm{d}M^n = \sqrt{g^s}\mathrm{d}s = \sqrt{g^\xi}\mathrm{d}\boldsymbol{\xi}$$

we also obtain the following form of the functional (5.29)

$$I[\boldsymbol{\xi}] = \int_{M^n} \mathrm{tr}(g_\xi^{ij})\mathrm{d}M^n$$

and

$$I[\boldsymbol{\xi}] = \int_{\Xi^n} \sqrt{g^\xi}\mathrm{tr}(g_\xi^{ij})\mathrm{d}\boldsymbol{\xi} .$$

Thus for $n = 1, 2,$ and 3 we have

$$I[\boldsymbol{\xi}] = \begin{cases} \displaystyle\int_{S^1} \sqrt{g^s}g_\xi^{11}\mathrm{d}s , & n = 1 , \\[1em] \displaystyle\int_{S^2} \sqrt{g^s}(g_\xi^{11} + g_\xi^{22})\mathrm{d}s^1\mathrm{d}s^2 , & n = 2 , \\[1em] \displaystyle\int_{S^3} \sqrt{g^s}(g_\xi^{11} + g_\xi^{22} + g_\xi^{33})\mathrm{d}s^1\mathrm{d}s^2\mathrm{d}s^3 , & n = 3 , \end{cases} \qquad(5.33)$$

with the corresponding contravariant metric elements g_ξ^{ij} and determinants g^s for each $n = 1, 2, 3$. Analogously, using suitable formulas for the elements of inverse matrices, we obtain the formulation of the functional with respect to an intermediate transformation $s(\boldsymbol{\xi})$ in terms of the covariant metric elements g_{ij}^ξ:

$$I[s] = \begin{cases} \displaystyle\int_{\Xi^1} \frac{1}{\sqrt{g^\xi}}\mathrm{d}\xi , & n = 1 , \\[1em] \displaystyle\int_{\Xi^2} \frac{1}{\sqrt{g^\xi}}(g_{11}^\xi + g_{22}^\xi)\mathrm{d}\xi^1\mathrm{d}\xi^2 , & n = 2 , \\[1em] \displaystyle\int_{\Xi^3} \frac{1}{\sqrt{g^\xi}}[g_{11}^\xi g_{22}^\xi + g_{11}^\xi g_{33}^\xi + g_{22}^\xi g_{33}^\xi \\ \qquad -(g_{12}^\xi)^2 - (g_{13}^\xi)^2 - (g_{23}^\xi)^2]\mathrm{d}\xi^1\mathrm{d}\xi^2\mathrm{d}\xi^3 , & n = 3 . \end{cases} \qquad(5.34)$$

Note that the functional $I[s]$, in this formulation, is defined on the set of invertible transformations $s(\boldsymbol{\xi}) \in C^2(\Xi^n)$.

When the manifold M^n is an n-dimensional domain X^n, the functional (5.29) is the functional referred to as a functional of grid smoothness on X^n,

$$I[\boldsymbol{\xi}] = \int_{X^n} \left(\sum_{i=1}^{n} g^{ii}_{\xi x} \right) d\boldsymbol{x} , \qquad (5.35)$$

where

$$g^{ij}_{\xi x} = \frac{\partial \xi^i}{\partial x^m} \frac{\partial \xi^j}{\partial x^m} , \quad i,j,m = 1,\ldots,n ,$$

introduced by Brackbill and Saltzman (1988). Therefore it is reasonable to call the functional (5.29), being the generalization of (5.35), as the functional of grid smoothness on the manifold M^n. It was shown by Liseikin (1999) that such a generalization of the functional (5.29) to n-dimensional regular surfaces preserves all salient features of grids obtained by applying the smoothness functional on domains.

5.2.2 Geometric Interpretation

This paragraph describes a geometric meaning of the smoothness functional (5.29) which justifies to some extent its expression for the generation of adaptive grids in domains and on surfaces. For this purpose we consider a monitor surface S^{rn} specified by (5.24) in the parametric coordinates for a monitor manifold over a physical geometry S^{xn}. The monitor surface is naturally parametrized in the grid coordinates $\xi^1, \ldots, \xi^n$ in the form

$$\boldsymbol{r}[s(\boldsymbol{\xi})] : \Xi^n \to R^{n+k+l} ,$$

while its metric in these coordinates is specified by

$$g^{\xi}_{ij} = \boldsymbol{r}_{\xi^i} \cdot \boldsymbol{r}_{\xi^j} , \quad i,j = 1,\ldots,n ,$$

where

$$\boldsymbol{r}_{\xi^i} = \frac{\partial}{\partial \xi_i} \boldsymbol{r}[s(\boldsymbol{\xi})] , \quad i = 1,\ldots,n$$

The functional of grid smoothness (5.29) with respect to the monitor surface is as follows:

$$I[\boldsymbol{\xi}] = \int_{S^{rn}} \left(\sum_{i=1}^{n} g^{ii}_{\xi} \right) dS^{rn} = \int_{S^{rn}} \mathrm{tr}(g^{ij}_{\boldsymbol{\xi}}) dS^{rn} . \qquad (5.36)$$

First, note that the trace of the contravariant n-dimensional tensor (g^{ij}_{ξ}) can be expressed through the invariants I_{n-1} and I_n of the orthogonal transforms of the covariant tensor (g^{ξ}_{ij}), namely

$$\operatorname{tr}(g_\xi^{ij}) = \frac{I_{n-1}}{I_n}, \qquad (5.37)$$

where I_i, $i = 1, \ldots, n$, is the sum of the principal minors of order i of the matrix (g_{ij}^ξ). Therefore the functional of smoothness (5.36) can also be expressed through these invariants:

$$I[\boldsymbol{\xi}] = \int_{S^{rn}} \left(\frac{I_{n-1}}{I_n}\right) dS^{rn}. \qquad (5.38)$$

The expression (5.38) of the smoothness functional through the invariants of orthogonal transformations was given by Liseikin (1991). A generalization of this expression by subsituting the quantity $(I_{n-1}/I_n)^q$, $0 \leq q \leq 1$ for I_{n-1}/I_n in (5.38) was suggested by Liseikin (1991, 1999) and Huang (2001).

Now, for the purpose of simplicity, we restrict our consideration to three dimensions. The functional (5.38) then has the form

$$I[\boldsymbol{\xi}] = \int_{S^{r3}} \left(\frac{I_2}{I_3}\right) dS^{r3}. \qquad (5.39)$$

In three dimensions the invariant I_3 of the covariant metric tensor (g_{ij}^ξ) is the Jacobian g^ξ of the matrix (g_{ij}^ξ) and it represents the volume V^3 of the three-dimensional basic parallelepiped $P^3 \subset R^{n+k+l}$ formed by the basic tangent vectors $\boldsymbol{r}_{\xi^i}$, $i = 1, 2, 3$. The invariant I_2 of the matrix (g_{ij}^ξ) is the sum of its principal minors of order 2. Every principal minor of order 2 equals the Jacobian of the two-dimensional matrix A^2 obtained from (g_{ij}^ξ) by crossing out a row and a column which intersect on the diagonal. Therefore each element of the matrix A^2 is a dot product of two tangential vectors of the basis $\boldsymbol{r}_{\xi^i}$, $i = 1, 2, 3$, and, consequently, the Jacobian of A^2 equals the square of the area of the parallelogram formed by these two vectors. So the invariants I_2, I_3 can be expressed as

$$I_2 = \sum_{m=1}^{3} \left(V_m^2\right)^2, \qquad I_3 = \left(V^3\right)^2, \qquad (5.40)$$

where V_m^2 is the area of the boundary segment of the basic parallelepiped P^3 formed by the basic tangent vectors $\boldsymbol{r}_{\xi^i}$, $i = 1, 2, 3$, except for $\boldsymbol{r}_{\xi^m}$, and V^3 is the volume of P^3. Therefore

$$\frac{I_2}{I_3} = \sum_{m=1}^{3} \left(V_m^2\right)^2 / \left(V^3\right)^2. \qquad (5.41)$$

It is obvious that

$$V^3 = d_m V_m^2, \qquad m = 1, 2, 3,$$

where d_m is the distance between the vertex of the vector $\boldsymbol{r}_{\xi^m}$ and the plane spanned by the vectors r_{ξ^i}, $i \neq m$. Hence, from (5.40) and (5.41)

$$\frac{I_2}{I_3} = \sum_{m=1}^{3} (1/d^m)^2 \ . \tag{5.42}$$

Now let us consider two grid hypersurfaces $\xi^m = c$ and $\xi^m = c + h$ in S^{r3} obtained by mapping a uniform rectangular grid with a step size h in the computational domain Ξ^3 onto the monitor surface S^{r3}. The distance l_m between a node on the surface $\xi^m = c$ and the nearest node on the surface $\xi^m = c + h$ equals $d_m h + O(h)^2$. Therefore (5.42) is equivalent to

$$\frac{I_2}{I_3} = \sum_{m=1}^{3} (h/l_m)^2 + O(h) \ .$$

The quantity $(h/l_m)^2$ increases as the nodes of a coordinate grid cluster in the direction normal to the hypersurface $\xi^m = c$, and therefore it can be considered as some measure of the grid concentration in this direction; consequently, the functional (5.39) defines an integral measure of the grid clustering in all directions. Therefore the problem of minimizing the functional of smoothness (5.36) for $n = 3$ can be interpreted as a problem of finding a grid with a minimum of nonuniform clustering, namely a quasiuniform grid on the monitor surface S^{r3}. Analogous interpretations are valid for arbitrary dimensions.

The interpretation of the smoothness functional considered above justifies, to some extent, its potential to generate adaptive grids in a domain or surface by projecting onto the domain or surface quasiuniform grids built on monitor surfaces (see Fig. 5.2) by the minimization of the functional.

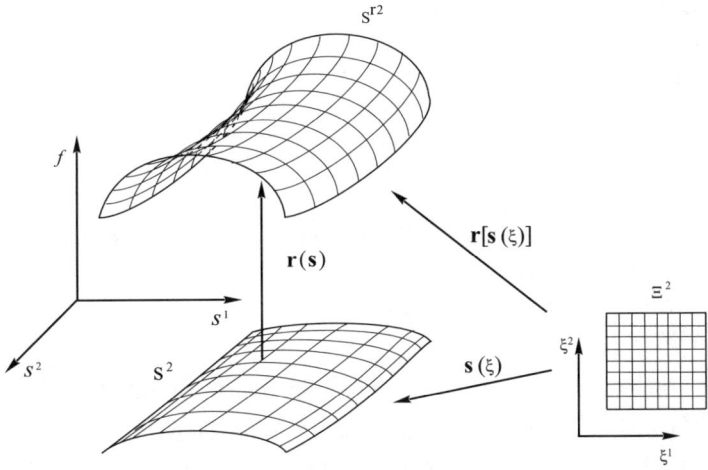

Fig. 5.2. Illustration for grid adaptation

5.2.3 Relation to Harmonic Functions

This paragraph discusses another interpretation of the smoothness functional (5.29), which is related to the harmonic-functions approach for generating adaptive grids. For this purpose we express the smoothness functional (5.29) in the following form

$$I[\boldsymbol{\xi}] = \int_{M^n} g_s^{ml} \frac{\partial \xi^i}{\partial s^m} \frac{\partial \xi^i}{\partial s^l} \, \mathrm{d}M^n, \qquad i, l, m = 1, \cdots, n, \qquad (5.43)$$

since

$$\sqrt{g^s}\mathrm{d}\boldsymbol{s} = \mathrm{d}M^n .$$

Now we describe the definition of a harmonic map between two general n-dimensional Riemannian manifolds M^n and Z^n with covariant metric tensors g_{ij} and D_{ij} in some local coordinates x^i, $i = 1, \cdots, n$, of M^n and z^i, $i = 1, \cdots, n$, of Z^n, respectively.

Every $C^1(M^n)$ map $\boldsymbol{z}(\boldsymbol{x}) : M^n \to Z^n$ defines an energy density by the following formula:

$$e(\boldsymbol{z}) = \frac{1}{2} g^{ij}(\boldsymbol{x}) D_{kl}(\boldsymbol{z}) \frac{\partial z^k}{\partial x^i} \frac{\partial z^l}{\partial x^j}, \qquad i, j, k, l = 1, \cdots, n, \qquad (5.44)$$

where (g^{ij}) is the contravariant metric tensor of M^n in the coordinates $x^1, \ldots, x^n$, i.e. $g_{ij} g^{jk} = \delta_i^k$. The total energy associated with the mapping $\boldsymbol{z}(\boldsymbol{x})$ is then defined as the integral of (5.44) over the manifold M^n:

$$E(\boldsymbol{z}) = \int_{M^n} e(\boldsymbol{z}) \mathrm{d}M^n . \qquad (5.45)$$

A transformation $\boldsymbol{z}(\boldsymbol{x})$ of class $C^2(M^n)$ is referred to as a harmonic mapping if it is a critical point of the functional of the total energy (5.45). The Euler–Lagrange equations whose solution minimizes the energy functional (5.45) are given by

$$\frac{\partial}{\partial x^k} \left(\sqrt{g} g^{kj} \frac{\partial z^l}{\partial x^j} \right) + \sqrt{g} g^{kj} \Gamma^l_{mp} \frac{\partial z^m}{\partial x^k} \frac{\partial z^p}{\partial x^j} = 0, \qquad (5.46)$$

where $g = \det(g_{ij})$ and Γ^l_{mp} are Christoffel symbols of the second kind on the manifold Z^n:

$$\Gamma^l_{mp} = \frac{1}{2} D^{lj} \left(\frac{\partial D_{jm}}{\partial z^p} + \frac{\partial D_{jp}}{\partial z^m} - \frac{\partial D_{mp}}{\partial z^j} \right). \qquad (5.47)$$

The following theorem guarantees the uniqueness of the harmonic mapping.

Theorem 5.2.1. *Let M^n, with metric g_{ij}, and Z^n, with metric D_{ij}, be two Riemannian manifolds with boundaries ∂M^n and ∂Z^n and let $\boldsymbol{\phi} : M^n \to Z^n$ be a diffeomorphism. If the curvature of Z^n is nonpositive and ∂Z^n is convex (with respect to the metric D_{ij}) then there exists a unique harmonic map*

$z(x) : M^n \to Z^n$ such that $z(x)$ is a homotopy equivalent to ϕ. In other words, one can deform z to ϕ by constructing a continuous family of maps $z_t : M^n \to Z^n$, $t \in [0,1]$, such that $z_0(x) = \phi(x)$, $z_1(x) = z(x)$, and $z_t(x) = z(x)$ for all $x \in \partial M^n$.

In applying the harmonic theory to grid generation, the manifold Z^n is assumed to correspond to the logical domain Ξ^n, with a Euclidean metric $D_{ij} = \delta^i_j$ in the coordinates $\xi^1, \ldots, \xi^n$. Since the Euclidean space Ξ^n is flat, i.e. it has zero curvature, and the domain Ξ^n is constructed by the user, both requirements of the above theorem can be satisfied. For the manifold M^n, one uses a set of points of a physical geometry S^{xn} with an introduced Riemannian metric g^x_{ij}. The functional of the total energy (5.45) has then the form

$$E(\boldsymbol{\xi}) = \frac{1}{2} \int_{M^n} \left(g^{kl}_x \frac{\partial \xi^i}{\partial x^k} \frac{\partial \xi^i}{\partial x^l} \right) \mathrm{d} M^n , \qquad i,k,l = 1, \cdots, n . \qquad (5.48)$$

Thus, assuming in (5.44) $\boldsymbol{s} = \boldsymbol{x}$, we obtain in the case of the Eucledian metric in Ξ^n that the integral (5.29) is twice the total energy associated with the mapping $\boldsymbol{\xi}(\boldsymbol{s}) : S^n \to \Xi^n$ representing a transformation between the manifold M^n, with its metric tensor g^s_{ij} in the coordinates s^i, and the computational domain Ξ^n, with the Cartesian coordinates ξ^i:

$$I[\boldsymbol{\xi}] = 2E[\boldsymbol{\xi}] .$$

As for the Euler–Lagrange equations (5.46) for the functional (5.48) we have

$$\frac{\partial}{\partial x^k} \left(\sqrt{g^x} g^{kj}_x \frac{\partial \xi^i}{\partial x^j} \right) = 0 , \qquad i,j,k = 1, \cdots, n , \qquad (5.49)$$

since, from (5.47), $\Upsilon^l_{mp} = 0$. So the equations (5.49) with $\boldsymbol{x} = \boldsymbol{s}$ are equivalent to the grid equations in (5.4).

Equations (5.49), in contrast to (5.46), are linear and of elliptic type, and have a conservative form. Therefore they satisfy the maximum principle, and the Dirichlet boundary value problem is a well-posed problem for this system of equations, i.e. the above theorem is obvious for the functional (5.48). Thus the functions $\xi^i(\boldsymbol{s})$ obtained from (5.4) compose a harmonic transformation

$$\boldsymbol{\xi}(\boldsymbol{s}) : S^n \to \Xi^n , \quad \boldsymbol{\xi}(\boldsymbol{s}) = [\xi^1(\boldsymbol{s}), \ldots, \xi^n(\boldsymbol{s})] .$$

Note the value of any grid generation method is commonly judged by its ability to rule out the construction of folded grids in domains or on surfaces with arbitrary geometry. In the case $n = 2$ the mathematical foundation of this requirement for the technique based on the generalized Laplace system in (5.4) is solid when the computation domain Ξ^2 is convex. It is founded on the following result, derived from a theorem of Rado.

Theorem 5.2.2. *Let M^2 be a simply connected bounded Riemannian manifold. In this case, the Jacobian of the transformation $\boldsymbol{\xi}(\boldsymbol{s})$ generated by the boundary value problem (5.4) does not vanish in the interior of S^2, if Ξ^2 is a convex domain and $\boldsymbol{\xi}(\boldsymbol{s}) : \partial S^2 \to \partial \Xi^2$ is a homeomorphism.*

5.2.4 Application to Adaptive Grid Generation

In accordance with a concept of Eiseman (1987) adaptive grids on a physical geometry $S^{xn} \subset R^{n+k}$ can be generated by projecting quasiuniform grids from a regular monitor surface defined as the graph of the values of some generally vector-valued monitor function over S^{xn}. This monitor function can be a solution to the problem of interest, a combination of its components or derivatives, or any other variable vector-valued quantity that suitably monitors the features of the physical solution and/or of the geometry of the physical domain or surface which significantly affect the accuracy of the calculations. The monitor functions provide an efficient opportunity to control the grid quality, in particular, the concentration of grid nodes and the size of angles between grid lines. The following paragraphs expound a relation of the monitor functions to adaptive grid generation in domains and on surfaces.

One of the techniques to generating quasiuniform grids on the monitor surface is based on the use of a smoothness functional proposed by Liseikin (1991) which generalizes the functional introduced by Brackbill and Saltzman (1982) for generating fixed grids in domains.

Generation of Adaptive Grids in Domains. In the case, important for the generation of adaptive grids in a physical domain $X^n \subset R^n$, the monitor manifold can be defined as an n-dimensional monitor surface S^{rn} formed by the values of some monitor vector-valued function

$$\boldsymbol{f}(\boldsymbol{x}) : X^n \to R^l, \quad \boldsymbol{x} = (x^1, \ldots, x^n), \quad \boldsymbol{f} = [f^1(\boldsymbol{x}), \ldots, f^l(\boldsymbol{x})], \quad (5.50)$$

over X^n. Thus the monitor surface S^{rn}, whose points are

$$(x^1, \ldots, x^n, f^1(\boldsymbol{x}), \ldots, f^l(\boldsymbol{x})), \quad \boldsymbol{x} = (x^1, \ldots, x^n) \in X^n,$$

is the subset of the $(n+l)$-dimensional space R^{n+l}. It is apparent that for the parametric domain S^n there may be taken the domain X^n and, consequently, the parametric mapping $\boldsymbol{r}(\boldsymbol{s}) : S^n \to R^{n+l}$ is defined as

$$\boldsymbol{r}(\boldsymbol{s}) = [\boldsymbol{s}, \boldsymbol{f}(\boldsymbol{s})] = [s^1, \ldots, s^n, f^1(\boldsymbol{s}), \ldots, f^l(\boldsymbol{s})], \quad \boldsymbol{s} = \boldsymbol{x}. \quad (5.51)$$

If S^{rn} is a monitor surface over the domain S^n formed by the values of a vector-valued function $\boldsymbol{f}(\boldsymbol{s})$ then it is obvious that the intermediate transformation $\boldsymbol{s}(\boldsymbol{\xi})$ that is the inverse of $\boldsymbol{\xi}(\boldsymbol{s})$ found by a solution of the problem (5.4) produces, in fact, the very adaptive grid in S^n determined by projecting the quasiuniform grid from S^{rn}. This adaptive grid provides node concentration in the zones of large variation of $\boldsymbol{f}(\boldsymbol{s})$ (Fig. 5.2).

In the case of a scalar monitor function $f(\boldsymbol{s})$ the covariant and contravariant metric elements of S^{rn} in the coordinates $s^1, \ldots, s^n$ are defined by (4.76). These expressions substituted in (5.4) give the following equations for determining the components $\xi^i(\boldsymbol{s})$ of the transformation $\boldsymbol{\xi}(\boldsymbol{s}) : S^n \to \Xi^n$:

5.2 Variational Formulation

$$\frac{1}{\sqrt{g^s}} \frac{\partial}{\partial s^j} \left[\sqrt{1+\nabla(f)} \left(\delta_k^j - \frac{1}{1+\nabla(f)} \frac{\partial f}{\partial s^j} \frac{\partial f}{\partial s^k} \right) \frac{\partial \xi^i}{\partial s^k} \right] = 0 , \qquad (5.52)$$

$$i, j, k = 1, \ldots, n .$$

The inverse mapping $s(\boldsymbol{\xi})$ forms an adaptive grid in S^n.

Analogously, the grid equations are written out with a vector-valued monitor function $\boldsymbol{f}(s)$ for generating adaptive grids.

Note the popular equations for generating adaptive grids in domains are based on the numerical solution of the Poisson system

$$\frac{\partial}{\partial x^j} \left(\frac{\partial \xi^i}{\partial x^j} \right) = P^i , \qquad i, j, = 1, \cdots, n , \qquad (5.53)$$

where the functions P^i are the control functions. These equations are not equivalent to the generalized Laplace equations in (5.4). In the case where S^{rn} is a monitor surface, the system in (5.4) derived by the minimization of the functional (5.29) can also be interpreted as a system of elliptic equations with a control function. The control function is the monitor mapping $\boldsymbol{f}(s)$ whose values over the physical domain or surface form the monitor hypersurface S^{rn}. The influence of the control function $\boldsymbol{f}(s)$ is realized through the magnitudes $\boldsymbol{f}_{s^i} \cdot \boldsymbol{f}_{s^j}$ in the terms g^s and g_s^{ml} in (5.29). These terms are determined by the tensor elements g_{ij}^s which define the covariant metric tensor of the surface S^{rn} in the coordinates s^i represented by the parametrization (5.51). The system in (5.4), in contrast to that of (5.53), has a divergent form and its solution is a harmonic function, as was mentioned above. Besides this, its solution is independent of parametrizations of X^n; as a rule, this is not valid for the system (5.53).

Generation of Adaptive Grids on Surfaces. When the monitor surface is formed by the values of the function $\boldsymbol{f}(\boldsymbol{x})$ over a general n-dimensional surface S^{xn} lying in the space R^{n+k} and represented by the parametrization

$$\boldsymbol{x}(s) : S^n \to R^{n+k} , \qquad \boldsymbol{x}(s) = [x^1(s), \ldots, x^{n+k}(s)]$$

from an n-dimensional parametric domain $S^n \in R^n$ then the monitor surface S^{rn} can be described by a parametrization from S^n in the form

$$\boldsymbol{r}(s) : S^n \to R^{n+l+k} , \qquad \boldsymbol{r}(s) = \{\boldsymbol{x}(s), \boldsymbol{f}[\boldsymbol{x}(s)]\} . \qquad (5.54)$$

In particular, a one-dimensional monitor surface S^{r1} over a curve S^{x1} lying in R^n and represented by $\boldsymbol{x}(s) : [a, b] \to R^n$, can be defined by the parametrization

$$\boldsymbol{r}(s) : [a, b] \to R^{n+l} , \qquad \boldsymbol{r}(s) = \{\boldsymbol{x}(s), \boldsymbol{f}[\boldsymbol{x}(s)]\} .$$

It is evident that the adaptive grid on the surface S^{xn} obtained by projecting the quasiuniform grid from S^{rn} is formed, in fact, by mapping a reference grid in Ξ^n with a composition of $\boldsymbol{x}(s)$ and the intermediate grid transformation $s(\boldsymbol{\xi})$, i.e. with $\boldsymbol{x}(s(\boldsymbol{\xi})) : \Xi^n \to S^{xn}$.

If the monitor surface over S^{xn} is formed by a scalar monitor function $f(s)$, then for the covariant metric tensor of S^{rn}, designated by g^s_{ij} in the coordinates $s^1, \ldots, s^n$, we have

$$g^s_{ij} = g^{xs}_{ij} + f_{s^i} f_{s^j}, \quad i,j = 1, \ldots, n,$$

where

$$g^{xs}_{ij} = \boldsymbol{x}_{s^i} \cdot \boldsymbol{x}_{s^j}, \quad i,j = 1, \ldots, n,$$

is the metric of S^{xn}. This monitor metric is a particular case of the metric (5.21). Using (5.21) for computing the elements g_s^{ij} of the contravariant metric tensor of S^{rn} and substituting them in (5.4) we obtain the following system for generating adaptive grids on surfaces

$$\frac{1}{\sqrt{g^s}} \frac{\partial}{\partial s^j} \left[\sqrt{g^{xs}[1 + \nabla(f)]} \right.$$

$$\left. \times \left(g^{jk}_{sx} - \frac{1}{1+\nabla(f)} g^{jl}_{sx} g^{km}_{sx} \frac{\partial f}{\partial s^l} \frac{\partial f}{\partial s^m} \right) \frac{\partial \xi^i}{\partial s^k} \right] = 0, \quad (5.55)$$

$$i,j,k,l,m = 1, \ldots, n.$$

The popular approach to generating adaptive grids on two-dimensional surfaces $S^{x2} \subset R^3$ uses a combination of the Beltramian operator and forcing terms:

$$\Delta_B[\xi^i] = P^i, \quad i = 1, 2. \qquad (5.56)$$

Here the Beltramian operator is defined through the metric of S^{x2}, i.e.

$$\Delta_B[\xi^i] = \frac{1}{\sqrt{g}} \frac{\partial}{\partial s^j} \left(\sqrt{g} g^{kj} \frac{\partial \xi^i}{\partial s^k} \right), \quad i,j = 1, 2, \qquad (5.57)$$

where the metric elements g^{kj} and g in the coordinates s^1, s^2 are defined as in (5.10).

If the forcing terms P^i, $i = 1, 2$, are specified as functions of the coordinates s^1, s^2 then the system (5.56) is independent of the parametrization of S^{x2}. However, the equations (5.56) are not the generalized Laplace equations and their solution $\boldsymbol{\xi}(s)$ is not a harmonic function. So the theorem of Rado is not valid and consequently $\boldsymbol{\xi}(s)$ may not be a one-to-one mapping. Thus the grid cells obtained from (5.56) may be folded.

The essential advantage of the comprehensive grid method is that it uses the generalized Laplace equations in (5.4) for the purpose of generating arbitrary grids on surfaces and in domains.

5.3 Conclusion

Here we itemize the features of the comprehensive grid generator formulated through the Dirichlet boundary value problem (5.4).

5.3 Conclusion

1. Equations in (5.4) are linear and elliptic therefore the Dirichlet boundary value problem for these equations is well-posed.
2. The generalized Laplace equations are invariant with respect to a choice of a coordinate system in S^{xn} which is a guarantee that the grid generated is the same for different parametrizations of S^{xn}. Note the grid obtained by the solution of both Poisson equations (5.53) and (5.56) and diffusion equations (5.9) for $n = 2$ may vary for different parametrizations of S^{xn}.
3. The elliptic system in (5.4) is of divergence form hence its solution is subject to the maximum principle. Therefore the grid nodes produced through the solution of (5.4) will be inside of S^{xn} if the computational domain Ξ^n is convex. Moreover, for $n = 2$ there is valid a general theorem of Rado from which a particular corollary follows that the transformation $\boldsymbol{\xi}(\boldsymbol{s})$ obtained by the solution of the Dirichlet boundary value problem for the system in (5.4) with an arbitrary metric is nondegenerate if Ξ^2 is convex and the boundary mapping determined by the Dirichlet condition is a one-to-one map between the boundaries of S^{x2} and Ξ^2. The properties mentioned substantiate the formulation of the Beltramian system in (5.4) with respect to the function $\boldsymbol{\xi}(\boldsymbol{s})$, though its formulation with respect to the intermediate transformation $\boldsymbol{s}(\boldsymbol{\xi})$ seamed to be more natural as such system would be linear and of divergence form and therefore more tractable for numerical implementations. However such divergent model, owing to the maximum principle, does not guarantee that all grid points will be inside of the physical geometry S^{xn} when the parametric domain S^n is not convex, let alone the grid cells may be folded. Note both the equations (5.56) for $P^i \neq 0$ and (5.9) for $n = 2$ are not the generalized Laplace equations therefore the theorem of a nondegenerate solution $\boldsymbol{\xi}(\boldsymbol{s})$ to these equations can not be applied.
4. Equations in (5.4) provide the generation of an arbitrary unfolded grid through the mapping approach. This follows from a result shown above in Sect. 5.2.5 that an arbitrary nondegenerate transformation (5.2) of the class $C^2[\Xi^n]$ is realized as the inverse to a solution of (5.4) with respect to the metric (5.14) formed through a corresponding specification of monitor functions.
5. Equations in (5.4) are equivalent to the Euler-Lagrange equations for the functional of energy

$$I[\boldsymbol{\xi}] = \frac{1}{2} \int_{S^n} \sqrt{g^s} g_s^{kl} \frac{\partial \xi^i}{\partial s^k} \frac{\partial \xi^i}{\partial s^l} d\boldsymbol{s}, \quad i, k, l = 1, \ldots, n. \tag{5.58}$$

Therefore grid generation on S^{xn} can be realized by the numerical solution of the problem of the minimization of the transformed functional (5.58) rewritten with respect to the intermediate transformation $\boldsymbol{s}(\boldsymbol{\xi})$:

$$I[\boldsymbol{s}] = \frac{1}{2} \int_{\Xi^n} \sqrt{g^\xi} \left(\sum_{i=1}^n g_\xi^{ii} \right) d\boldsymbol{\xi}, \quad i = 1, \ldots, n, \tag{5.59}$$

where $g^\xi = \det\left(g_{ij}^\xi\right)$. Since

$$\sqrt{g^\xi}d\boldsymbol{\xi} = dM^n ,$$

we obtain from (5.59)

$$I[\boldsymbol{s}] = \frac{1}{2}\int_{M^n} \operatorname{tr}(g_\xi^{ij})dM^n = \frac{1}{2}\int_{M^n} \frac{I_{n-1}}{I_n}dM^n , \quad i,j = 1,\ldots,n ,\quad(5.60)$$

where I_{n-1} is the $(n-1)$th invariant of the matrix $\left(g_{ij}^\xi\right)$, $I_n = \det\left(g_{ij}^\xi\right)$. Analogously to the case expounded for the natural metric of the monitor surface S^{rn} the expression I_{n-1}/I_n is a measure of local grid spacing in M^n, while (5.60) is an integral measure of grid spacing in M^n. So in the process of the minimization of (5.60) and consequently (5.58) and (5.59) the grid aspires to become uniform in M^n.

6. The systems of equations in (5.4) allows one to generate grids on surfaces or in domains of an arbitrary dimension $n > 0$, and hence they can be applied to obtain grids in n-dimensional blocks by means of the successive uniform generation of grids on curvilinear edges, faces, parallelepipeds, etc., using the solution at a step $i < n$ as the Dirichlet boundary condition for the following step $i+1 \leq n$. Thus both the interior and the boundary grid points of a domain or surface can be calculated by the similar elliptic solvers.

So the requirements 1–5 outlined in Chap. 1 are held by the mathematical model (5.4). Therefore the model is really a promising one for the development of comprehensive grid generation codes.

Thus the Beltramian operator is a universal tool and its implementation in grid technology will allow one to generate in a unified manner a grid both in an arbitrary domain and on its boundary segments.

6. Relations to Monitor Manifolds

An arbitrary physical geometry is considered in modern grid technology as an n-dimensional surface S^{xn}, in particular, a domain (curve or line in the one-dimensional case) whose mathematical representation is formulated locally in a general parametric form

$$\boldsymbol{x}(\boldsymbol{s}) : S^n \to R^{n+n_0}, \quad \boldsymbol{s} = (s^1, \ldots, s^n), \quad \boldsymbol{x} = (x^1, \ldots, x^{n+n_0}), \quad (6.1)$$

where $\boldsymbol{x}(\boldsymbol{s})$ is a smooth transformation of rank n, while $S^n \subset R^n$ is a parametric domain.

The parametrization (6.1) yields a formula for the elements of the covariant metric tensor of the regular surface S^{xn} designated by g_{ij}^{xs} in the coordinates $s^1, \ldots, s^n$:

$$g_{ij}^{xs} = \boldsymbol{x}_{s^i} \cdot \boldsymbol{x}_{s^j}, \quad i, j = 1, \ldots, n. \quad (6.2)$$

The matrix inverse to (g_{ij}^{xs}) is referred to as the contravariant metric tensor of S^{xn} in the coordinates $s^1, \ldots, s^n$. Its elements are designated by g_{sx}^{ij}.

A monitor manifold over S^{xn} introduced for providing grid control through the comprehensive grid equations reviewed in Chap. 5 is formed by imposing on S^{xn} a new monitor metric g_{ij}^s in the coordinates $s^1, \ldots, s^n$ by the following relations

$$g_{ij}^s = z(\boldsymbol{s}) g_{ij}^{xs} + v^k(\boldsymbol{s}) f_{s^i}^k(\boldsymbol{s}) f_{s^j}^k(\boldsymbol{s}), \quad i, j = 1, \ldots, n, \quad k = 1, \ldots, l, \quad (6.3)$$

where $z(\boldsymbol{s}) > 0$ and $v^k(\boldsymbol{s}) \geq 0$ are weight functions, while $f^k(\boldsymbol{s})$, $k = 1, \ldots, l$, are monitor functions. We designate by M^n the monitor manifold over S^{xn} whose metric is specified by (6.3). Particular cases of the monitor manifolds are the monitor surfaces S^{rn} over S^{xn} ($z(\boldsymbol{s}) = v^k(\boldsymbol{s}) = 1$, $k = 1, \ldots, l$), represented by the parametrizations (5.51) and (5.54). Another example of a monitor manifold was formulated by the metric (4.21).

The influence of the monitor metric (6.3) on the grid properties is realized through the coefficients in the comprehensive grid equations in the boundary value problem (5.4). These coefficients are the elements of the contravariant metric tensor of the monitor manifold M^n and the Jacobian of the monitor metric tensor. However the geometric characteristics of grids and grid cells are also depended on the characteristics of the physical geometry S^{xn}. In particular, the measure of grid spacing (4.62) and clustering (4.64) in the physical geometry is expressed through invariants of the Beltrami's differential parameters and the mean curvature of a grid hypersurface with respect to the metric of S^{xn} (see Chap. 8). Therefore for analyzing and transforming the grid equations in a suitable form and for efficient controlling the grid behavior with the monitor and weight functions in (6.3) it is necessary to compute the geometric characteristics of the monitor manifold M^n through the geometric characteristics of the physical geometry S^{xn} and the weight and monitor functions forming the monitor metric.

In this chapter we establish some relations between certain geometric characteristics of original physical geometries and monitor surfaces and manifolds over them. These relations provide efficient means to control grid properties by monitor functions.

6.1 Computation of Geometric Characteristics

This section presents a sequence of recursive formulas aimed at computation of geometric characteristics of monitor manifolds whose metrics are specified in the form (6.3).

6.1.1 Recursive Representation of the Monitor Metric

The metric (6.3) can be formulated in a recursive form through a sequence of simpler monitor metrics designated in the coordinates $s^1, \ldots, s^n$ as g_{ij}^{ks}, $k = 0, 1, \ldots, l$, where

$$g_{ij}^{0s} = z(\boldsymbol{s}) g_{ij}^{xs},$$

$$g_{ij}^{ks} = g_{ij}^{(k-1)s} + v_k(\boldsymbol{s}) f_{s^i}^k f_{s^j}^k, \qquad (6.4)$$

$$i, j = 1, \ldots, n, \quad k = 1, \ldots, l, \quad k \text{ fixed}.$$

Each of these metrics determines a Riemannian manifold M_k^n, $k = 0, \ldots, l$, in addition M_k^n, $k > 0$ is a monitor manifold over M_{k-1}^n, while $M_l^n = M^n$.

The recursive formula (6.4) gives one an opportunity for computing the geometric characteristics of M^n using a sequence of the relations which connect the geometric characteristics of M_k^n and M_{k-1}^n. For establishing these relations we consider a standard monitor manifold M^s with a simply prescribed monitor metric g_{ij}^s over some original manifold M having a metric g_{ij}^{xs} in the coordinates $s^1, \ldots, s^n$ such that:

$$g_{ij}^s = g_{ij}^{xs} + v(\boldsymbol{s}) f_{s^i} f_{s^j}, \quad i, j = 1, \ldots, n, \qquad (6.5)$$

where $v(\boldsymbol{s}) > 0$ and $f(\boldsymbol{s})$ are two scalar-valued functions. In particular when M is a regular surface S^{xn} and $v(\boldsymbol{s}) \equiv 1$ then M^s is a monitor surface over S^{xn} formed by the scalar-valued monitor function $f(\boldsymbol{s})$.

Let us designate by g^s, g_s^{ij}, $\nabla^s(\varphi,\psi)$, $\nabla^s(\varphi)$, $[ij,l]^s$, ${}^s\Upsilon_{ij}^l$, $\nabla_{ij}^s(\varphi)$, $\Delta_B^s[\varphi]$ and g^{xs}, g_{sx}^{ij}, $\nabla(\varphi,\psi)$, $\nabla(\varphi)$, $[ij,l]$, Υ_{ij}^l, $\nabla_{ij}(\varphi)$, $\Delta_B[\varphi]$ the Jacobians of the covariant metric tensors, the elements of the contravariant metric tensors, Beltrami's mixed and first differential parameters, Christoffel symbols, mixed derivatives and Beltrami's second differential parameter with respect to the metric g_{ij}^s and g_{ij}^{xs}, respectively, in the coordinates $s^1,\dots,s^n$. Now, assuming that these geometric quantities with respect to the metric of M are known, we establish formulas for the same geometric characteristics with respect to the metric (6.5) of the standard monitor manifold M^s.

6.1.2 Jacobian of the Covariant Metric Tensor

For the Jacobian of the metric tensor g_{ij}^s defined by (6.5) we have

$$g^s = g^{xs}\left(1 + v g_{sx}^{ij}\frac{\partial f}{\partial s^i}\frac{\partial f}{\partial s^j}\right) = g^{xs}[1 + v\nabla(f)], \qquad i,j = 1,\dots,n, \quad (6.6)$$

where

$$g^s = \det(g_{ij}^s), \quad g^{xs} = \det(g_{ij}^{xs}),$$

$$\nabla(f) \equiv \nabla(f,f) = g_{sx}^{ij}\frac{\partial f}{\partial s^i}\frac{\partial f}{\partial s^j}, \qquad i,j = 1,\dots,n.$$

When $n = 2$ then this equation is readily proved using (4.8) by the following consideration

$$\begin{aligned}g^s &= g^{xs} + v\left[g_{11}^{xs}\left(\frac{\partial f}{\partial s^2}\right)^2 - 2g_{12}^{xs}\frac{\partial f}{\partial s^1}\frac{\partial f}{\partial s^2} + g_{22}^{xs}\left(\frac{\partial f}{\partial s^1}\right)^2\right]\\ &= g^{xs}\left(1 + vg_{sx}^{ij}\frac{\partial f}{\partial s^i}\frac{\partial f}{\partial s^j}\right) = g^{xs}[1 + v\nabla(f)], \qquad i,j = 1,2.\end{aligned}$$

In order to prove (6.6) for an arbitrary $n > 1$ we note that the rank of the matrix $\left(\dfrac{\partial f}{\partial s^i}\dfrac{\partial f}{\partial s^j}\right)$ does not exceed 1. Therefore

$$g^s = g^{xs} + \sum_{i=1}^n \det(g_{km}^i),$$

where (g_{km}^i), $k,m = 1,\dots,n$, is the matrix obtained from the matrix (g_{km}^{xs}) by substituting the vector

$$v\left(\frac{\partial f}{\partial s^i}\frac{\partial f}{\partial s^1},\dots,\frac{\partial f}{\partial s^i}\frac{\partial f}{\partial s^n}\right) = v\frac{\partial f}{\partial s^i}\mathrm{grad}f$$

for its ith row, i.e.

$$(g^i_{km}) = \begin{pmatrix} g^{xs}_{11} & \cdots & g^{xs}_{1n} \\ \cdots & \cdots & \cdots \\ g^{xs}_{i-11} & \cdots & g^{xs}_{i-1n} \\ v\dfrac{\partial f}{\partial s^i}\dfrac{\partial f}{\partial s^1} & \cdots & v\dfrac{\partial f}{\partial s^i}\dfrac{\partial f}{\partial s^n} \\ g^{xs}_{i+11} & \cdots & g^{xs}_{i+1n} \\ \cdots & \cdots & \cdots \\ g^{xs}_{n1} & \cdots & g^{xs}_{nn} \end{pmatrix}.$$

Since

$$\det(g^i_{km}) = v\frac{\partial f}{\partial s^i}\frac{\partial f}{\partial s^k}G^{ik} = vg^{xs}g^{ik}_{sx}\frac{\partial f}{\partial s^i}\frac{\partial f}{\partial s^k}, \quad i,k,m = 1,\ldots,n, \ i \text{ fixed},$$

where G^{ik} is the (ik)th cofactor of the matrix (g^{xs}_{ij}), for which

$$G^{ik} = g^{xs}g^{ik}_{sx}, \quad i,k = 1,\ldots,n,$$

we readily find that the formula (6.6) is valid for an arbitrary $n > 1$. For $n = 1$ the form (6.6) is obvious.

6.1.3 Contravariant Metric Tensor

Elements with Respect to Parametric Coordinates. The elements of the contravariant metric tensor of the standard manifold M^s in the parametric coordinates $s^1, \ldots, s^n$ are as follows:

$$g^{ij}_s = g^{ij}_{sx} - dg^{ik}_{sx}g^{jm}_{sx}\frac{\partial f}{\partial s^k}\frac{\partial f}{\partial s^m} = g^{ij}_{sx} - d\nabla(f,s^i)\nabla(f,s^j), \quad (6.7)$$

$$i,j,m = 1,\ldots,n,$$

where

$$d = \frac{v}{1+v\nabla(f)}.$$

Indeed

$$g^{ij}_s g^s_{jp} = \left(g^{ij}_{sx} - dg^{ik}_{sx}g^{jm}_{sx}\frac{\partial f}{\partial s^k}\frac{\partial f}{\partial s^m}\right)\left(g^{xs}_{jp} + v\frac{\partial f}{\partial s^j}\frac{\partial f}{\partial s^p}\right)$$

$$= \delta^i_p - dg^{ik}_{sx}\frac{\partial f}{\partial s^k}\frac{\partial f}{\partial s^p} + vg^{ij}_{sx}\frac{\partial f}{\partial s^j}\frac{\partial f}{\partial s^p}$$

$$- dvg^{jm}_{sx}\frac{\partial f}{\partial s^j}\frac{\partial f}{\partial s^m}g^{ik}_{sx}\frac{\partial f}{\partial s^k}\frac{\partial f}{\partial s^p}$$

$$= \delta^i_p - g^{ik}_{sx}\frac{\partial f}{\partial s^k}\frac{\partial f}{\partial s^p}[d - v + vd\nabla(f)] = \delta^i_p, \quad i,j,k,l,m = 1,\ldots,n,$$

6.1 Computation of Geometric Characteristics

since

$$d - v + vd\nabla(f) = \frac{v}{1+v\nabla(f)} - v + \frac{v^2\nabla(f)}{1+v\nabla(f)} = 0 \ .$$

Thus the matrix (g_s^{ij}) whose elements are defined by (6.7) is the inverse of the matrix (g_{ij}^s) and, consequently, its elements comprise the contravariant metric tensor of M^s in the coordinates $s^1, \ldots, s^n$.

In particular, if in (6.5)

$$g_{ij}^{xs} = \delta_j^i \ , \quad i,j = 1, \ldots, n \ , \quad \text{and} \quad v(\boldsymbol{s}) = 1 \ ,$$

i.e. M^s is a monitor surface over the parametric domain $S^n \subset R^n$ with a scalar-valued monitor function $f(\boldsymbol{s})$ then, from (6.7),

$$g_s^{ij} = \delta_j^i - \frac{1}{1+\nabla^E(f)} f_{s^i} f_{s^j} \ , \quad i,j = 1, \ldots, n \ , \tag{6.8}$$

where

$$\nabla^E(f) = f_{s^i} f_{s^i} \ , \quad i = 1, \ldots, n \ ,$$

which corresponds to (4.76).

Elements with Respect to Grid Coordinates. The intermediate mapping $\boldsymbol{s}(\boldsymbol{\xi})$ in (5.2) introduced to generate a grid on both S^{xn} and M^s determines a new coordinate system ξ^i, $i = 1, \ldots, n$, on S^{xn} and this manifold. It is also used to specify the values $(g_{ij}^{x\xi})$ and (g_{ij}^{ξ}), $(g_{\xi x}^{ij})$ and (g_{ξ}^{ij}), $i,j = 1, \ldots, n$, of the covariant and contravariant metric tensor of S^{xn} and M^s, respectively, in these grid coordinates. Note the elements of the covariant and contravariant tensors in the coordinates s^i and ξ^i are connected by the corresponding tensor relations (4.18) and (4.19). Thus if the original metric of M^s is specified by (6.5) then we obtain, using (4.18), the following formula in the grid coordinates ξ^i, $i = 1, \ldots, n$, for the elements of the covariant metric tensor of the monitor manifold M^s

$$g_{ij}^{\xi} = g_{kl}^s \frac{\partial s^k}{\partial \xi^i} \frac{\partial s^l}{\partial \xi^j} = g_{ij}^{x\xi} + v[\boldsymbol{s}(\boldsymbol{\xi})] \frac{\partial f}{\partial \xi^i} \frac{\partial f}{\partial \xi^j} \ , \quad i,j,k,l = 1, \ldots, n \ , \tag{6.9}$$

where

$$\frac{\partial f}{\partial \xi^i} = \frac{\partial f[\boldsymbol{s}(\boldsymbol{\xi})]}{\partial \xi^i} \ , \quad i = 1, \ldots, n \ .$$

The elements of the contravariant metric tensor of M^s in the grid coordinates ξ^i, $i = 1, \ldots, n$, are computed, in accordance with (4.19) and (6.7), as follows:

$$\begin{aligned} g_\xi^{ij} &= g_s^{kl} \frac{\partial \xi^i}{\partial s^k} \frac{\partial \xi^j}{\partial s^l} = g_{\xi x}^{ij} - dg_{sx}^{pk} g_{sx}^{lm} \frac{\partial f}{\partial s^k} \frac{\partial f}{\partial s^m} \frac{\partial \xi^i}{\partial s^p} \frac{\partial \xi^j}{\partial s^l} \\ &= g_{\xi x}^{ij} - dg_{\xi x}^{ik} g_{\xi x}^{jm} \frac{\partial f}{\partial \xi^k} \frac{\partial f}{\partial \xi^m} = g_{\xi x}^{ij} - d\nabla(f, \xi^i)\nabla(f, \xi^j) \ , \end{aligned} \tag{6.10}$$

$$i,j,k,m,p = 1, \ldots, n \ .$$

In particular, when M^s is a monitor surface over the domain $S^n \subset R^n$, i.e. $v(s) \equiv 1$ in (6.5) then we find, from (6.9) and (6.10)

$$g_{ij}^{\xi} = g_{ij}^{s\xi} + \frac{\partial f}{\partial \xi^i}\frac{\partial f}{\partial \xi^j} \,,$$

$$g_{\xi}^{ij} = g_{\xi s}^{ij} - \frac{1}{1+\nabla^E(f)}\nabla^E(f,\xi^i)\nabla^E(f,\xi^j) \,, \qquad (6.11)$$

$$i,j = 1,\ldots,n \,,$$

where

$$g_{ij}^{s\xi} = \frac{\partial \boldsymbol{s}}{\partial \xi^i} \cdot \frac{\partial \boldsymbol{s}}{\partial \xi^j} \,, \quad g_{\xi s}^{ij} = \frac{\partial \boldsymbol{\xi}}{\partial s^i} \cdot \frac{\partial \boldsymbol{\xi}}{\partial s^j} \,, \quad i,j = 1,\ldots,n \,, \qquad (6.12)$$

$$\nabla^E(f,\varphi) = f_{s^i}\varphi_{s^i} \,, \quad i = 1,\ldots,n \,.$$

We also have the relations

$$g_s^{ii} = g_{\xi}^{ml}g_{ml}^{s\xi} \,, \qquad g^{\xi} = g^s g^{s\xi} \,, \qquad i,l,m = 1,\ldots,n \,,$$

where

$$g^{s\xi} = \det(g_{ij}^{s\xi}) \,, \quad i,j = 1,\ldots,n \,.$$

6.1.4 Beltrami's Mixed and First Differential Parameters

The Beltrami's mixed differential parameter of any invariants φ and ψ with respect to the monitor metric (6.5) is readily computed, using (4.51) and (6.7). Namely,

$$\nabla^s(\varphi,\psi) = \varphi_{s^i}\psi_{s^j}g_s^{ij} = \varphi_{s^i}\psi_{s^j}g_{sx}^{ij} - dg_{sx}^{ik}\varphi_{s^i}f_{s^k}g_{sx}^{jm}\psi_{s^j}f_{s^m}$$
$$= \nabla(\varphi,\psi) - d\nabla(\varphi,f)\nabla(\psi,f) \,, \quad i,j = 1,\ldots,n \,, \qquad (6.13)$$

This equation yields the following formula for the Beltrami's first differential parameter of φ

$$\nabla^s(\varphi) = \nabla(\varphi) - d[\nabla(\varphi,f)]^2 \,. \qquad (6.14)$$

6.1.5 Christoffel Symbols

By virtue of the general formula (4.22) for the elements of the Christoffel symbols of the first kind in the coordinates $s^1,\ldots,s^n$ we have, taking advantage of (6.5),

$$[ij,k]^s = \frac{1}{2}\left(\frac{\partial g_{ik}^s}{\partial s^j} + \frac{\partial g_{jk}^s}{\partial s^i} - \frac{\partial g_{ij}^s}{\partial s^k}\right)$$
$$= [ij,k] + vf_{s^is^j}f_{s^k} + \frac{1}{2}[(v_{s^i}f_{s^j} + v_{s^j}f_{s^i})f_{s^k} - f_{s^i}f_{s^j}v_{s^k}] \,, \qquad (6.15)$$
$$i,j,k = 1,\ldots,n \,.$$

This formula, (4.24), and (6.7) yield the following expression for the elements of the Christoffel symbols of the second kind in the coordinates $s^1, \ldots, s^n$

$$^s\Upsilon_{ij}^k = g_s^{kp}[ij, p]^s = [g_{sx}^{kp} - d\nabla(f, s^k)\nabla(f, s^p)]$$

$$\times \{[ij, p] + vf_{s^i s^j} f_{s^p} + \frac{1}{2}[(v_{s^i} f_{s^j} + v_{s^j} f_{s^i})f_{s^p} - f_{s^i} f_{s^j} v_{s^p}]\}$$

$$= \Upsilon_{ij}^k + v\nabla(f, s^k)f_{s^i s^j}$$

$$+ \frac{1}{2}[\nabla(f, s^k)(v_{s^i} f_{s^j} + v_{s^j} f_{s^i}) - \nabla(v, s^k) f_{s^i} f_{s^j}] \quad (6.16)$$

$$- d\nabla(f, s^k)\{f_{s^m}\Upsilon_{ij}^m + v\nabla(f)f_{s^i s^j}$$

$$+ \frac{1}{2}[\nabla(f)(v_{s^i} f_{s^j} + v_{s^j} f_{s^i}) - \nabla(v, f) f_{s^i} f_{s^j}]\}$$

$$= \Upsilon_{ij}^k + d\nabla(f, s^k)\nabla_{ij}(f) + D_{ij}^k, \quad i, j, k, m, p = 1, \ldots, n ,$$

where

$$D_{ij}^k = \frac{1}{2}\{d\nabla(f, s^k)(v_{s^i} f_{s^j} + v_{s^j} f_{s^i}) + [d\nabla(f, s^k)\nabla(v, f) - \nabla(v, s^k)]f_{s^i} f_{s^j}\}.$$

Note when $v(s) \equiv const = c$, in particular if M^s is a monitor surface over S^{xn}, then $v_{s^i} = 0$, $i = 1, \ldots, n$, and consequently formulas (6.15) and (6.16) are as follows:

$$[ij, k]^s = [ij, k] + cf_{s^i s^j} f_{s^k} ,$$
$$^s\Upsilon_{ij}^k = \Upsilon_{ij}^k + d\nabla(f, s^k)\nabla_{ij}(f), \quad i, j, k = 1, \ldots, n , \quad (6.17)$$

where $d = c/[1 + c\nabla(f)]$.

6.1.6 Mixed Derivatives

For the components of the tensor of the covariant mixed derivatives $\nabla_{ij}^s(\varphi)$ we have from (4.39)

$$\nabla_{ij}^s(\varphi) = \varphi_{s^i s^j} - \varphi_{s^k}\,^s\Upsilon_{ij}^k , \quad i, j, k = 1, \ldots, n ,$$

and using (6.16) yields

$$\nabla_{ij}^s(\varphi) = \nabla_{ij}(\varphi) - d\nabla(f, \varphi)\nabla_{ij}(f) + B_{ij}, \quad i, j = 1, \ldots, n , \quad (6.18)$$

where

$$B_{ij} = -\varphi_{s^k} D_{ij}^k = -\frac{1}{2}\{d\nabla(f, \varphi)(v_{s^i} f_{s^j} + v_{s^j} f_{s^i})$$

$$+ [d\nabla(f, \varphi)\nabla(v, f) - \nabla(v, \varphi)]f_{s^i} f_{s^j}\}, \quad i, j = 1, \ldots, n .$$

As $B_{ij} = 0$, $i, j = 1, \ldots, n$, when $v(s) \equiv const$ so, in this case,

$$\nabla_{ij}^s(\varphi) = \nabla_{ij}(\varphi) - d\nabla(f, \varphi)\nabla_{ij}(f), \quad i, j = 1, \ldots, n , \quad (6.19)$$

with the invariant d defined as in (6.17).

6.1.7 Beltrami's Second Differential Parameter

The Beltrami's second differential parameter $\Delta_B^s[\varphi]$ of a function $\varphi(s)$ is computed, in accordance with (4.56), as follows:

$$\Delta_B^s[\varphi] = g_s^{ij}\nabla_{ij}^s(\varphi), \quad i,j = 1,\ldots,n.$$

Therefore by virtue of (6.7) and (6.18) we find

$$\begin{aligned}\Delta_B^s[\varphi] &= [g_{sx}^{ij} - d\nabla(f,s^i)\nabla(f,s^j)]\\ &\quad \times [\nabla_{ij}(\varphi) - d\nabla(f,\varphi)\nabla_{ij}(f) + B_{ij}]\\ &= \Delta_B[\varphi] - d\nabla(f,\varphi)\Delta_B[f] - d\nabla(f,s^i)\nabla(f,s^j)\\ &\quad \times [\nabla_{ij}(\varphi) - d\nabla(f,\varphi)\nabla_{ij}(f)] + C, \quad i,j=1,\ldots,n,\end{aligned} \quad (6.20)$$

where

$$C = g_s^{ij}B_{ij} = \frac{d}{v}\left\{\frac{1}{2}\nabla(f)\nabla(v,\varphi) - d\nabla(f,\varphi)\nabla(f,v)\left[1 + \frac{1}{2}\nabla(f)\right]\right\},$$
$$i,j = 1,\ldots,n.$$

Since (4.108) with $g_s^{ij} \equiv g_{sr}^{ij}$ and $f(s) \equiv \varphi(s)$, we have

$$\nabla(f,s^i)\nabla(f,s^j)\nabla_{ij}(f) = \frac{1}{2}g_{sx}^{jm}f_{s^m}\frac{\partial}{\partial s^j}\nabla(f) = \frac{1}{2}\nabla(\nabla(f),f),$$
$$i,j,m = 1,\ldots,n,$$

and consequently (6.20) also has the following form

$$\begin{aligned}\Delta_B^s[\varphi] &= \Delta_B[\varphi] - d\nabla(f,\varphi)\Delta_B[f] + \frac{1}{2}d^2\nabla(f,\varphi)\nabla(\nabla(f),f)\\ &\quad - d\nabla(f,s^i)\nabla(f,s^j)\nabla_{ij}(\varphi) + C, \quad i,j=1,\ldots,n.\end{aligned} \quad (6.21)$$

If $v(s) \equiv const$ then $C = 0$ so, in this case,

$$\begin{aligned}\Delta_B^s[\varphi] &= \Delta_B[\varphi] - d\nabla(f,\varphi)\Delta_B[f] + \frac{1}{2}d^2\nabla(f,\varphi)\nabla(\nabla(f),f)\\ &\quad - d\nabla(f,s^i)\nabla(f,s^j)\nabla_{ij}(\varphi), \quad i,j=1,\ldots,n.\end{aligned} \quad (6.22)$$

6.1.8 Geometric Characteristics for a Spherical Monitor Metric

Now we compute the geometric quantities of M^s if its metric is specified in the coordinates $s^1,\ldots,s^n$ by

$$g_{ij}^s = z(s)g_{ij}^{xs}, \quad z(s) > 0, \quad i,j = 1,\ldots,n, \quad (6.23)$$

where g_{ij}^{xs} is a metric of the original manifold M. We readily find, using the original definition of the geometric characteristics described in Chap. 4,

$$g_s^{ij} = \frac{1}{z}g_{sx}^{ij}, \quad i,j = 1,\ldots,n, \quad (6.24)$$

$$\nabla^s(\varphi,\psi) = \frac{1}{z}\nabla(\varphi,\psi)\,, \quad \nabla^s(\varphi) = \frac{1}{z}\nabla(\varphi)\,, \tag{6.25}$$

$$[ij,l]^s = z[ij,l] + \frac{1}{2}(z_{s^i}g_{jl}^{xs} + z_{s^j}g_{il}^{xs} - z_{s^l}g_{ij}^{xs})\,, \quad i,j,l = 1,\ldots,n\,, \tag{6.26}$$

$$^s\Upsilon_{ij}^k = \Upsilon_{ij}^k + \frac{1}{2z}[z_{s^i}\delta_k^j + z_{s^j}\delta_k^i - \nabla(z,s^k)g_{ij}^{xs}]\,, \quad i,j,k = 1,\ldots,n\,, \tag{6.27}$$

$$\nabla_{ij}^s(\varphi) = \nabla_{ij}(\varphi) - \frac{1}{2z}[z_{s^i}\varphi_{s^j} + z_{s^j}\varphi_{s^i} - \nabla(z,\varphi)g_{ij}^{xs}]\,, \quad i,j = 1,\ldots,n\,, \tag{6.28}$$

$$\Delta_B^s[\varphi] = \frac{1}{z}\Delta_B[\varphi] + \frac{n-2}{2z^2}\nabla(z,\varphi)\,, \tag{6.29}$$

$$v^s(\varphi) = \nabla^s\left(\frac{1}{\sqrt{z}}\sqrt{\frac{\nabla E(\varphi)}{\nabla(\varphi)}},\varphi\right)$$
$$= \frac{1}{z^{3/2}}v(\varphi) - \frac{1}{2z^{3/2}}\sqrt{\frac{\nabla E(\varphi)}{\nabla(\varphi)}}\nabla(z,\varphi)\,. \tag{6.30}$$

6.1.9 Computation of Geometric Characteristics of a Monitor Manifold

The formulas (6.6–6.30) give one an opportunity to compute successively the geometric characteristics described above of the monitor manifold M^n over S^{xn}, whose metric is specified by (6.3), through the geometric characteristics of S^{xn} and derivatives of the monitor and weight functions.

Let, for example, the metric elements of a monitor manifold M^s over S^{xn} be formulated in the coordinates $s^1,\ldots,s^n$ as follows:

$$g_{ij}^s = z(\boldsymbol{s})g_{ij}^{xs} + f_{s^i}f_{s^j}\,, \quad i,j = 1,\ldots,n\,. \tag{6.31}$$

If we assume by M^z a monitor manifold with a spherical monitor metric whose covariant metric elements designated in the coordinates $s^1,\ldots,s^n$ by g_{ij}^{zs} are defined as

$$g_{ij}^{zs} = z(\boldsymbol{s})g_{ij}^{xs}\,, \quad i,j = 1,\ldots,n\,, \tag{6.32}$$

then M^s is a monitor manifold over M^z formed by a scalar-valued monitor function $f(\boldsymbol{s})$, with the following metric elements in the coordinates $s^1,\ldots,s^n$

$$g_{ij}^s = g_{ij}^{zs} + f_{s^i}f_{s^j}\,, \quad i,j = 1,\ldots,n\,.$$

The geometric characteristics discussed above of M^z are computed by the formulas (6.24 – 6.30), while the same geometric characteristics of M^s can be computed by the formulas (6.7–6.22) in which g_{ij}^{xs} is identified with g_{ij}^{zs}. Thus using these relations we can express the geometric characteristics of M^s through the geometric characteristics of S^{xn} and derivatives of the functions $z(\boldsymbol{s})$ and $f(\boldsymbol{s})$.

We shall designate by $\nabla^z(\varphi,\psi)$, $\nabla^z(\varphi,[ij,l]^z$, $^z\Upsilon_{ij}^k$, $\nabla_{ij}^z(\varphi)$, $\Delta_B^z[\varphi]$ the geometric characteristics in the metric (6.32) which are computed by the corresponding formulas (6.25–6.29).

Contravariant Metric Tensor. For the elements of the contravariant metric tensor of M^s in the coordinates $s^1, \ldots, s^n$ we readily find using (6.7), for $v(s) \equiv 1$, and (6.24)

$$\begin{aligned} g_s^{ij} &= g_{zs}^{ij} - \frac{1}{1+\nabla^z(f)} g_{zs}^{ik} f_{s^k} g_{zs}^{jm} f_{s^m} \\ &= \frac{1}{z} g_{sx}^{ij} - d_3 \nabla(f, s^i) \nabla(f, s^j) \,, \quad i,j,k,m = 1,\ldots,n \,, \end{aligned} \tag{6.33}$$

where

$$d_3 = \frac{1}{z[z+\nabla(f)]} \,.$$

Beltrami's Mixed and First Differential Parameters. Formulas (4.50) and (6.33) give rize to the following expression for the Beltrami's mixed differential parameter of functions $\varphi(s)$ and $\psi(s)$

$$\nabla^s(\varphi, \psi) = \frac{1}{z} \nabla(\varphi, \psi) - d_3 \nabla(f, \varphi) \nabla(f, \psi) \,. \tag{6.34}$$

So the Beltrami's first differential parameter of a function $\varphi(s)$ is expressed as follows:

$$\nabla^s(\varphi) = \frac{1}{z} \nabla(\varphi) - d_3 [\nabla(f, \varphi)]^2 \,. \tag{6.35}$$

Christoffel Symbols. For the Christoffel symbols of the first kind we have, taking advantage of (4.22), (6.26), and (6.15) with $v(s) \equiv 1$

$$\begin{aligned}{} [ij,k]^s &= [ij,k]^z + f_{s^i s^j} f_{s^k} \\ &= z[ij,k] + f_{s^i s^j} f_{s^k} + \frac{1}{2}(z_{s^i} g_{jk}^{xs} + z_{s^j} g_{ik}^{xs} - z_{s^k} g_{ij}^{xs}) \,, \end{aligned} \tag{6.36}$$

$i,j,k = 1,\ldots,n$.

Using (6.17) with $v(s) \equiv 1$ gives

$${}^s\Upsilon_{ij}^k = {}^z\Upsilon^k{}_{ij} + \frac{1}{1+\nabla^z(f)} \nabla^z(f, s^k) \nabla_{ij}^z(f) \,, \quad i,j,k = 1,\ldots,n \,.$$

Taking advantage in this equation of (6.25), (6.27), and (6.28) yields

$$\begin{aligned} {}^s\Upsilon_{ij}^m &= \Upsilon_{ij}^m + d_3 \nabla(f, s^m) \nabla_{ij}(f) \\ &\quad + \frac{1}{2z}[z_{s^i} \delta_m^j + z_{s^j} \delta_m^i - \nabla(z, s^m) g_{ij}^{xs}] \\ &\quad - \frac{1}{2} d_3 \nabla(f, s^m)[z_{s^i} f_{s^j} + z_{s^j} f_{s^i} - \nabla(f, z) g_{ij}^{xs}] \,, \end{aligned} \tag{6.37}$$

$i,j,k,m = 1,\ldots,n$.

Tensor of Mixed Derivatives. For the tensor of mixed derivatives the formula (6.19) produces

$$\nabla^s_{ij}(\varphi) = \nabla^z_{ij}(\varphi) - d_3 \nabla^z(f,\varphi) \nabla^z_{ij}(f) , \quad i,j = 1,\ldots,n .$$

Substituting into this formula the expressions (6.25) and (6.28) yields

$$\nabla^s_{ij}(\varphi) = \nabla_{ij}(\varphi) - zd_3 \nabla(f,\varphi) \nabla_{ij}(f)$$

$$- \frac{1}{2z}[z_{s^i}\varphi_{s^j} + z_{s^j}\varphi_{s^i} - \nabla(z,\varphi)g^{xs}_{ij}] \quad (6.38)$$

$$+ \frac{1}{2}d_3[z_{s^i}f_{s^j} + z_{s^j}f_{s^i} - \nabla(z,f)g^{xs}_{ij}] , \quad i,j = 1,\ldots,n .$$

Beltrami's Second Differential Parameter. The Beltrami's second differential parameter of a function $\varphi(s)$ is computed, in accordance with (6.22) for $v(s) \equiv 1$, as follows:

$$\Delta^s_B[\varphi] = \Delta^z_B[\varphi] - \frac{z}{z+\nabla(f)}\nabla^z(f,\varphi)\Delta^z_B[f]$$

$$+ \frac{1}{2}\left[\frac{z}{z+\nabla(f)}\right]^2 \nabla^z(f,\varphi)\nabla^z(\nabla^z(f),f) \quad (6.39)$$

$$- \frac{z}{z+\nabla(f)}\nabla^z(f,s^i)\nabla^z(f,s^j)\nabla^z_{ij}(\varphi) , \quad i,j=1,\ldots,n .$$

So using (6.25), (6.28), and (6.29) we obtain

$$\Delta^s_B[\varphi] = \frac{1}{z}\Delta_B[\varphi] + \frac{n-2}{2z^2}\nabla(z,\varphi)$$

$$- \frac{1}{z+\nabla(f)}\nabla(f,\varphi)\left[\frac{1}{z}\Delta_B[f] + \frac{n-2}{2z^2}\nabla(z,f)\right]$$

$$+ \frac{1}{2[z+\nabla(f)]^2}\nabla(f,\varphi)\nabla\left(\frac{1}{z}\nabla(f),f\right) \quad (6.40)$$

$$- d_3\left[\nabla(f,s^i)\nabla(f,s^j)\nabla_{ij}(\varphi) - \frac{1}{z}\nabla(f,z)\nabla(f,\varphi)\right.$$

$$\left. + \frac{1}{2z}\nabla(z,\varphi)\nabla(f)\right] , \quad i,j=1,\ldots,n .$$

6.2 Geometric Characteristics of Monitor Surfaces

Let the monitor manifold over a regular surface S^{xn} represented by (6.1) is a monitor surface S^{rn} formed in the coordinates $s^1,\ldots,s^n$ by a vector-valued monitor function $\boldsymbol{f}(s) = [f^1(s),\ldots,f^l(s)]$, i.e. S^{rn} is parametrized by

$$\boldsymbol{r}(s) : S^n \to R^{n+n_0+l} , \quad \boldsymbol{r}(s) = [\boldsymbol{x}(s), \boldsymbol{f}(s)] . \quad (6.41)$$

For the elements g^s_{ij} of the covariant metric tensor of S^{rn} in the coordinates $s^1,\ldots,s^n$ we have from (6.41)

$$g^s_{ij} = \mathbf{r}_{s^i} \cdot \mathbf{r}_{s^j} = g^{xs}_{ij} + \mathbf{f}_{s^i} \cdot \mathbf{f}_{s^j}, \quad i,j = 1,\ldots,n.$$

Though equations (6.6 – 6.22) are proved in the case of a scalar-valued monitor function $f(\mathbf{s})$, nevertheless, they allow one to compute successively the elements of the contravariant metric tensor (g^{ij}_s) and invariants of the monitor surface S^{rn} formed by a vector-valued monitor function $\mathbf{f}(\mathbf{s})$ over S^{xn} (i.e. represented by (6.41)) through the metric elements of the surface S^{xn} and the first and second derivatives of $f^i(\mathbf{s})$, $i = 1,\ldots,l$.

6.2.1 Computation of the Elements of the Contravariant Metric Tensor

Two-Dimensional Monitor Function. Let first $\mathbf{f}(\mathbf{s}) = [f^1(\mathbf{s}), f^2(\mathbf{s})]$. Then assuming S^{yn} as the monitor surface formed by the scalar-valued monitor function $f^1(\mathbf{s})$ over S^{xn} and represented correspondently by the parametrization

$$\mathbf{y}(\mathbf{s}) : S^n \to R^{n+n_0+1}, \quad \mathbf{y}(\mathbf{s}) = [\mathbf{x}(\mathbf{s}), f^1(\mathbf{s})], \quad (6.42)$$

we get that S^{rn} is the monitor surface over S^{yn} formed by the second scalar-valued monitor function $f^2(\mathbf{s})$ and having the corresponding parametrization

$$\mathbf{r}(\mathbf{s}) : S^n \to R^{n+n_0+2}, \quad \mathbf{r}(\mathbf{s}) = [\mathbf{y}(\mathbf{s}), f^2(\mathbf{s})]. \quad (6.43)$$

Now, introducing designations g^1_{ij}, g^{ij}_1, g_1 and g^2_{ij}, g^{ij}_2, g_2, $i,j = 1,\ldots,n$, for the covariant and contravariant metric tensors and their determinants of S^{yn} and S^{rn}, respectively, we find, in accordance with (6.6) and (6.7) for $v(\mathbf{s}) \equiv 1$,

$$g_1 = \det(g^1_{ij}) = g^{xs}[1 + \nabla(f^1)], \quad i,j = 1,\ldots,n,$$

$$g^1_{ij} = g^{xs}_{ij} + \frac{\partial f^1}{\partial s^i}\frac{\partial f^1}{\partial s^j}, \quad i,j = 1,\ldots,n,$$

$$g^{ij}_1 = g^{ij}_{sx} - \frac{g^{xs}}{g^1}g^{ik}_{sx}g^{jm}_{sx}\frac{\partial f^1}{\partial s^k}\frac{\partial f^1}{\partial s^m}, \quad i,j,k,m = 1,\ldots,n,$$

$$g_2 = g^s = \det(g^2_{ij}) = g_1\left(1 + g^{ij}_1\frac{\partial f^2}{\partial s^i}\frac{\partial f^2}{\partial s^j}\right), \quad i,j = 1,\ldots,n, \quad (6.44)$$

$$g^2_{ij} = g^s_{ij} = g^1_{ij} + \frac{\partial f^2}{\partial s^i}\frac{\partial f^2}{\partial s^j} = g^{xs}_{ij} + \frac{\partial \mathbf{f}}{\partial s^i}\cdot\frac{\partial \mathbf{f}}{\partial s^j}, \quad i,j = 1,\ldots,n,$$

$$g^{ij}_2 = g^{ij}_s = g^{ij}_1 - \frac{g_1}{g^s}g^{ik}_1 g^{jm}_1\frac{\partial f^2}{\partial s^k}\frac{\partial f^2}{\partial s^m}, \quad i,j,k,m = 1,\ldots,n.$$

Thus substituting g^{ij}_1 and g_1 in g^{ij}_2 and g_2 we obtain the expressions for g^s and g^{ij}_s through the elements g^{ij}_{sx} of the surface S^{xn} and the first derivatives of the monitor functions $f^1(\mathbf{s})$ and $f^2(\mathbf{s})$:

6.2 Geometric Characteristics of Monitor Surfaces

$$g_1 = g^{xs}[1 + \nabla(f^1)], \qquad (6.45)$$

$$g^s = g_1 \left(1 + g_{sx}^{ij} \frac{\partial f^2}{\partial s^i} \frac{\partial f^2}{\partial s^j} - \frac{g^{xs}}{g_1} g_{sx}^{ik} \frac{\partial f^1}{\partial s^k} \frac{\partial f^2}{\partial s^i} g_{sx}^{jm} \frac{\partial f^1}{\partial s^m} \frac{\partial f^2}{\partial s^j}\right)$$

$$= g^{xs}\{[1 + \nabla(f^1)][1 + \nabla(f^2)] - [\nabla(f^1, f^2)]^2\}, \qquad (6.46)$$

$$i, j, k, m = 1, \ldots, n,$$

$$g_s^{ij} = g_{sx}^{ij} - \frac{1}{1 + \nabla(f^1)} g_{sx}^{ik} g_{sx}^{jm} \frac{\partial f^1}{\partial s^k} \frac{\partial f^1}{\partial s^m}$$

$$- d[1 + \nabla(f^1)] \left[g_{sx}^{ik} \frac{\partial f^2}{\partial s^k} - \frac{1}{1 + \nabla(f^1)} g_{sx}^{ik} \frac{\partial f^1}{\partial s^k} \nabla(f^1 f^2)\right]$$

$$\times \left[g_{sx}^{jm} \frac{\partial f^2}{\partial s^m} - \frac{1}{1 + \nabla(f^1)} g_{sx}^{jm} \frac{\partial f^1}{\partial s^m} \nabla(f^1 f^2)\right] \qquad (6.47)$$

$$= g_{sx}^{ij} - d g_{sx}^{ik} g_{sx}^{jm} \left\{[1 + \nabla(f^1)] \frac{\partial f^2}{\partial s^k} \frac{\partial f^2}{\partial s^m} + [1 + \nabla(f^2)] \frac{\partial f^1}{\partial s^k} \frac{\partial f^1}{\partial s^m}\right.$$

$$\left. - \nabla(f^1 f^2) \left(\frac{\partial f^1}{\partial s^k} \frac{\partial f^2}{\partial s^m} + \frac{\partial f^2}{\partial s^l} \frac{\partial f^1}{\partial s^m}\right)\right\}, \quad i, j, k, m = 1, \ldots, n,$$

where

$$d = 1/\{[1 + \nabla(f^1)][1 + \nabla(f^2)] - [\nabla(f^1, f^2)]^2\} = g^{xs}/g^s.$$

In particular, when $g_{ij}^{sx} = \delta_i^j$, $i, j = 1, \ldots, n$, then $g_{sx}^{ij} = \delta_j^i$, $i, j = 1, \ldots, n$, therefore from (6.47)

$$g_s^{ij} = \delta_j^i - d\left\{[1 + \nabla^E(f^1)] \frac{\partial f^2}{\partial s^i} \frac{\partial f^2}{\partial s^j} + [1 + \nabla^E(f^2)] \frac{\partial f^1}{\partial s^i} \frac{\partial f^1}{\partial s^j}\right.$$

$$\left. - \nabla^E(f^1 f^2) \left(\frac{\partial f^1}{\partial s^i} \frac{\partial f^2}{\partial s^j} + \frac{\partial f^2}{\partial s^j} \frac{\partial f^1}{\partial s^i}\right)\right\}, \quad i, j = 1, \ldots, n, \qquad (6.48)$$

where

$$\nabla^E(f) = \frac{\partial f}{\partial s^i} \frac{\partial f}{\partial s^i}, \quad \nabla^E(f^1, f^2) = \frac{\partial f^1}{\partial s^i} \frac{\partial f^2}{\partial s^i}, \quad i = 1, \ldots, n,$$

$$d = 1/\{[1 + \nabla^E(f^1)][1 + \nabla^E(f^2)] - [\nabla^E(f^1, f^2)]^2\}.$$

If the monitor function $\boldsymbol{f}(s)$ is a vector-valued function $\boldsymbol{f}(s) = [f^1(s), f^2(s)]$ then the elements of the contravariant metric tensor g_ξ^{ij} of the monitor surface S^{rn} over a domain S^n in the grid coordinates $\xi^1, \ldots, \xi^n$ have, in accordance with (6.48) and (4.15), the following form

$$g_\xi^{ij} = g_{\xi s}^{ij} - d g_{\xi s}^{ik} g_{\xi s}^{jm} \left\{ [1 + \nabla^E(f^1)] \frac{\partial f^2}{\partial \xi^k} \frac{\partial f^2}{\partial \xi^m} \right.$$
$$\left. + [1 + \nabla^E(f^2)] \frac{\partial f^1}{\partial \xi^k} \frac{\partial f^1}{\partial \xi^m} - \nabla^E(f^1 f^2) \left(\frac{\partial f^1}{\partial \xi^k} \frac{\partial f^2}{\partial \xi^m} + \frac{\partial f^2}{\partial \xi^k} \frac{\partial f^1}{\partial \xi^m} \right) \right\}, \quad (6.49)$$
$$i, j, k, m = 1, \ldots, n,$$

where the elements $g_{\xi s}^{jm}$ of the contravariant metric tensor of S^n in the grid coordinates are defined by (6.12). Note the constant d in this formula is an invariant of parametrizations hence it is the same as in (6.48).

Similarly, if $\boldsymbol{f}(\boldsymbol{s}) = [f^1(\boldsymbol{s}), f^2(\boldsymbol{s})]$ is a monitor function over a surface S^{xn} we get from (6.47) and (4.15) for the elements of the contravariant metric tensor of S^{rn} represented by (6.41) in the grid coordinates $\xi^1, \ldots, \xi^n$

$$g_\xi^{ij} g_{\xi x}^{ij} - d g_{\xi x}^{ik} g_{\xi x}^{jm} \left\{ [1 + \nabla(f^1)] \frac{\partial f^2}{\partial \xi^k} \frac{\partial f^2}{\partial \xi^m} + [1 + \nabla(f^2)] \frac{\partial f^1}{\partial \xi^k} \frac{\partial f^1}{\partial \xi^m} \right.$$
$$\left. - \nabla(f^1 f^2) \left(\frac{\partial f^1}{\partial \xi^k} \frac{\partial f^2}{\partial \xi^m} + \frac{\partial f^2}{\partial \xi^k} \frac{\partial f^1}{\partial \xi^m} \right) \right\}, \quad i, j, k, m = 1, \ldots, n. \quad (6.50)$$

Here the constant d being an invariant of parametrizations is the same as in (6.47).

Arbitrary Vector-Valued Monitor Functions. Formula (6.47) prompts one on how the elements of the contravariant metric tensor of the monitor surface S^{rn} represented by (6.41) are expressed in the case of an arbitrary vector-valued monitor function $\boldsymbol{f}(\boldsymbol{s}) = [f^1(\boldsymbol{s}), \ldots, f^l(\boldsymbol{s})]$ in the parametric coordinates $s^1, \ldots, s^n$. For this purpose we designate by (d_{ab}) and (d^{ab}), $a, b = 1, \ldots, l$, two mutually inverse $(l \times l)$ matrices, where

$$d^{ab} = \delta_b^a + \nabla(f^a, f^b), \quad a, b = 1, \ldots, l. \quad (6.51)$$

Theorem 6.2.1. *The elements of the contravariant metric tensor of the monitor surface S^{rn} represented by (6.41) are expressed in the coordinates $s^1, \ldots, s^n$ by the following formula*

$$g_s^{ij} = g_{sx}^{ij} - g_{sx}^{ik} g_{sx}^{jm} d_{ab} \frac{\partial f^a}{\partial s^k} \frac{\partial f^b}{\partial s^m}$$
$$= g_{sx}^{ij} - d_{ab} \nabla(f^a, s^i) \nabla(f^b, s^j), \quad (6.52)$$
$$i, j, k, m = 1, \ldots, n, \quad a, b = 1, \ldots, l.$$

Proof. Equation (6.52) is valid if we shall show that the elements g_s^{ij} defined by (6.52) are subject to the relations

$$g_{kj}^s g_s^{ji} = \delta_k^i, \quad i, j, k = 1, \ldots, n, \quad (6.53)$$

where g_{kj}^s is the (kj)th element of the covariant metric tensor of S^{rn} in the coordinates $s^1, \ldots, s^n$. As

6.2 Geometric Characteristics of Monitor Surfaces

$$g^s_{kj} = g^{xs}_{kj} + \frac{\partial f^a}{\partial s^k}\frac{\partial f^a}{\partial s^j}, \quad j, k = 1, \ldots, n, \quad a = 1, \ldots, l, \tag{6.54}$$

we have

$$\begin{aligned}
g^s_{kj}g^{ji}_s &= \left(g^{xs}_{kj} + \frac{\partial f^a}{\partial s^k}\frac{\partial f^a}{\partial s^j}\right)\left(g^{ji}_{sx} - g^{jm}_{sx}g^{ip}_{sx}d_{bc}\frac{\partial f^b}{\partial s^m}\frac{\partial f^c}{\partial s^p}\right) \\
&= \delta^k_i + \frac{\partial f^a}{\partial s^j}g^{ji}_{sx}\frac{\partial f^a}{\partial s^k} - \frac{\partial f^c}{\partial s^p}g^{ip}_{sx}d_{bc}\frac{\partial f^b}{\partial s^k} \\
&\quad - \frac{\partial f^c}{\partial s^p}g^{ip}_{sx}d_{bc}\nabla(f^a, f^b)\frac{\partial f^a}{\partial s^k} \\
&= \delta^k_i + \frac{\partial f^a}{\partial s^j}g^{ji}_{sx}b^a_k,
\end{aligned} \tag{6.55}$$

$$i, j, k, m, p = 1, \ldots, n, \quad a, b, c = 1, \ldots, l,$$

where

$$b^a_k = \frac{\partial f^a}{\partial s^k} - d_{ab}\frac{\partial f^b}{\partial s^k} - d_{ab}\nabla(f^c, f^b)\frac{\partial f^c}{\partial s^k}, \quad k = 1, \ldots, n, \quad a, b, c = 1, \ldots, l.$$

Since

$$\nabla(f^c, f^b) = d^{cb} - \delta^c_b, \quad b, c = 1, \ldots, l, \tag{6.56}$$

we find

$$b^a_k = \frac{\partial f^a}{\partial s^k} - d_{ab}\frac{\partial f^b}{\partial s^k} - d_{ab}(d^{cb} - \delta^c_b)\frac{\partial f^c}{\partial s^k} = 0,$$

$$k = 1, \ldots, n, \quad a, b, c = 1, \ldots, l,$$

therefore (6.55) yields that the equation (6.53) is valid.

In particular, for $l = 1$ and $f^a = f$, we find

$$d^{aa} = 1 + \nabla(f), \quad d_{aa} = \frac{1}{1 + \nabla(f)}, \quad a = 1,$$

hence (6.52) in this particular case is as follows:

$$g^{ij}_s = g^{ij}_{sx} - \frac{1}{1 + \nabla(f)}\nabla(f^a, s^i)\nabla(f^a, s^j), \quad i, j = 1, \ldots, n, \quad a = 1,$$

which coincides with (6.7) for $v(s) \equiv 1$.

If $\boldsymbol{f}(s) = [f^1(s), f^2(s)]$ then for the matrices (d^{ab}) and (d_{ab}) we have

$$d^{ab} = \begin{pmatrix} 1 + \nabla(f^1, f^1) & \nabla(f^1, f^2) \\ \nabla(f^1, f^2) & 1 + \nabla(f^2, f^2) \end{pmatrix},$$

$$d_{ab} = d\begin{pmatrix} 1 + \nabla(f^2, f^2) & -\nabla(f^1, f^2) \\ -\nabla(f^1, f^2) & 1 + \nabla(f^1, f^1) \end{pmatrix},$$

6. Relations to Monitor Manifolds

where

$$d = \frac{1}{\det(d^{ab})} = \frac{1}{[1 + \nabla(f^1)][1 + \nabla(f^2)] - [\nabla(f^1, f^2)]^2} \,.$$

Thus for the elements d_{ab}, $a, b = 1, 2$, there is valid the following formula

$$\begin{aligned}d_{ab} &= (-1)^{a+b} d d^{3-a 3-b} \\ &= (-1)^{a+b} d[\delta^a_b + \nabla(f^{3-a}, f^{3-b})]\,, \quad a, b = 1, 2\,, \quad a, b \text{ fixed}\,.\end{aligned} \qquad (6.57)$$

Therefore from (6.52), for $l = 2$, we find

$$g^{ij}_s = g^{ij}_{sx} - d[\delta^a_b + (-1)^{a+b} \nabla(f^{3-a}, f^{3-b})] \nabla(f^a, s^i) \nabla(f^b, s^j)\,,$$

$$i, j = 1, \ldots, n\,, \quad a, b = 1, 2\,, \quad a, b \text{ fixed}\,.$$

Note this formula coincides with (6.47) obtained by the successive application of formula (6.7).

Special Monitor Function.

Multidimensional Case. Let in (6.41) $l = n$ and $f^a(s) = s^a$, $a = 1, \ldots, n$. Then for the elements of the covariant metric tensor of $S^{\tau n}$ in the coordinates $s^1, \ldots, s^n$ we obtain

$$g^s_{ij} = g^{xs}_{ij} + \delta^i_j\,, \quad i, j = 1, \ldots, n\,. \qquad (6.58)$$

As for this monitor function $\boldsymbol{f}(\boldsymbol{s}) \equiv \boldsymbol{s}$

$$\nabla(f^a, s^i) = \nabla(f^a, f^i) = g^{ai}_{sx}\,, \quad a, i = 1, \ldots, n\,,$$

so, availing us of (6.51), we find

$$d^{ab} = \delta^a_b + g^{ab}_{sx}\,, \quad a, b = 1, \ldots, n\,. \qquad (6.59)$$

Therefore

$$\begin{aligned}d_{ab} \nabla(f^a, s^i) \nabla(f^b, s^j) &= d_{ab}(d^{ai} - \delta^a_i)(d^{bj} - \delta^b_j) \\ &= d^{ij} - \delta^i_j - \delta^i_j + d_{ij} = g^{ij}_{sx} - \delta^i_j + d_{ij}\,, \quad a, b, i, j = 1, \ldots, n\,.\end{aligned}$$

Thus, substituting these relations in (6.52) yields

$$g^{ij}_s = \delta^i_j - d_{ij}\,, \quad i, j = 1, \ldots, n\,. \qquad (6.60)$$

Two-Dimensional Case. For $n = 2$ we find, using (6.59),

$$d_{ij} = (-1)^{i+j} \frac{1}{\det(d^{ab})} d^{3-i 3-j} = w\left(\delta^i_j + \frac{1}{g^{xs}} g^{xs}_{ij}\right)\,, \qquad (6.61)$$

$$a, b, i, j = 1, 2\,, \quad i, j \text{ fixed}\,,$$

where

$$w = \frac{1}{\det(d^{ab})} = \frac{g^{xs}}{1 + g^{xs}_{11} + g^{xs}_{22} + g^{xs}}\,, \quad g^{xs} = \det(g^{xs}_{ij})\,.$$

6.2 Geometric Characteristics of Monitor Surfaces

In order to distinguish the geometric characteristics of S^{r2} in any coordinates v^1, v^2 with the metric (6.58) for $n = 2$ we shall further substitute in them the index $2v$ for v. So for the elements g_{ij}^{2s} and g_{2s}^{ij} of the corresponding covariant and contravariant metric tensor of S^{r2} in the coordinates s^1, s^2 we find, from (6.60) for $n = 2$ and (6.61),

$$g_{2s}^{ij} = (1-w)\delta_j^i - \frac{w}{g^{xs}}g_{ij}^{xs}, \quad i,j = 1, 2. \tag{6.62}$$

Using (4.15) and (6.62) yields the following formula for the elements of the contravariant metric tensor of S^{r2} in the grid coordinates ξ^1, ξ^2

$$g_{2\xi}^{ij} = (1-w)g_{\xi s}^{ij} - \frac{w}{g^{xs}}g_{km}^{x\xi}g_{\xi s}^{ki}g_{\xi s}^{mj}, \quad i,j,k,m = 1,2,$$

where

$$g_{\xi s}^{ij} = \frac{\partial \xi^i}{\partial s^k}\frac{\partial \xi^j}{\partial s^k}, \quad i,j,k = 1,2.$$

Since (6.58) and (4.8) we also find

$$g_{2s}^{ij} = \frac{w}{g^{xs}}(\delta_j^i + g^{xs}g_{sx}^{ij}), \quad i,j = 1,2. \tag{6.63}$$

As

$$g_{ij}^{xs} = g^{xs}[(g_{sx}^{11} + g_{sx}^{22})\delta_j^i - g_{sx}^{ij}], \quad i,j = 1,2, \tag{6.64}$$

and

$$g_{sx}^{ik}g_{sx}^{jk} = g_{sx}^{ij}(g_{sx}^{11} + g_{sx}^{22}) - \frac{1}{g^{xs}}\delta_j^i, \quad i,j,k = 1,2, \tag{6.65}$$

hence (6.61) and (6.63) also have the following corresponding forms

$$d_{ij} = w[(1 + g_{sx}^{11} + g_{sx}^{22})\delta_j^i - g_{sx}^{ij}], \quad i,j = 1,2, \tag{6.66}$$

$$g_{2s}^{ij} = w[(1 + g_{sx}^{11} + g_{sx}^{22})g_{sx}^{ij} - g_{sx}^{ik}g_{sx}^{jk}], \quad i,j,k = 1,2. \tag{6.67}$$

6.2.2 Beltrami's Mixed and First Differential Parameters

General Formulas. For the Beltrami's mixed differential parameter with respect to the metric (6.54) we obtain, using (4.51) and (6.52),

$$\nabla^s(\varphi, \psi) = \varphi_{s^i}\psi_{s^j}g_s^{ij} = \varphi_{s^i}\psi_{s^j}g_{sx}^{ij} - d_{ab}\varphi_{s^i}\frac{\partial f^a}{\partial s^m}g_{sx}^{im}\psi_{s^j}\frac{\partial f^b}{\partial s^k}g_{sx}^{jk}$$

$$= \nabla(\varphi,\psi) - d_{ab}\nabla(\varphi, f^a)\nabla(\psi, f^b), \tag{6.68}$$

$$i,j,k,m = 1,\ldots,n, \quad a,b = 1,\ldots l.$$

As to the Beltrami's first differential parameter of $\varphi(s)$ we find from (6.68)

$$\nabla^s(\varphi) = \nabla^s(\varphi,\varphi) = \nabla(\varphi) - d_{ab}\nabla(\varphi, f^a)\nabla(\varphi, f^b) \,,$$
$$a, b = 1, \ldots l \,. \tag{6.69}$$

We see that these formulas in the particular case ($l = 1$) give the corresponding expressions (6.13) and (6.14) for $v(s) \equiv 1$.

Special Two-Dimensional Case. Using (4.51) and (6.63) yields for the metric (6.58) with $n = 2$

$$\nabla^{2s}(\varphi, \psi) = \frac{w}{g^{xs}}[\nabla^E(\varphi, \psi) + g^{xs}\nabla(\varphi, \psi)] \,,$$
$$\nabla^{2s}(\varphi) = \frac{w}{g^{xs}}[\nabla^E(\varphi) + g^{xs}\nabla(\varphi)] \,. \tag{6.70}$$

Analogous using (4.51) and (6.67) for the metric (6.58) with $n = 2$ gives

$$\nabla^{2s}(\varphi, \psi) = w[(1 + g_{sx}^{11} + g_{sx}^{22})\nabla(\varphi, \psi) - \nabla(\varphi, s^k)\nabla(\psi, s^k)] \,,$$
$$\nabla^{2s}(\varphi) = w[(1 + g_{sx}^{11} + g_{sx}^{22})\nabla(\varphi) - \nabla(\varphi, s^k)\nabla(\varphi, s^k)] \,, \quad k = 1, 2 \,. \tag{6.71}$$

6.2.3 Christoffel Symbols

General Formula. The elements of the Christoffel symbols of the first kind of the monitor surface S^{rn} represented by (6.41) in the coordinates $s^1, \ldots, s^n$ are readily obtained using (4.27) and (6.41), namely,

$$[ij, m]^s = [ij, m] + \boldsymbol{f}_{s^i s^j} \cdot \boldsymbol{f}_{s^m} \,, \quad i, j, m = 1, \ldots, n \,. \tag{6.72}$$

Taking advantage of this formula and (6.52) we find the following formula for the elements of the Christoffel symbols of the second kind of S^{rn} in the coordinates $s^1, \ldots, s^n$

$$\begin{aligned}
{}^s\Upsilon_{ij}^k &= g_s^{km}[ij,m]^s = \left(g_{sx}^{km} - g_{sx}^{kp}g_{sx}^{mt}d_{ab}\frac{\partial f^a}{\partial s^p}\frac{\partial f^b}{\partial s^t}\right) \\
&\quad \times ([ij,m] + \boldsymbol{f}_{s^i s^j}\cdot\boldsymbol{f}_{s^m}) = \Upsilon_{ij}^k - g_{sx}^{kp}d_{ab}\frac{\partial f^a}{\partial s^p}\frac{\partial f^b}{\partial s^t}\Upsilon_{ij}^t \\
&\quad + g_{sx}^{km}\boldsymbol{f}_{s^i s^j}\cdot\boldsymbol{f}_{s^m} - \frac{\partial^2 f^b}{\partial s^i \partial s^j}\nabla(f^b, f^c)g_{sx}^{kp}d_{ac}\frac{\partial f^a}{\partial s^p} \\
&= \Upsilon_{ij}^k + g_{sx}^{kp}d_{ab}\frac{\partial f^a}{\partial s^p}\left(\frac{\partial^2 f^b}{\partial s^i \partial s^j} - \frac{\partial f^b}{\partial s^t}\Upsilon_{ij}^t\right) + \frac{\partial^2 f^b}{\partial s^i \partial s^j}g_{sx}^{kp}D_p^b \,, \\
&= \Upsilon_{ij}^k + d_{ab}\nabla(f^a, s^k)\nabla_{ij}(f^b) + \frac{\partial^2 f^b}{\partial s^i \partial s^j}g_{sx}^{kp}D_p^b \,,
\end{aligned} \tag{6.73}$$

$$i, j, k, m, p, t = 1, \ldots, n \,, \quad a, b, c = 1, \ldots l \,,$$

where

$$D_p^b = \frac{\partial f^b}{\partial s^p} - d_{ab}\frac{\partial f^a}{\partial s^p} - \nabla(f^b, f^c)d_{ac}\frac{\partial f^a}{\partial s^p},$$

$$p = 1, \ldots, n, \quad a, b, c = 1, \ldots, l.$$

Since (6.56) we find

$$D_p^b = \frac{\partial f^b}{\partial s^p} - d_{ab}\frac{\partial f^a}{\partial s^p} - \frac{\partial f^b}{\partial s^p} + d_{ab}\frac{\partial f^a}{\partial s^p} = 0,$$

$$p = 1, \ldots, n, \quad a, b = 1, \ldots, l,$$

therefore (6.73) yields

$$\begin{aligned}{}^s\Upsilon_{ij}^k &= \Upsilon_{ij}^k + g_{sx}^{kp}d_{ab}\frac{\partial f^a}{\partial s^p}\nabla_{ij}(f^b) \\ &= \Upsilon_{ij}^k + d_{ab}\nabla(f^a, s^k)\nabla_{ij}(f^b),\end{aligned} \quad (6.74)$$

$$i, j, k = 1, \ldots, n, \quad a, b = 1, \ldots l.$$

So the formula in the second line of (6.17) for $v(s) \equiv 1$ is a particular case of (6.74) for $l = 1$.

Special Two-Dimensional Case. Using (6.56) and (6.63) in (6.74) gives for the metric (6.58) with $n = 2$

$$\begin{aligned}{}^{2s}\Upsilon_{ij}^k &= \Upsilon_{ij}^k + d_{ab}(d^{ak} - \delta_k^a)\nabla_{ij}(s^b) = \Upsilon_{ij}^k + (\delta_k^b - d_{bk})\nabla_{ij}(s^b), \\ &= \Upsilon_{ij}^k + \frac{w}{g^{xs}}(\delta_k^b + g^{xs}g_{sx}^{bk})\nabla_{ij}(s^b), \quad a, b, i, j, k = 1, 2.\end{aligned} \quad (6.75)$$

While application of (6.60) and (6.67) to (6.74) produces

$$^{2s}\Upsilon_{ij}^k = \Upsilon_{ij}^k + w[(1 + g_{sx}^{11} + g_{sx}^{22})g_{sx}^{bk} - g_{sx}^{bm}g_{sx}^{km}]\nabla_{ij}(s^b), \quad (6.76)$$

$$a, b, i, j, k, m = 1, 2.$$

6.2.4 Mixed Derivatives

General Formula. In accordance with (4.39) and using (6.74) we obtain the following formula for the components of the covariant tensor of mixed derivatives with respect to the metric (6.54) of the monitor surface S^{rn} in the coordinates $s^1, \ldots, s^n$:

$$\begin{aligned}\nabla_{ij}^s(\varphi) &= \varphi_{s^is^j} - \varphi_{s^k}{}^s\Upsilon_{ij}^k \\ &= \varphi_{s^is^j} - \varphi_{s^k}\Upsilon_{ij}^k - d_{ab}\nabla(f^a, \varphi)\nabla_{ij}(f^b) \\ &= \nabla_{ij}(\varphi) - d_{ab}\nabla(f^a, \varphi)\nabla_{ij}(f^b),\end{aligned} \quad (6.77)$$

$$i, j, k = 1, \ldots, n, \quad a, b = 1, \ldots, l.$$

156   6. Relations to Monitor Manifolds

Special Two-Dimensional Case. For the metric (6.58) with $n = 2$ we find from (6.61) and (6.77)

$$\nabla_{ij}^{2s}(\varphi) = \nabla_{ij}(\varphi) - w\nabla(s^k, \varphi)\nabla_{ij}(s^k) - \frac{w}{g^{xs}}\varphi_{s^k}\nabla_{ij}(s^k)$$
$$= \nabla_{ij}(\varphi) - wL^k[\varphi]\nabla_{ij}(s^k), \quad i, j, k = 1, 2,$$
(6.78)

where

$$L^k[\varphi] = \nabla(s^k, \varphi) + \frac{\varphi_{s^k}}{g^{xs}}.$$

While application of (6.76) to the second line of (6.77) yields

$$\nabla_{ij}^{2s}(\varphi) = \nabla_{ij}(\varphi) - w[(1 + g_{sx}^{11} + g_{sx}^{22})\nabla(s^k, \varphi)$$
$$- g_{sx}^{km}\nabla(s^m, \varphi)]\nabla_{ij}(s^k), \quad i, j, k, m = 1, 2.$$
(6.79)

6.2.5 Beltrami's Second Differential Parameter

General Formula. Using (4.56), (6.52), and (6.77) we have for the invariant $\Delta_B^s[\varphi]$:

$$\Delta_B^s[\varphi] = g_s^{ij}\nabla_{ij}^s(\varphi)$$
$$= [g_{sx}^{ij} - d_{ab}\nabla(f^a, s^i)\nabla(f^b, s^j)][\nabla_{ij}(\varphi) - d_{ch}\nabla(f^c, \varphi)\nabla_{ij}(f^h)]$$
$$= \Delta_B[\varphi] - d_{ab}\nabla(f^a, \varphi)\Delta_B[f^b] - L,$$
(6.80)

$$i, j, k, m = 1, \ldots, n, \quad a, b, c, h = 1, \ldots, l,$$

where

$$L = d_{ab}\nabla(f^a, s^i)\nabla(f^b, s^j)[\nabla_{ij}(\varphi) - d_{ch}\nabla(f^c, \varphi)\nabla_{ij}(f^h)],$$

$$i, j = 1, \ldots, n, \quad a, b, c, h = 1, \ldots, l.$$

Note L is an invariant of parametrizations of S^{xn} therefore it can be computed in an arbitrary coordinate system.

Taking advantage of (4.25) and (6.51) gives

$$\frac{\partial}{\partial s^k}d_{ab} = d_{hb}d^{ch}\frac{\partial}{\partial s^k}d_{ac} = -d_{hb}d_{ac}\frac{\partial}{\partial s^k}d^{ch}$$

$$= -d_{hb}d_{ac}\frac{\partial}{\partial s^k}\left(g_{sx}^{ij}\frac{\partial f^c}{\partial s^i}\frac{\partial f^h}{\partial s^j}\right)$$

$$= -d_{hb}d_{ac}\left[g_{sx}^{ij}\left(\frac{\partial f^h}{\partial s^j}\frac{\partial^2 f^c}{\partial s^i \partial s^k} + \frac{\partial f^c}{\partial s^i}\frac{\partial^2 f^h}{\partial s^j \partial s^k}\right)\right.$$
$$\left. - \frac{\partial f^c}{\partial s^i}\frac{\partial f^h}{\partial s^j}(g_{sx}^{im}\Upsilon_{km}^j + g_{sx}^{pj}\Upsilon_{pk}^i)\right],$$
$$i, j, k, m, p = 1, \ldots, n, \quad a, b, c, h = 1, \ldots, l.$$

So using here the formula (4.39) yields

$$\frac{\partial}{\partial s^k} d_{ab} = -2 d_{hb} d_{ac} \nabla(f^h, s^i) \nabla_{ik}(f^c) \,, \tag{6.81}$$
$$i, k = 1, \ldots, n \,, \quad a, b, c, h = 1, \ldots, l \,,$$

therefore

$$d_{ab} d_{ch} \nabla(f^a, s^i) \nabla(f^b, s^j) \nabla(f^c, \varphi) \nabla_{ij}(f^h)]$$
$$= -\frac{1}{2} \nabla(f^c, \varphi) \nabla(f^b, s^j) \frac{\partial}{\partial s^j} d_{bc} = -\frac{1}{2} \nabla(f^c, \varphi) \nabla(d_{bc}, f^b) \,, \tag{6.82}$$
$$i, j, = 1, \ldots, n \,, \quad a, b, c, h = 1, \ldots, l \,.$$

Thus

$$L = d_{ab} \nabla(f^a, s^i) \nabla(f^b, s^j) \nabla_{ij}(\varphi) + \frac{1}{2} \nabla(f^c, \varphi) \nabla(d_{bc}, f^b) \,,$$
$$i, j = 1, \ldots, n \,, \quad a, b, c = 1, \ldots, l \,,$$

and consequently (6.80) also has the following form

$$\Delta_B^s[\varphi] = \Delta_B[\varphi] - d_{ab} \nabla(f^a, \varphi) \Delta_B[f^b] - \frac{1}{2} \nabla(f^c, \varphi) \nabla(d_{bc}, f^b)$$
$$- d_{ab} \nabla(f^a, s^i) \nabla(f^b, s^j) \nabla_{ij}(\varphi) \tag{6.83}$$
$$i, j = 1, \ldots, n \,, \quad a, b, c = 1, \ldots, l \,.$$

Note this formula is a generalization of the formula (6.22) for $v(s) \equiv 1$, $C = 0$.

Special Two-Dimensional Case. For the metric (6.58) with $n = 2$ we obtain, by virtue of (6.61) and (6.65),

$$d_{ab} \nabla(f^a, s^i) \nabla(f^b, s^j) = w \left(\delta_b^a + \frac{1}{g^{xs}} g_{ab}^{xs} \right) g_{sx}^{ai} g_{sx}^{bj}$$
$$= w \left(g_{sx}^{ik} g_{sx}^{jk} + \frac{1}{g^{xs}} g_{sx}^{ij} \right)$$
$$= \frac{w}{g^{xs}} [(1 + g_{11}^{xs} + g_{22}^{xs}) g_{sx}^{ij} - \delta_j^i] \,, \tag{6.84}$$
$$a, b, i, j, k = 1, 2 \,,$$

$$d_{ab} \nabla(f^a, \varphi) = w \left(\delta_b^a + \frac{1}{g^{xs}} g_{ab}^{xs} \right) g_{sx}^{am} \varphi_{sm} = w L^b[\varphi] \,,$$
$$a, b, m = 1, 2,$$

therefore (6.83) in this case has the following form

$$\Delta_B^{2s}[\varphi] = w\left\{\Delta_B[\varphi] - wL^k[\varphi]\Delta_B[s^k]\right\}$$
$$+ \frac{w}{g^{xs}}\left\{\nabla_{mm}(\varphi) - wL^k[\varphi]\nabla_{pp}(s^k)\right\}, \quad (6.85)$$
$$k, m, p = 1, 2.$$

Application of (6.67) and (6.78) gives, in accordance with (4.56),

$$\Delta_B^{2s}[\varphi] = w(d^1 g_{sx}^{ij} - g_{sx}^{ik} g_{sx}^{jk})\left\{\nabla_{ij}(\varphi) - wL^k[\varphi]\nabla_{ij}(s^k)\right\}$$
$$= d_1 w\{\Delta_B[\varphi] - wL^k[\varphi]\Delta_B[s^k]\} \quad (6.86)$$
$$- w g_{sx}^{ik} g_{sx}^{jk}\left\{\nabla_{ij}(\varphi) - wL^k[\varphi]\nabla_{ij}(s^k)\right\}, \quad i, j, k = 1, 2,$$

where

$$d_1 = 1 + g_{sx}^{11} + g_{sx}^{22}.$$

Other equivalent expressions for $\Delta_B^{2s}[\varphi]$ can be obtained by application of (6.63) and (6.66) and (6.79) to (4.56).

6.3 Particular Two-Dimensional Case

In the technology for generating grids on two-dimensional surfaces S^{x2} locally parametrized by

$$\boldsymbol{x}(\boldsymbol{s}) : S^2 \to R^3, \quad \boldsymbol{x}(\boldsymbol{s}) = [x^1(\boldsymbol{s}), x^2(\boldsymbol{s}), x^3(\boldsymbol{s})], \quad (6.87)$$

and with the corresponding covariant metric tensor

$$g_{ij}^{xs} = \boldsymbol{x}_{s^i} \cdot \boldsymbol{x}_{s^j}, \quad i, j = 1, 2, \quad (6.88)$$

in the coordinates s^1, s^2 there is an important monitor surface S^{r2} over S^{x2} formed by a specific vector-valued monitor function of the kind

$$\boldsymbol{F}(\boldsymbol{s}) = [s^1, s^2, f(\boldsymbol{s})].$$

It appears that the transformed grid equations with respect to the monitor metric formed by this vector-valued function are very convenient for the implementation into numerical codes (see Sect. 7.2.7).

The monitor surface S^{r2} can be locally parametrized in the coordinates s^1, s^2 by

$$\boldsymbol{r}(\boldsymbol{s}) : S^2 \to R^6, \quad \boldsymbol{r}(\boldsymbol{s}) = [\boldsymbol{x}(\boldsymbol{s}), \boldsymbol{F}(\boldsymbol{s})], \quad (6.89)$$

and consequently the elements g_{ij}^s of the covariant metric tensor of S^{r2} in the coordinates s^1, s^2 are as follows:

$$g_{ij}^s = \boldsymbol{r}_{s^i} \cdot \boldsymbol{r}_{s^j} = g_{ij}^{xs} + \delta_j^i + f_{s^i} f_{s^j}, \quad i, j = 1, 2. \quad (6.90)$$

Thus the monitor surface S^{r2} also is a monitor surface over the regular surface with the metric (6.58).

The geometric characteristics of S^{r2} can be computed in two ways: by using the values of the elements of the matrix d_{ij}, $i,j = 1,2,3$, in the corresponding formulas (6.52), (6.68–6.69), (6.74), (6.77), and (6.83) with $f^1 = s^1$, $f^2 = s^2$, $f^3 = f$; or by virtue of the formulas (6.7), (6.13), (6.14), (6.17), (6.19) and (6.22) in which $v(s) = 1$ and the metric g_{ij}^{xs} in the coordinates s^1, s^2 is identified with the special metric $g_{ij}^{2s} = g_{ij}^{xs} + \delta_j^i$.

In this section we establish some relations between the geometric characteristics of S^{r2} and S^{x2} by using individually one of these two ways.

6.3.1 Preliminary Results

In this subsection we find a formula for the elements of the matrix d_{ij}, $i,j = 1,2,3$, when $f^1(s) = s^1$, $f^2(s) = s^2$, $f^3(s) = f(s)$. For this purpose we formulate a readily verified lemma.

Lemma 6.3.1. *Let (g^{ij}) and (a_{ij}), $i,j = 1,2$, be some symmetric 2×2 matrix. Then*

$$g^{ik}g^{jm}a_{km} = g^{ij}g^{km}a_{km} - (-1)^{i+j}ga_{3-i\,3-j}\,, \tag{6.91}$$

$$i,j,k,m,p = 1,2\,, \quad i,j \text{ fixed}\,,$$

where

$$g = \det(g^{ij})\,.$$

Proof. Let $i = j = 1$. Then

$$g^{1k}g^{1m}a_{km} = g^{11}g^{1m}a_{1m} + g^{12}g^{1m}a_{2m}$$

$$= g^{11}g^{km}a_{km} + (g^{12}g^{1m} - g^{11}g^{2m})a_{2m}$$

$$= g^{11}g^{km}a_{km} - ga_{22}\,, \quad k,m = 1,2\,.$$

Analogously, if $i = j = 2$, then

$$g^{2k}g^{2m}a_{km} = g^{22}g^{km}a_{km} - ga_{11}\,, \quad k,m = 1,2\,.$$

Let now $i = 1$, $j = 2$ or $i = 2$, $j = 1$. In this case we find

$$g^{1k}g^{2m}a_{km} = g^{11}g^{2m}a_{1m} + g^{12}g^{2m}a_{2m}$$

$$= g^{12}g^{km}a_{km} + (g^{11}g^{2m} - g^{12}g^{1m})a_{1m}$$

$$= g^{12}g^{km}a_{km} + ga_{12}\,, \quad k,m = 1,2\,.$$

Uniting all the relations written above gives the equation (6.91), and proof of the lemma.

6. Relations to Monitor Manifolds

Now we proceed to the major result.

Theorem 6.3.1. *The elements of the matrix d_{ij}, $i,j = 1,2,3$, are computed by the followig formulas*

$$d_{ij} = \frac{1}{B}\{g^{xs}_{ij} + f_{s^i}f_{s^j} + g^{xs}[1+\nabla(f)]\delta^i_j\}, \quad i,j = 1,2,$$

$$d_{3i} = d_{i3} = -\frac{1}{B}[g^{xs}\nabla(s^i, f) + f_{s^i}] = -\frac{g^{xs}}{B}L^i[f], \quad i,j = 1,2, \qquad (6.92)$$

$$d_{33} = \frac{1}{B}(1 + g^{xs}_{11} + g^{xs}_{22} + g^{xs}) = \frac{g^{xs}}{Bw},$$

where

$$B = g^{xs}\det(d^{ij}) = 1 + g^{xs}_{11} + g^{xs}_{22} + g^{xs}[1+\nabla(f)] + f_{s^m}f_{s^m}$$

$$= g^{xs}[d_1 + \nabla(f)] + 1 + \nabla^E(f)$$

$$= g^{xs}\left[\frac{1}{w} + L^m[f]f_{s^m}\right],$$

$$i,j = 1,2,3, \quad m = 1,2, \qquad (6.93)$$

$$L^k[y] = \nabla(s^k, y) + \frac{y_{s^k}}{g^{xs}}, \quad k = 1,2,$$

$$d_1 = 1 + g^{11}_{sx} + g^{22}_{sx}, \quad \frac{1}{w} = d_1 + \frac{1}{g^{xs}}.$$

Proof. The theorem will be proved if we show that

$$d^{ij}d_{jk} = \delta^i_k, \quad i,j,k = 1,2,3. \qquad (6.94)$$

For this purpose we consider four cases: 1) $i \neq 3$ and $k \neq 3$; 2) $i \neq 3$ and $k = 3$; 3) $i = 3$ and $k \neq 3$; 4) $i = 3$ and $k = 3$.

Let first both $i \neq 3$ and $k \neq 3$. Then, using (6.92) and the relations

$$d^{ij} = \delta^i_j + \nabla(f^i, f^j), \quad i,j = 1,2,3,$$

we find

$$d^{ij}d_{jk} = \frac{1}{B}[\delta^i_m + \nabla(s^i, s^m)]\{g^{xs}_{mk} + f_{s^m}f_{s^k} + g^{xs}[1+\nabla(f)]\delta^m_k\}$$

$$- \frac{1}{B}\nabla(s^i, f)[g^{xs}\nabla(s^k, f) + f_{s^k}]$$

$$= \frac{1}{B}\{g^{xs}_{ik} + f_{s^i}f_{s^k} + g^{xs}[1+\nabla(f)]\delta^i_k + \delta^i_k + g^{xs}g^{ik}_{sx}[1+\nabla(f)]$$

$$- g^{xs}\nabla(s^i, f)\nabla(s^k, f)\}, \quad i,k,m = 1,2, \quad j = 1,2,3.$$

$$(6.95)$$

Since (6.91), for $g^{ij}_{sx} = g^{ij}$, $f_{s^i}f_{s^j} = a_{ij}$,

6.3 Particular Two-Dimensional Case

$$g^{xs}\nabla(s^i,f)\nabla(s^k,f) = g^{xs}g^{ik}_{sx}\nabla(f) - (-1)^{i+k}f_{s^{3-i}}f_{s^{3-k}}\,, \tag{6.96}$$

$$i,k = 1,2\,,\quad i,k \text{ fixed}\,,$$

therefore, from (6.95),

$$d^{ij}d_{jk} = \frac{1}{B}\{g^{xs}_{ik} + f_{s^i}f_{s^k} + g^{xs}[1+\nabla(f)]\delta^i_k + \delta^i_k$$
$$+ g^{xs}g^{ik}_{sx} + (-1)^{i+k}f_{s^{3-i}}f_{s^{3-k}}\}\,, \tag{6.97}$$

$$i,k = 1,2\,,\quad j = 1,2,3\,,\quad i,k \text{ fixed}\,.$$

Note

$$g^{xs}_{ik} + g^{xs}g^{ik}_{sx} = (g^{xs}_{11}+g^{xs}_{22})\delta^i_k\,,\quad i,k=1,2\,,$$

$$f_{s^i}f_{s^k} + (-1)^{i+k}f_{s^{3-i}}f_{s^{3-k}} = f_{s^m}f_{s^m}\delta^i_k = \nabla^E(f)\delta^i_k\,,$$

$$i,k,m = 1,2\,,\quad i,k \text{ fixed}\,,$$

so, availing us of these relations and (6.83) in (6.97), we obtain

$$d^{ij}d_{jk} = \frac{1}{B}\{1 + g^{xs}_{11} + g^{xs}_{22} + f_{s^m}f_{s^m} + g^{xs}[1+\nabla(f)]\}\delta^i_k = \delta^i_k \tag{6.98}$$

$$i,k,m = 1,2\,,\quad j = 1,2,3\,.$$

Let now in (6.94) $i \neq 3$ and $k = 3$. Then, using (6.92),

$$d^{ij}d_{j3} = -\frac{1}{B}[\delta^i_m + \nabla(s^i,s^m)][g^{xs}\nabla(s^m,f) + f_{s^m}]$$
$$+ \frac{1}{B}\nabla(s^i,f)(1 + g^{xs}_{11} + g^{xs}_{22} + g^{xs})\,, \tag{6.99}$$

$$i,m = 1,2\,,\ j = 1,2,3\,.$$

Since (6.65)

$$g^{xs}\nabla(s^i,s^m)\nabla(s^m,f) = g^{xs}g^{im}_{sx}g^{mk}_{sx}f_{s^k}$$
$$= \nabla(s^i,f)(g^{xs}_{11}+g^{xs}_{22}) - f_{s^i}\,, \tag{6.100}$$

$$i,m,k = 1,2\,,$$

therefore we find from (6.99)

$$d^{ij}d_{j3} = \frac{1}{B}[\nabla(s^i,f)(g^{xs}_{11}+g^{xs}_{22}) - f_{s^i}$$
$$- \nabla(s^i,f)(g^{xs}_{11}+g^{xs}_{22}) + f_{s^i}] = 0\,, \tag{6.101}$$

$$i = 1,2\,.$$

Now let $k \neq 3$ and $i = 3$. In this case, availing us of (6.92),

$$\begin{aligned}d^{3j}d_{jk} &= \frac{1}{B}\nabla(s^m, f)\{g^{xs}_{mk} + f_{s^m}f_{s^k} + g^{xs}[1 + \nabla(f)]\delta^m_k\} \\ &\quad - \frac{1}{B}[1 + \nabla(f)][g^{xs}\nabla(s^k, f) + f_{s^k}] \\ &= \frac{1}{B}\{f_{s^k} + \nabla(f)f_{s^k} + g^{xs}[1 + \nabla(f)]\nabla(s^k, f) \\ &\quad - [1 + \nabla(f)][g^{xs}\nabla(s^k, f) + f_{s^k}]\} = 0 \, , \\ k, m &= 1,2 \, , \ j = 1,2,3 \, .\end{aligned} \quad (6.102)$$

Let us consider the final case $i = 3$ and $k = 3$. We have then, taking advantage of (6.92),

$$\begin{aligned}d^{3j}d_{j3} &= -\frac{1}{B}\nabla(s^m, f)[g^{xs}\nabla(s^m, f) + f_{s^m}] \\ &\quad + \frac{1}{B}[1 + \nabla(f)](1 + g^{xs}_{11} + g^{xs}_{22} + g^{xs}) \, , \\ m &= 1,2 \, , \ j = 1,2,3 \, .\end{aligned} \quad (6.103)$$

Since (6.96)

$$g^{xs}\nabla(s^m, f)\nabla(s^m, f) = \nabla(f)(g^{xs}_{11} + g^{xs}_{22}) - f_{s^m}f_{s^m} \, , \ m = 1,2 \, ,$$

hence (6.103) gives

$$\begin{aligned}d^{3j}d_{j3} &= \frac{1}{B}\{-\nabla(f)(1 + g^{xs}_{11} + g^{xs}_{22}) + f_{s^m}f_{s^m} \\ &\quad + [1 + \nabla(f)](1 + g^{xs}_{11} + g^{xs}_{22} + g^{xs})\} = 1 \, , \\ m &= 1,2 \, , \ j = 1,2,3 \, .\end{aligned} \quad (6.104)$$

Thus the formulas (6.98), (6.101–6.102), and (6.104) confirm that the relations (6.94) are held. This proves the theorem.

6.3.2 Contravariant Metric Tensor

The elements g^{ij}_s of the contravariant metric tensor of S^{r2} parametrized in the coordinates s^1, s^2 by (6.89) can be computed by the formula (6.7) in which $v = 1$, while the metric g^{xs}_{ij} is identified with (6.58) for $n = 2$, i.e.

$$g^{ij}_s = g^{ij}_{2s} - d_2 g^{ik}_{2s} f_{s^k} g^{jm}_{2s} f_{s^m} \, , \quad i,j,k,m = 1,2 \, , \quad (6.105)$$

where

$$d_2 = \frac{1}{1 + \nabla^{2s}(f)} \, .$$

Here, in accordance with the second line of (6.70),

$$\nabla^{2s}(f) = w\left[\nabla(f) + \frac{1}{g^{xs}}\nabla^E(f)\right] = wL^i[f]f_{s^i}, \quad i = 1, 2,$$

where

$$L^k[\varphi] = \nabla(s^k, \varphi) + \frac{\varphi_{s^k}}{g^{xs}}.$$

Thus, using (6.62), we obtain for the elements g_s^{ij} of S^{r2}

$$\begin{aligned}g_s^{ij} &= (1-w)\delta_j^i - \frac{w}{g^{xs}}g_{ij}^{xs} - d_2[(1-w)\delta_k^i - \frac{w}{g^{xs}}g_{ik}^{xs}]f_{s^k} \\ &\times [(1-w)\delta_m^j - \frac{w}{g^{xs}}g_{jm}^{xs}]f_{s^m}, \quad i,j,k,m = 1,2.\end{aligned} \quad (6.106)$$

To transform (6.105) we shall take advantage of the lemma 6.3.1 with $g_{2s}^{ij} = g^{ij}$, $f_{s^i}f_{s^j} = a_{ij}$. By virtue of (6.91) we obtain

$$g_{2s}^{ik}f_{s^k}g_{2s}^{jl}f_{s^l} = g_{2s}^{ij}\nabla^{2s}(f) - (-1)^{i+j}(g^{2s})^{-1}f_{s^{3-i}}f_{s^{3-j}}, \quad (6.107)$$

$$i, j, k, l = 1, 2, \quad i, j \text{ fixed},$$

where

$$g^{2s} = \det(g_{ij}^{2s}) = g^{xs}/w, \quad i,j = 1, 2.$$

Analogously, using (6.91), we find

$$g_{sx}^{ik}f_{s^k}g_{sx}^{jl}f_{s^l} = g_{sx}^{ij}\nabla(f) - (-1)^{i+j}(g^{xs})^{-1}f_{s^{3-i}}f_{s^{3-j}},$$

$$i, j, k, l = 1, 2, \quad i, j \text{ fixed}.$$

So substituting $(-1)^{i+j}f_{s^{3-i}}f_{s^{3-j}}$, i,j fixed, from this equation into (6.107) gives

$$\begin{aligned}g_{2s}^{ik}f_{s^k}g_{2s}^{jl}f_{s^l} &= g_{2s}^{ij}\nabla^{2s}(f) - w[\nabla(f)g_{sx}^{ij} - g_{sx}^{ik}f_{s^k}g_{sx}^{jl}f_{s^l}] \\ &= g_{2s}^{ij}\nabla^{2s}(f) - w[\nabla(f)g_{sx}^{ij} - \nabla(f, s^i)\nabla(f, s^j)], \quad (6.108)\end{aligned}$$

$$i, j, k, l = 1, 2.$$

Using this formula in (6.105) yields

$$g_s^{ij} = d_2\{g_{2s}^{ij} + w[\nabla(f)g_{sx}^{ij} - \nabla(f, s^i)\nabla(f, s^j)]\}, \quad i,j = 1,2. \quad (6.109)$$

After substituting the expression for g_{2s}^{ij} from (6.67) in (6.109) we obtain the following formula of g_s^{ij}

$$\begin{aligned}g_s^{ij} &= wd_2\{[d_1 + \nabla(f)]g_{sx}^{ij} - \nabla(s^i, s^k)\nabla(s^j, s^k) \\ &- \nabla(s^i, f)\nabla(s^j, f)\}, \quad i, j, k = 1, 2,\end{aligned} \quad (6.110)$$

where

$$d_1 = 1 + g_{sx}^{11} + g_{sx}^{22}.$$

From (6.64) we have

$$\nabla(s^i, s^k)\nabla(s^j, s^k) = g_{sx}^{ij}(g_{sx}^{11} + g_{sx}^{22}) - \frac{1}{g^{xs}}\delta_j^i, \quad i,j,k = 1,2,$$

therefore (6.110) also has the following form

$$g_s^{ij} = wd_2\{[1 + \nabla(f)]g_{sx}^{ij} + \frac{1}{g^{xs}}\delta_j^i - \nabla(s^i, f)\nabla(s^j, f)\}, \quad (6.111)$$

$$i,j = 1,2.$$

Note this formula can also be obtained by substituting in (6.109) the expression for g_{2s}^{ij} from (6.63).

Analogously to (6.108) we find

$$\nabla(s^i, f)\nabla(s^j, f)\} = \nabla(f)g_{sx}^{ij} - \frac{1}{g^{xs}}[\nabla^E(f)\delta_j^i - f_{s^i}f_{s^j}], \quad i,j = 1,2,$$

therefore equations (6.111) also are as follows:

$$g_s^{ij} = wd_2\left\{g_{sx}^{ij} + \frac{1}{g^{xs}}[1 + \nabla^E(f)]\delta_j^i - \frac{1}{g^{xs}}f_{s^i}f_{s^j}\right\}, \quad i,j = 1,2. \quad (6.112)$$

6.3.3 Beltrami's Mixed and First Differential Parameters

For the Beltrami's mixed differential parameter we obtain, using (4.51) and (6.110)

$$\nabla^s(\varphi, \psi) = \varphi_{s^i}\psi_{s^j}g_s^{ij}$$

$$= wd_2\Big\{[d_1 + \nabla(f)]\nabla(\varphi, \psi) - \nabla(\varphi, s^k)\nabla(\psi, s^k) \quad (6.113)$$

$$- \nabla(\varphi, f)\nabla(\psi, f)\Big\}, \quad i,j = 1,2,$$

while the application of (6.111) gives

$$\nabla^s(\varphi, \psi) = wd_2\Big\{[1 + \nabla(f)]\nabla(\varphi, \psi)$$

$$+ \frac{1}{g^{xs}}\nabla^E(\varphi, \psi) - \nabla(\varphi, f)\nabla(\psi, f)\Big\}. \quad (6.114)$$

Thus the Beltrami's first differential parameter is expressed, in accordance with (6.113), by

$$\nabla^s(\varphi) = wd_2\Big\{[d_1 + \nabla(f)]\nabla(\varphi)$$

$$- \nabla(\varphi, s^k)\nabla(\varphi, s^k) - [\nabla(\varphi, f)]^2\Big\}, \quad (6.115)$$

6.3 Particular Two-Dimensional Case

while the formula (6.114) yields

$$\nabla^s(\varphi) = wd_2 \left\{ [1 + \nabla(f)]\nabla(\varphi) + \frac{1}{g^{xs}}\nabla^E(\varphi) - [\nabla(\varphi, f)]^2 \right\}. \qquad (6.116)$$

Christoffel Symbols. A formula for the Christoffel symbols of the first kind in the coordinates s^1, s^2 follows readily from (4.27) and (6.89)

$$[ij, k]^s = [ij, k] + f_{s^i s^j} f_{s^k}, \quad i, , j, k = 1, 2. \qquad (6.117)$$

This formula, (4.24), and (6.110) give the expression for the Christoffel symbols of the second kind in the coordinates s^1, s^2

$$\begin{aligned}
{}^s\Upsilon_{ij}^p &= g_s^{kp}[ij, k]^s \\
&= wd_2\{[d_1 + \nabla(f)]g_{sx}^{kp} - \nabla(s^k, s^m)\nabla(s^p, s^m) \\
&\quad - \nabla(s^k, f)\nabla(s^p, f)\}\{[ij, k] + f_{s^i s^j}f_{s^k}\} \\
&= wd_2\{[d_1 + \nabla(f)][\Upsilon_{ij}^p + f_{s^i s^j}\nabla(s^p, f)] \\
&\quad - \nabla(s^p, s^m)\Upsilon_{ij}^m - \nabla(s^p, s^m)\nabla(f, s^m)f_{s^i s^j} \\
&\quad - \nabla(s^p, f)f_{s^m}\Upsilon_{ij}^m - \nabla(s^p, f)\nabla(f)f_{s^i s^j}\} \\
&= wd_2\{[d_1 + \nabla(f)]\Upsilon_{ij}^p + [d_1\nabla(s^p, f) - \nabla(s^p, s^m)\nabla(s^m, f)]f_{s^i s^j} \\
&\quad - [\nabla(s^p, s^m) + \nabla(s^p, f)f_{s^m}]\Upsilon_{ij}^m\}, \quad i, j, k, m, p = 1, 2.
\end{aligned} \qquad (6.118)$$

Availing us of (6.100) and (6.93) in (6.118) we obtain

$$\begin{aligned}
{}^s\Upsilon_{ij}^p &= wd_2\Big\{[d_1 + \nabla(f)]\Upsilon_{ij}^p + L^p[f](f_{s^i s^j} - f_{s^m}\Upsilon_{ij}^m) \\
&\quad - \left[\nabla(s^p, s^m) - \frac{1}{g^{xs}}f_{s^p}f_{s^m}\right]\Upsilon_{ij}^m\Big\} \\
&= wd_2\Big\{[d_1 + \nabla(f)]\Upsilon_{ij}^p - \left[\nabla(s^p, s^m) - \frac{1}{g^{xs}}f_{s^p}f_{s^m}\right]\Upsilon_{ij}^m \\
&\quad + L^p[f]\nabla_{ij}(f)\Big\}, \quad i, j, m, p = 1, 2.
\end{aligned} \qquad (6.119)$$

Using (4.24), (6.111), and (6.117) also yields the following equivalent formula of the Christoffel symbols of the second kind in the coordinates s^1, s^2

$$^s\Upsilon^p_{ij} = g^{kp}_s[ij,k]^s$$

$$= wd_2\{[1+\nabla(f)]g^{kp}_{sx} + \frac{1}{g^{xs}}\delta^k_p$$

$$- \nabla(s^k,f)\nabla(s^p,f)\}\{[ij,k] + f_{s^is^j}f_{s^k}\}$$

$$= wd_2\{\Upsilon^p_{ij} + \frac{1}{g^{xs}}[ij,p] - \nabla(s^p,f)f_{s^m}\Upsilon^m_{ij} \qquad (6.120)$$

$$+ [1+\nabla(f)]\nabla(s^p,f)f_{s^is^j} + \frac{1}{g^{xs}}f_{s^is^j}f_{s^p} - \nabla(f)\nabla(s^p,f)f_{s^is^j}\}$$

$$= wd_2\{\Upsilon^p_{ij} + \nabla(s^p,f)\nabla_{ij}(f) + \frac{1}{g^{xs}}([ij,p] + f_{s^is^j}f_{s^p})\},$$

$$i,j,k,p = 1,2.$$

Tensor of Mixed Derivatives. Formulas (6.19), (6.70), (6.78), and (6.93) give rize to to the following expression for the elements of the tensor of mixed derivatives in the coordinates s^1, s^2

$$\nabla^s_{ij}(\varphi) = \nabla^{2s}_{ij}(\varphi) - d_2\nabla^{2s}(f,\varphi)\nabla^{2s}_{ij}(f)$$

$$= \nabla_{ij}(\varphi) - wL^k[\varphi]\nabla_{ij}(s^k)$$

$$- d_2 w\left[\nabla(f,\varphi) + \frac{1}{g^{xs}}\nabla^E(f,\varphi)\right]$$

$$\times \left\{\nabla_{ij}(f) - wL^k[f]\nabla_{ij}(s^k)\right\} \qquad (6.121)$$

$$= \nabla_{ij}(\varphi) - d_2 w L^m[f]\varphi_{s^m}\nabla_{ij}(f)$$

$$+ w\{d_2 w L^m[f]\varphi_{s^m}L^k[f] - L^k[\varphi]\}\nabla_{ij}(s^k),$$

$$i,j,k,m = 1,2.$$

Beltrami's Second Differential Parameter. To compute the Beltrami's second differential parameter of $\varphi(s)$ we use the formulas (6.80) with $l=3$, $f^1(s) = s^1$, $f^2(s) = s^2$, $f^3(s) = f(s)$ and (6.92).

Taking advantage of (6.93) we obtain

$$d_{ab}\nabla(f^a,\varphi)\Delta_B[f^b] = d_{ij}\nabla(s^i,\varphi)\Delta_B[s^j] + d_{i3}\{\nabla(s^i,\varphi)\Delta_B[f]$$
$$+ \nabla(f,\varphi)\Delta_B[s^i]\} + d_{33}\nabla(f,\varphi)\Delta_B[f]$$
$$= [d_{i3}\nabla(s^i,\varphi) + d_{33}\nabla(f,\varphi)]\Delta_B[f]$$
$$+ [d_{ij}\nabla(s^i,\varphi) + d_{j3}\nabla(f,\varphi)]\Delta_B[s^j] \qquad (6.122)$$
$$= \frac{1}{B}\{(1 + g_{11}^{xs} + g_{22}^{xs} + g^{xs})\nabla(f,\varphi) - g^{xs}L^i[f]\nabla(s^i,\varphi)\}\Delta_B[f]$$
$$+ \frac{1}{B}\{[g_{ij}^{xs} + f_{s^i}f_{s^j}]\nabla(s^i,\varphi) + g^{xs}[1 + \nabla(f)]\nabla(s^j,\varphi)$$
$$- g^{xs}L^j[f]\nabla(f,\varphi)\}\Delta_B[s^j] \ , \ i,j = 1,2 \ , \ a,b = 1,2,3 \ .$$

Multiplying (6.100) by φ_{s^i} we find

$$g^{xs}\nabla(s^i,\varphi)\nabla(s^i,f) = \nabla(f,\varphi)(g_{11}^{xs} + g_{22}^{xs}) - \nabla^E(f,\varphi) \ , \ i = 1,2 \ , \quad (6.123)$$

therefore

$$g^{xs}L^i[f]\nabla(s^i,\varphi) = g^{xs}\left[\nabla(s^i,f) + \frac{1}{g^{xs}}f_{s^i}\right]\nabla(s^i,\varphi)$$
$$= \nabla(f,\varphi)(g_{11}^{xs} + g_{22}^{xs}) - \nabla^E(f,\varphi) + \nabla(f,\varphi) \ , \ i = 1,2 \ .$$

Besides this

$$(g_{ij}^{xs} + f_{s^i}f_{s^j})\nabla(s^i,\varphi) + g^{xs}[1 + \nabla(f)]\nabla(s^i,\varphi) - g^{xs}L^j[f]\nabla(f,\varphi)$$
$$= \varphi_{s^j} + g^{xs}\{[1 + \nabla(f)]\nabla(s^j,\varphi) - \nabla(f,\varphi)\nabla(s^j,f)\}$$
$$= b_j(\varphi) \ , \quad i,j = 1,2 \ ,$$

where

$$b_j(\varphi) = g^{xs}\{L^j[\varphi] + \nabla(f)\nabla(s^j,\varphi) - \nabla(f,\varphi)\nabla(s^j,f)\} \ , \ j = 1,2 \ .$$

So (6.122) is as follows:

$$d_{ab}\nabla(f^a,\varphi)\Delta_B[f^b] = \frac{1}{B}\{[g^{xs}\nabla(f,\varphi)$$
$$+ \nabla^E(f,\varphi)]\Delta_B[f] + b_j(\varphi)\Delta_B[s^j]\} \ , \qquad (6.124)$$
$$j = 1,2 \ , \ a,b = 1,2,3 \ .$$

In the same way

$$d_{ch}\nabla(f^c,\varphi)\nabla_{ij}(f^h) = \frac{1}{B}\{[g^{xs}\nabla(f,\varphi) + \nabla^E(f,\varphi)]\nabla_{ij}(f)$$
$$+ b_k(\varphi)\nabla_{ij}(s^k)\} \ , \qquad (6.125)$$
$$i,j,k = 1,2 \ , \quad c,h = 1,2,3 \ .$$

Taking advantage of (6.52), (6.64), and (6.92) we also obtain

$$d_{ab}\nabla(f^a, s^i)\nabla(f^b, s^j) = d_{ab}(d^{ai} - \delta_i^a)(d^{bj} - \delta_j^b)$$

$$= d^{ij} + d_{ij} - 2\delta_j^i = \nabla(s^i, s^j) + d_{ij} - \delta_j^i$$

$$= \nabla(s^i, s^j) + \frac{1}{B}\{g_{ij}^{xs} + f_{s^i}f_{s^j} + g^{xs}[1 + \nabla(f)]\delta_j^i\} - \delta_j^i \tag{6.126}$$

$$= \frac{1}{B}\{[1 + g^{xs}(d_1 + \nabla(f)) + \nabla^E(f)]g_{sx}^{ij} + f_{s^i}f_{s^j}$$

$$- [1 + \nabla^E(f)]\delta_j^i\}, \quad i, j = 1, 2, \quad a, b = 1, 2, 3.$$

Substituting (6.124–6.126) in (6.80) produces

$$\Delta_B^s[\varphi] = \frac{1}{B}\bigg\{\Delta_B[\varphi] - \frac{1}{B}[g^{xs}\nabla(f, \varphi) + \nabla^E(f, \varphi)]\Delta_B[f]$$

$$- \frac{1}{B}b_k(\varphi)\Delta_B[s^k]\bigg\} - \frac{1}{B}\{f_{s^i}f_{s^j} - [1 + \nabla^E(f)]\delta_j^i\} \tag{6.127}$$

$$\times \{\nabla_{ij}(\varphi) + [g^{xs}\nabla(f, \varphi) + \nabla^E(f, \varphi)]\nabla_{ij}(f) + b_m(\varphi)\nabla_{ij}(s^m)\},$$

$$i, j, k, m = 1, 2.$$

7. Grid Equations with Respect to Intermediate Transformations

Here we establish some equivalent forms of the comprehensive grid equations specified in (5.4) with respect to the components $s^i(\boldsymbol{\xi})$ of intermediate transformations (5.2) using for this purpose the relations outlined in Chaps. 4–6.

7.1 Relations to Comprehensive Equations

Let s^i, $i = 1,\ldots,n$, be a local coordinate system in a Riemannian manifold M^n. Expanding the differentiation in $\Delta_B[s^i]$ (formula (5.3)), we have

$$\Delta_B[s^i] = \frac{1}{\sqrt{g^s}} \frac{\partial}{\partial s^j}(\sqrt{g^s} g_s^{ji})$$

$$= \frac{g_s^{ji}}{\sqrt{g^s}} \frac{\partial}{\partial s^j} \sqrt{g^s} + \frac{\partial}{\partial s^j} g_s^{ji}, \quad i,j = 1,\ldots,n. \tag{7.1}$$

The application of (4.26) to the first item in the second line of this system of equations yields

$$\frac{g_s^{ji}}{\sqrt{g^s}} \frac{\partial}{\partial s^j} \sqrt{g^s} = g_s^{ji} \Upsilon_{kj}^k, \quad i,j,k = 1,\ldots,n,$$

while for the second item we find from (4.25)

$$\frac{\partial}{\partial s^j} g_s^{ji} = -g_s^{ik} \Upsilon_{kj}^j - g_s^{lj} \Upsilon_{lj}^i, \quad i,j,k,l = 1,\ldots,n.$$

Substituting these relations in (7.1) gives the following expression of $\Delta_B[s^i]$

$$\Delta_B[s^i] \equiv -g_s^{kj} \Upsilon_{kj}^i, \quad i,j,k = 1,\ldots,n. \tag{7.2}$$

Thus, assuming $\boldsymbol{\xi} = \boldsymbol{s}$ in (7.2), the equations in (5.4) formulated for generating grids are as follows:

$$\Delta_B[\xi^i] \equiv -g_\xi^{kj} \Upsilon_{kj}^i = 0, \quad i,j,k = 1,\ldots,n. \tag{7.3}$$

Remind, the quantities in this formula are in the grid coordinates ξ^i, $i = 1,\ldots,n$, in particular, the Christoffel symbols of the second kind are, in accordance with (4.24), as

$$\Upsilon_{kj}^i =^\xi \Upsilon_{kj}^i = g_\xi^{im}[kj,m]^\xi, \quad i,j,k,m = 1,\ldots,n.$$

Thus the grid equations (7.3) also have the following equivalent form

$$g_\xi^{kj} g_\xi^{im}[kj,m]^\xi = 0, \quad i,j,k,m = 1,\ldots,n.$$

Multiplying these equations by g_{ip}^ξ gives the following grid system

$$g_\xi^{kj}[kj,p]^\xi = 0, \quad j,k,p = 1,\ldots,n. \tag{7.4}$$

Let M^n be a regular surface S^{rn} represented by (4.1) then using in (7.4) the relations (4.27) with $s = \xi$ we find that the comprehensive grid equations in (5.4) with respect to the metric (4.6) are equivalent to

$$g_\xi^{kj}(\boldsymbol{r}_{\xi^k \xi^j} \cdot \boldsymbol{r}_{\xi^i}) = 0, \quad i,j,k = 1,\ldots,n, \tag{7.5}$$

where

$$\boldsymbol{r}_{\xi^k \xi^j} = \frac{\partial^2 \boldsymbol{r}[\boldsymbol{s}(\boldsymbol{\xi})]}{\partial \xi^k \partial \xi^j}, \quad j,k = 1,\ldots,n, \quad \boldsymbol{r}_{\xi^i} = \frac{\partial \boldsymbol{r}[\boldsymbol{s}(\boldsymbol{\xi})]}{\partial \xi^i}, \quad i = 1,\ldots,n.$$

Note equations (7.5) mean that in the grid coordinates $\xi^1, \ldots, \xi^n$ the vector

$$g_\xi^{ij} \boldsymbol{r}_{\xi^i \xi^j}, \quad i,j = 1,\ldots,n,$$

is orthogonal to the n-dimensional plane formed by the basic tangent vectors $\boldsymbol{r}_{s^i}$, $i = 1,\ldots,n$, consequently, it is orthogonal to the surface $S^{rn} \subset R^{n+l}$. As

$$g_\xi^{ij} \boldsymbol{r}_{\xi^i \xi^j} + \frac{1}{\sqrt{g^\xi}} \frac{\partial}{\partial \xi^j}(\sqrt{g^\xi} g_\xi^{ij}) \boldsymbol{r}_{\xi^i}$$
$$= \frac{1}{\sqrt{g^\xi}} \frac{\partial}{\partial \xi^j}(\sqrt{g^\xi} g_\xi^{ij} \boldsymbol{r}_{\xi^i}) = \Delta_B[\boldsymbol{r}], \quad i,j = 1,\ldots,n,$$

we obtain, taking advantage of (5.5) that in the grid coordinates $\xi^1, \ldots, \xi^n$

$$g_\xi^{ij} \boldsymbol{r}_{\xi^i \xi^j} = \Delta_B[\boldsymbol{r}], \quad i,j = 1,\ldots,n.$$

If $S^{rn} \subset R^{n+1}$ then there is defined the mean curvature of S^{rn} by the formula

$$K_m = \frac{1}{n} g_\xi^{ij} \boldsymbol{r}_{\xi^i \xi^j} \cdot \boldsymbol{n}, \quad i,j = 1,\ldots,n,$$

where $\boldsymbol{n}$ is the unit normal to S^{rn} (see (4.66)). Since

$$\boldsymbol{n} = \frac{1}{\|\Delta_B[\boldsymbol{r}]\|} \Delta_B[\boldsymbol{r}], \quad \|\Delta_B[\boldsymbol{r}]\| = \sqrt{\Delta_B[\boldsymbol{r}] \cdot \Delta_B[\boldsymbol{r}]},$$

we obtain

$$K_m = \frac{1}{n\|\Delta_B[\boldsymbol{r}]\|} \Delta_B[\boldsymbol{r}] \cdot \Delta_B[\boldsymbol{r}] = \frac{1}{n} \|\Delta_B[\boldsymbol{r}]\|. \tag{7.6}$$

As $\Delta_B[\boldsymbol{r}]$ is an invariant of parametrization of a regular surface the formula (7.6) allows one to formulate the mean curvature for a surface $S^{rn} \subset R^{n+l}$ with an arbitrary whole number $l \geq 1$, let alone $l = 1$.

7.2 Resolved Grid Equations

In general the systems (7.4) and (7.5) are not resolved with respect to the components $s^i(\boldsymbol{\xi})$ of the intermediate transformation $\boldsymbol{s}(\boldsymbol{\xi})$. In this section we establish some resolved forms of grid equations convenient for the implementation into numerical codes. Crucial for this purpose is a basic elliptic operator.

7.2.1 Basic Elliptic Operator

The basic elliptic grid operator designated in local coordinates $\boldsymbol{v} = (v^1, \ldots, v^n)$ of the Riemannian manifold M^n by L^v is specified at an arbitrary twice-differentiable function $y(\boldsymbol{v})$ as

$$L^v[y] = g_v^{ij} \frac{\partial^2 y}{\partial v^i \partial v^j}, \quad i,j = 1, \ldots, n, \tag{7.7}$$

where g_v^{ij} is the (ij)th element of the contravariant metric tensor of M^n in the coordinates $v^1, \ldots, v^n$. Note the value the operator at $y(\boldsymbol{v})$ is not an invariant of parametrizations of M^n. Equation (4.58) yields the following relation between the basic elliptic and Beltrami operators:

$$L^v[y] = \Delta_B[y] + y_{v^k} g_v^{ij} \Upsilon_{ij}^k, \quad i,j,k = 1, \ldots, n,$$

while the application of (7.2) to this formula gives

$$L^v[y] = \Delta_B[y] - y_{v^k} \Delta_B[v^k], \quad k = 1, \ldots, n.$$

Assuming here that the coordinates $v^1, \ldots, v^n$ are the grid coordinates $\xi^1, \ldots, \xi^n$ and taking advantage of (5.4) we find

$$L^\xi[y] = \Delta_B[y], \tag{7.8}$$

i.e. in the grid coordinates the value of the basic elliptic operator at an arbitrary function $y(\boldsymbol{\xi})$ coincides with Beltrami's second differential parameter of this function.

7.2.2 General Grid Equations

A general form of the grid equations resolved with respect to the intermediate transformation $\boldsymbol{s}(\boldsymbol{\xi})$ and an arbitrary metric in a Riemannian manifold M^n is obtained after substituting in (7.8) $s^i(\boldsymbol{\xi})$ for $y(\boldsymbol{\xi})$. Thus we have the following system of the comprehensive grid equations with respect to the components $s^i(\boldsymbol{\xi})$, $i = 1, \ldots, n$,

$$L^\xi[s^i] = \Delta_B[s^i], \quad i = 1, \ldots, n. \tag{7.9}$$

As the value of the Beltramian operator is independent of the choice of parametrizations we also find from (7.9)

$$L^\xi[s^i] = \frac{1}{\sqrt{g^\xi}} \frac{\partial}{\partial \xi^k}\left(\sqrt{g^\xi} g_\xi^{kp} \frac{\partial s^i}{\partial \xi^p}\right), \quad i,k,p = 1,\ldots,n. \quad (7.10)$$

Another inference of equations (7.9) is found after multiplying the system in (5.4) by $\partial s^l/\partial \xi^i$. Indeed this multiplication yields

$$\frac{1}{\sqrt{g^s}} \frac{\partial}{\partial s^j}\left(\sqrt{g^s} g_s^{jk} \frac{\partial \xi^i}{\partial s^k}\right) \frac{\partial s^l}{\partial \xi^i}$$

$$= \frac{1}{\sqrt{g^s}}\left(\frac{\partial}{\partial s^j}(\sqrt{g^s} g_s^{jl}) - \sqrt{g^s} g_s^{jk} \frac{\partial^2 s^l}{\partial \xi^i \partial \xi^m} \frac{\partial \xi^i}{\partial s^k} \frac{\partial \xi^m}{\partial s^j}\right)$$

$$= \Delta_B[s^l] - g_\xi^{im} \frac{\partial^2 s^l}{\partial \xi^i \partial \xi^m} = 0, \quad i,j,k,l,m = 1,\ldots,n,$$

since (5.4), (4.15) and (4.18). We see that the equations in the last line are identical to (7.9).

7.2.3 Equations for a Spherical Monitor Metric

Let the monitor metric g_{ij}^s of a monitor manifold M^n over a domain S^n be a spherical one, i.e. in the Cartesian coordinates $s^1,\ldots,s^n$ of S^n it is of the form

$$g_{ij}^s = v(s)\delta_j^i, \quad i,j = 1,\ldots,n, \quad v(s) > 0. \quad (7.11)$$

Then

$$g_s^{ij} = \frac{1}{v(s)}\delta_j^i, \quad i,j = 1,\ldots,n,$$

so, in accordance with (7.2) and (6.27), for $f(s) = s^k$,

$$\Delta_B[s^k] = -\frac{1}{2(v(s))^2}\delta_j^i\left(\delta_j^k \frac{\partial v}{\partial s^i} + \delta_i^k \frac{\partial v}{\partial s^j} - \delta_j^i \frac{\partial v}{\partial s^k}\right)$$

$$= \frac{n-2}{2(v(s))^2} \frac{\partial v}{\partial s^k}, \quad i,j,k = 1,\ldots,n.$$

Therefore the equations (7.9), in the spherical metric (7.11), have the following form

$$L^\xi[s^i] = \frac{n-2}{2[v(s)]^2} \frac{\partial v}{\partial s^k}, \quad k = 1,\ldots,n. \quad (7.12)$$

Since the rule (4.15)

$$g_\xi^{im} = g_s^{kj} \frac{\partial \xi^i}{\partial s^k} \frac{\partial \xi^m}{\partial s^j} = \frac{1}{v(s)} g_{\xi s}^{im}, \quad i,j,k,m = 1,\ldots,n,$$

where the terms

$$g_{\xi s}^{im} = \frac{\partial \xi^i}{\partial s^k} \frac{\partial \xi^m}{\partial s^k}, \quad i,k,m = 1,\ldots,n,$$

are the elements of the contravariant Eucledian metric tensor of S^n in the grid coordinates $\xi^1,\ldots,\xi^n$. Therefore we readily see that for the basic operator specified by (7.7) in the metric (7.11) there is valid the relation

$$L^\xi[y] = \frac{1}{v(s)} g_{\xi s}^{im} \frac{\partial^2 y}{\partial \xi^i \partial \xi^m}, \quad i,m = 1,\ldots,n.$$

So (7.12) is also expressed as follows:

$$g_{\xi s}^{im} \frac{\partial^2 s^k}{\partial \xi^i \partial \xi^m} = \frac{n-2}{2v(s)} \frac{\partial v}{\partial s^k}, \quad i,k,m = 1,\ldots,n. \tag{7.13}$$

As

$$\frac{n-2}{2v(s)} \frac{\partial v}{\partial s^i} = \frac{1}{w(s)} \frac{\partial w}{\partial s^i}, \quad i = 1,\ldots,n,$$

where

$$w(s) = [v(s)]^{(n-2)/2},$$

we find, using (7.13),

$$g_{\xi s}^{ij} \frac{\partial^2 s^k}{\partial \xi^i \partial \xi^j} - \frac{1}{w} \frac{\partial w}{\partial s^k} = 0, \quad i,j,k = 1,\cdots,n. \tag{7.14}$$

Availing us of the relation (2.24), we have

$$\frac{\partial w}{\partial s^k} = \frac{\partial w}{\partial \xi^i} \frac{\partial \xi^i}{\partial s^k} = g_{\xi s}^{ij} \frac{\partial w}{\partial \xi^i} \frac{\partial s^k}{\partial \xi^j}.$$

Therefore we conclude, from the system (7.14),

$$g_{\xi s}^{ij} \left(\frac{\partial^2 s^k}{\partial \xi^i \partial \xi^j} - \frac{1}{w} \frac{\partial w}{\partial \xi^i} \frac{\partial s^k}{\partial \xi^j} \right) = w g_{\xi s}^{ij} \frac{\partial}{\partial \xi^i} \left(\frac{1}{w} \frac{\partial s^k}{\partial \xi^j} \right) = 0.$$

Thus we obtain one more compact form of the grid equations in the spherical metric (7.11):

$$g_{\xi s}^{ij} \frac{\partial}{\partial \xi^i} \left(\frac{1}{w} \frac{\partial s^k}{\partial \xi^j} \right) = 0, \quad i,j,k = 1,\ldots,n. \tag{7.15}$$

Note $w(s) \equiv 1$ for n=2 hence these equation for n=2 are equivalent to the inverted Laplace equations (5.6) regardless of the form of $v(s)$.

7.2.4 Equations for a Spherical Monitor Metric Over a Surface

Grid equations analogous to (7.15) are also readily obtained for the spherical metric

$$g_{ij}^s = v(s) g_{ij}^{xs}, \quad i,j = 1,\ldots,n, \quad v(s) > 0, \tag{7.16}$$

of a monitor manifold M^n over an arbitrary physical geometry S^{xn} with its covariant and contravariant metric tensors designated by g_{ij}^{xs} and g_{sx}^{ij}, respectively, in the coordinates $s^1,\ldots,s^n$. Since

$$g_s^{ij} = \frac{1}{v(s)} g_{sx}^{ij}, \quad i,j = 1,\ldots,n,$$

$$g^s = [v(s)]^n g^{xs}, \quad g^{xs} = \det(g_{ij}^{xs}),$$

the original grid equations $\Delta_B[\xi^i] = 0$ in (5.4) with respect to $\xi^i(s)$, $i = 1,\ldots,n$, in the metric (7.16) have the following form

$$\Delta_B[\xi^i] \equiv \frac{1}{g^s} \frac{\partial}{\partial s^j} \left\{ \sqrt{g^{xs}} [v(s)]^{(n-2)/2} g_{xs}^{jk} \frac{\partial \xi^i}{\partial s^j} \right\} = 0, \quad i,j,k = 1,\ldots,n. \quad (7.17)$$

As $g^s > 0$ these equations for $n = 2$ are equivalent to the Beltrami equations with respect to the original metric g_{ij}^{xs} regardless of $v(s)$.

In accordance with (6.29) we find

$$\Delta_B[s^i] = \frac{1}{v(s)} \Delta_B^{xs}[s^i] + \frac{n-2}{2[v(s)]^2} \frac{\partial v}{\partial s^k} g_{sx}^{ki}, \quad i,k = 1,\ldots,n,$$

where Δ_B^{xs} is the operator of Beltrami with respect to the metric g_{ij}^{xs} of the geometry S^{xn}. Similarly to (7.14) we obtain, using (7.9), the following system of grid equations resolved with respect to $s^i(\boldsymbol{\xi})$

$$g_{\xi x}^{km} \frac{\partial^2 s^i}{\partial \xi^k \partial \xi^m} = \Delta_B^{xs}[s^i] + \frac{1}{w(s)} \frac{\partial w}{\partial s^k} g_{sx}^{ki}$$

$$= \Delta_B^{xs}[s^i] + \frac{1}{w(s)} \frac{\partial v}{\partial \xi^k} g_{\xi x}^{km} \frac{\partial s^i}{\partial \xi^m}, \quad i,k,m = 1,\cdots,n, \quad (7.18)$$

where

$$w(s) = [v(s)]^{(n-2)/2}.$$

In the same way as it was made for (7.15) we find a more compact form of the grid equations (7.18)

$$g_{\xi x}^{km} \frac{\partial}{\partial \xi^k} \left(\frac{1}{w(s)} \frac{\partial s^i}{\partial \xi^m} \right) = \frac{1}{w(s)} \Delta_B^{xs}[s^i], \quad i,k,m = 1,\cdots,n. \quad (7.19)$$

7.2.5 Domain Grid Equations with Respect to a Monitor Surface

Let the physical domain X^n be identified with the parametric domain S^n while the intermediate transformation

$$s(\boldsymbol{\xi}): \Xi^n \to S^n$$

for generating a grid in X^n be determined as the inverse of the map

$$\boldsymbol{\xi}(s): S^n \to \Xi^n$$

satisfying (5.4) with respect to the metric of a monitor surface S^{rn} over the domain S^n. If the monitor surface S^{rn} over the domain X^n is formed by a vector-valued monitor function $\boldsymbol{f}(\boldsymbol{x}) = [f^1(\boldsymbol{x}),\ldots,f^l(\boldsymbol{x})]$ then, assuming $\boldsymbol{s} = \boldsymbol{x}$, the parametrization of S^{rn} is determined by

7.2 Resolved Grid Equations

$$r(s) : S^n \to R^{n+l}, \quad r(s) = [s, f(s)].$$

Hence in the grid coordinates $\xi^1, \ldots, \xi^n$ we obtain

$$g^\xi_{ij} = s_{\xi^i} \cdot s_{\xi^j} + f_{\xi^i} \cdot f_{\xi^j}, \quad i,j = 1, \ldots, n,$$

$$r_{\xi^m \xi^j} = (s_{\xi^m \xi^j}, f_{\xi^m \xi^j}), \quad j, m = 1, \ldots, n,$$

$$r_{\xi^k} = (s_{\xi^k}, f_{\xi^k}), \quad k = 1, \ldots, n,$$

where for a function $v(s)$

$$v_{\xi^i} = \frac{\partial v[s(\xi)]}{\partial \xi^i}, \quad v_{\xi^m \xi^j} = \frac{\partial^2 v[s(\xi)]}{\partial \xi^m \partial \xi^j}, \quad j, m = 1, \ldots, n.$$

Therefore, in this case, the grid system (7.5) is as follows:

$$g^{mj}_\xi (s_{\xi^m \xi^j} \cdot s_{\xi^k} + f_{\xi^m \xi^j} \cdot f_{\xi^k}) = 0, \quad j, k, m = 1, \ldots, n, \tag{7.20}$$

and the multiplication of this system by $\partial \xi^k / \partial s^i$ yields the following grid system, with respect to $s^i(\xi)$, $i = 1, \ldots, n$,

$$g^{mj}_\xi \left(\frac{\partial^2 s^i}{\partial \xi^m \partial \xi^j} + f_{\xi^m \xi^j} \cdot f_{s^i} \right) = 0, \quad i, j, m = 1, \ldots, n, \tag{7.21}$$

where

$$f_{s^i} = \frac{\partial f[s]}{\partial s^i}, \quad i = 1, \ldots, n.$$

Using the definition of the basic operator L^ξ specified by (7.7), this system is rewritten in the form

$$L^\xi[s^i] + f^k_{s^i} L^\xi[f^k] = 0, \quad i = 1, \ldots, n, \quad k = 1, \ldots, l. \tag{7.22}$$

Note, if $\xi^1, \ldots, \xi^n$ are the coordinates satisfying (5.4) and consequently (5.5) then for a function $v(\xi)$

$$L^\xi[v] \equiv g^{mj}_\xi \frac{\partial^2 v}{\partial \xi^m \partial \xi^j} = \frac{1}{\sqrt{g^\xi}} \frac{\partial}{\partial \xi^j} \left(\sqrt{g^\xi} g^{jm}_\xi \frac{\partial v}{\partial \xi^m} \right) = \Delta_B[v],$$

$$j, m = 1, \ldots, n,$$

so the system (7.21) also has the following equivalent forms

$$\Delta_B[s^i] + f^p_{s^i} \Delta_B[f^p] = 0, \quad i = 1, \ldots, n, \quad p = 1, \ldots, l,$$

$$L^\xi[s^i] + f^k_{s^i} \Delta_B[f^k] = 0, \quad i = 1, \ldots, n, \quad k = 1, \ldots, l. \tag{7.23}$$

Remind $\Delta_B(f^p)$ is independent of a parametrization of S^{rn} over S^n therefore it can be computed in an arbitrary coordinate system, in particular, in the parametric coordinates $s^1, \ldots, s^n$.

If we consider in S^n new curvelinear coordinates $v = (v^1, \ldots, v^n)$, $v \in V^n$ connected with the Cartesian coordinates $s = (s^1, \ldots, s^n)$ by the relations

$$v(s) : S^n \to V^n, \quad s(v) : V^n \to S^n,$$

then the parametrization of the monitor surface S^{rn} in the coordinates $v^1, \ldots, v^n$ is as follows:

$$\boldsymbol{r}_1(\boldsymbol{v}) : V^n \to R^{n+l}, \quad \boldsymbol{r}_1(\boldsymbol{v}) = \{\boldsymbol{s}(\boldsymbol{v}), \boldsymbol{f}[\boldsymbol{s}(\boldsymbol{v})]\} .$$

Therefore the grid equations equivalent to (7.21) and resolved with respect to the functions $v^i(\boldsymbol{\xi})$ can be obtained from (7.9). Using (7.2), (4.24), and (4.27) in (7.9) gives these equations the form

$$g_\xi^{ij} \frac{\partial^2 v^p}{\partial \xi^i \partial \xi^j} = -g_v^{ij} g_v^{pm}(\boldsymbol{s}_{v^i v^j} \cdot \boldsymbol{s}_{v^m} + \boldsymbol{f}_{v^i v^j} \cdot \boldsymbol{f}_{v^m}) , \qquad (7.24)$$

$$i, j, m, p = 1, \ldots, n ,$$

where

$$\boldsymbol{f}_{v^i v^j} = \frac{\partial^2 \boldsymbol{f}[\boldsymbol{s}(\boldsymbol{v})]}{\partial v^i \partial v^j} , \quad \boldsymbol{f}_{v^m} = \frac{\partial \boldsymbol{f}[\boldsymbol{s}(\boldsymbol{v})]}{\partial v^m} .$$

Availing us of the notion of the basic elliptic operator (7.7) this system is transformed to

$$L^\xi[v^p] = -g_v^{mp}(s_{v^m}^i L^v[s^i] + f_{v^m}^k L^v[f^k]) ,$$

$$i, m, p = 1, \ldots, n , \quad k = 1, \ldots, l .$$

Note, in general equations (7.24) are more complicated than (7.21) and (7.22).

7.2.6 Domain Grid Equations with Respect to a Monitor Metric

General Case. Let the grid equations in a domain S^n be formulated with respect to a monitor metric g_{ij}^s over this domain. The domain S^n with the monitor metric is a monitor manifold over S^n. We designate it by M^n. Let us assume that in local coordinates $s^1, \ldots, s^n$ the monitor metric of the manifold M^n be specified in the form (5.23), namely,

$$g_{ij}^s = z(\boldsymbol{s})\delta_j^i + v^k(\boldsymbol{s})f_{s^i}^k(\boldsymbol{s})f_{s^j}^k(\boldsymbol{s}) ,$$

$$= z(\boldsymbol{s})\boldsymbol{s}_{s^i} \cdot \boldsymbol{s}_{s^j} + v^k(\boldsymbol{s})f_{s^i}^k(\boldsymbol{s})f_{s^j}^k(\boldsymbol{s}) , \qquad (7.25)$$

$$i, j = 1, \ldots, n , \quad k = 1, \ldots, l ,$$

where $z(\boldsymbol{s}) > 0$ and $v^k(\boldsymbol{s}) \geq 0$, $k = 1, \ldots, l$, are weight functions while $f^k(\boldsymbol{s})$, $k = 1, \ldots, l$, are monitor functions. Then we have, similarly to (6.15) and (6.26), the following expressions for the Christoffel symbols of the first kind in the grid coordinates $\xi^1, \ldots, \xi^n$

$$[ij, m] = z\boldsymbol{s}_{\xi^i \xi^j} \cdot \boldsymbol{s}_{\xi^m} + \frac{1}{2}(z_{\xi^i} g_{jm}^{s\xi} + z_{\xi^j} g_{im}^{s\xi} - z_{\xi^m} g_{ij}^{s\xi})$$

$$+ v^k f_{\xi^i \xi^j}^k f_{\xi^m}^k + \frac{1}{2}[f_{\xi^m}^k(v_{\xi^i}^k f_{\xi^j}^k + v_{\xi^j}^k f_{\xi^i}^k) - v_{\xi^m}^k f_{\xi^i}^k f_{\xi^j}^k] , \qquad (7.26)$$

$$i, j, m = 1, \ldots, n , \quad k = 1, \ldots, l ,$$

7.2 Resolved Grid Equations

where

$$g_{ij}^{s\xi} = s_{\xi^i} \cdot s_{\xi^j}, \quad i,j = 1,\ldots,n,$$

are the elements of the Eucledian covariant metric tensor of the domain S^n in the grid coordinates $\xi^1, \ldots, \xi^n$. Hence the grid equations (7.4) in the metric (7.25) are as follows:

$$g_{\boldsymbol{\xi}}^{ij}[ij,m] \equiv z g_{\boldsymbol{\xi}}^{ij} s_{\xi^i \xi^j} \cdot s_{\xi^m} + v^k g_{\boldsymbol{\xi}}^{ij} f_{\xi^i \xi^j}^k f_{\xi^m}^k$$

$$+ z_{\xi^i} g_{\boldsymbol{\xi}}^{ij} g_{jm}^{s\xi} - \frac{1}{2} z_{\xi^m} g_{\boldsymbol{\xi}}^{ij} g_{ij}^{s\xi} + f_{\xi^m}^k \nabla(v^k, f^k) \quad (7.27)$$

$$- \frac{1}{2} v_{\xi^m}^k \nabla(f^k) = 0,$$

$$i,j,m = 1,\ldots,n, \quad k = 1,\ldots,l,$$

where

$$\nabla(\varphi,\psi) = g_s^{ij} \varphi_{s^i} \psi_{s^j}, \quad i,j = 1,\ldots,n, \quad \nabla(f) = \nabla(f,f).$$

Multiplying this system by $(\partial \xi^m / \partial s^p)/z$ yields the following equivalent grid system resolved with respect to $s^p(\boldsymbol{\xi})$, $p = 1,\ldots,n$,

$$L[s^p] + \frac{1}{z}\left\{ v^k f_{s^p}^k L[f^k] + z_{\xi^i} g_{\boldsymbol{\xi}}^{ij} g_{jm}^{s\xi} \frac{\partial \xi^m}{\partial s^p} \right.$$

$$\left. - \frac{1}{2} z_{s^p} g_{\boldsymbol{\xi}}^{ij} g_{ij}^{s\xi} + f_{s^p}^k \nabla(v^k, f^k) - \frac{1}{2} v_{s^p}^k \nabla(f^k) \right\} = 0, \quad (7.28)$$

$$i,j,m,p = 1,\ldots,n, \quad k = 1,\ldots,l,$$

where $L^{\boldsymbol{\xi}}$ is the basic elliptic operator defined by (7.7).

As

$$g_{jm}^{s\xi} \frac{\partial \xi^m}{\partial s^p} = \frac{\partial s^p}{\partial \xi^j}, \quad j,m,p = 1,\ldots,n,$$

we obtain

$$z_{\xi^i} g_{\boldsymbol{\xi}}^{ij} g_{jm}^{s\xi} \frac{\partial \xi^m}{\partial s^p} = z_{\xi^i} g_{\boldsymbol{\xi}}^{ij} \frac{\partial s^p}{\partial \xi^j} = \nabla(z, s^p), \quad i,j,m,p = 1,\ldots,n.$$

Analogously

$$g_{\boldsymbol{\xi}}^{ij} g_{ij}^{s\xi} = \nabla(s^m, s^m), \quad i,j,m = 1,\ldots,n.$$

Availing us of these relations in (7.28) we find the following expression of the grid system in the metric (7.25)

$$L^{\boldsymbol{\xi}}[s^p] + \frac{1}{z}\left\{ v^k f_{s^p}^k L[f^k] + \nabla(z, s^p) - \frac{1}{2} z_{s^p} \nabla(s^i, s^i) \right.$$

$$\left. + f_{s^p}^k \nabla(v^k, f^k) - \frac{1}{2} v_{s^p}^k \nabla(f^k) \right\} = 0, \quad i,p = 1,\ldots,n, \quad k = 1,\ldots,l. \quad (7.29)$$

In particular, when $z(s) = v^k(s) = 1$, $k = 1, \ldots, l$, this system coincides with (7.21).

Note

$$zL^\xi[s^p] + \nabla(z, s^p) = zg_\xi^{ij}\frac{\partial^2 s^p}{\partial \xi^i \partial \xi^j} + g_\xi^{ij}\frac{\partial z}{\partial \xi^i}\frac{\partial s^p}{\partial \xi^j} = L_z^\xi[s^p], \quad i, j, p = 1, \ldots, n,$$

where

$$L_z^\xi[y] = g_\xi^{ij}\frac{\partial}{\partial \xi^i}\left(z\frac{\partial y}{\partial \xi^j}\right), \quad i, j = 1, \ldots, n.$$

Therefore the grid system (7.29) has also the following equivalent form

$$L_z^\xi[s^p] + v^k f_{s^p}^k L[f^k] - \frac{1}{2}z_{s^p}\nabla(s^i, s^i) + f_{s^p}^k \nabla(v^k, f^k)$$
$$- \frac{1}{2}v_{s^p}^k \nabla(f^k) = 0, \quad i, p = 1, \ldots, n, \quad k = 1, \ldots, l. \tag{7.30}$$

The formulas given for the elliptic operators $L^\xi[\ ]$ and $L_z^\xi[\ ]$ require the knowledge of the elements g_ξ^{ij} of the contravariant metric tensor of M^n in the coordinates $\xi^1, \ldots, \xi^n$. These elements can be computed by the successive application of the formulas (6.9), (6.24), and (6.25).

Particular Case. In the particular case $l = 1$ in (7.25) we obtain in the coordinates $s^1, \ldots, s^n$, assuming in (7.25) $v^1(s) = v(s)$, $f^1(s) = f(s)$, and introducing an intermediate spherical monitor metric

$$g_{ij}^{1s} = z(s)\delta_j^i, \quad i, j = 1, \ldots, n,$$

the following formula for the elements of the contravariant metric tensor

$$g_{ij}^s = g_{ij}^{1s} + v(s)f_{s^i}f_{s^j}, \quad i, j = 1, \ldots, n, \tag{7.31}$$

The relations (6.24) and (6.25) give expressions for the elements g_{1s}^{ij} of the contravariant metric tensor and the Beltrami mixed parameter $\nabla^1(\varphi, \psi)$ with respect to the intermediate metric g_{ij}^{1s}

$$g_{1s}^{ij} = \frac{1}{z}\delta_j^i, \quad i, j = 1, \ldots, n,$$
$$\nabla^1(\varphi, \psi) = \frac{1}{z}\nabla^E(\varphi, \psi), \tag{7.32}$$

where

$$\nabla^E(\varphi, \psi) = \varphi_{s^i}\psi_{s^i}, \quad i = 1, \ldots, n.$$

Equations (6.7), (6.13), and (7.32) also yield

$$g_s^{ij} = g_{1s}^{ij} - d_1 g_{1s}^{ik} g_{1s}^{jm} f_{s^k} f_{s^m}$$
$$= \frac{1}{z}\delta_j^i - \frac{d_1}{z^2}f_{s^i}f_{s^j}, \quad i, j = 1, \ldots, n, \tag{7.33}$$
$$\nabla(\varphi, \psi) = \nabla^1(\varphi, \psi) - d_1\nabla(\varphi, f)\nabla(\psi, f),$$

where

$$d_1 = \frac{v}{1 + v\nabla^1(f)} = \frac{zv}{z + v\nabla^E(f)},$$
$$\nabla(\varphi, \psi) = g_s^{ij}\varphi_{s^i}\psi_{s^j}, \quad i,j = 1,\ldots,n.$$

Thus we find from (7.32) and (7.33)

$$g_s^{ij} = \frac{1}{z}\left(\delta_j^i - \frac{v}{z + v\nabla^E(f)}f_{s^i}f_{s^j}\right), \quad i,j = 1,\ldots,n,$$
$$\nabla(\varphi, \psi) = \frac{1}{z}\left\{\nabla^E(\varphi, \psi) - \frac{v}{z + v\nabla^E(f)}\nabla^E(f, \varphi)\nabla^E(f, \psi)\right\}.$$
(7.34)

Consequently in the grid coordinates $\xi^1, \ldots, \xi^n$ we obtain, using the relations (4.19),

$$g_\xi^{ij} = g_s^{km}\frac{\partial \xi^i}{\partial s^k}\frac{\partial \xi^j}{\partial s^m} = \frac{1}{z}\left(g_{\xi s}^{ij} - \frac{v}{z + v\nabla^E(f)}g_{\xi s}^{ik}f_{\xi^k}g_{\xi s}^{jm}f_{\xi^m}\right),$$
$$i,j,k,m = 1,\ldots,n,$$
(7.35)

where

$$g_{\xi s}^{ij} = \boldsymbol{\xi}_{s^i} \cdot \boldsymbol{\xi}_{s^j}, \quad i,j = 1,\ldots,n,$$

are the elements of the Eucledian contravariant metric tensor of the parametric domain S^n in the grid coordinates $\xi^1, \ldots, \xi^n$.

Using (7.34) yields

$$\nabla(s^i, s^i) = \frac{1}{z}\left[n - \frac{v}{z + v\nabla^E(f)}\nabla^E(f)\right], \quad i = 1,\ldots,n,$$
$$\nabla(v, f) = \frac{1}{z}\left[\nabla^E(v, f) - \frac{v}{z + v\nabla^E(f)}\nabla^E(f, v)\right] = \frac{\nabla^E(f, v)}{z + v\nabla^E(f)}, \quad (7.36)$$
$$\nabla(f) = \frac{\nabla^E(f)}{z + v\nabla^E(f)}.$$

Therefore the grid equations (7.30) in this particular case are as follows:

$$L^z[s^p] + vf_{s^p}L[f] - \frac{1}{2z}\left[n - \frac{v}{z + v\nabla^E(f)}\nabla^E(f)\right]z_{s^p}$$
$$+ \frac{\nabla^E(v, f)}{z + v\nabla^E(f)}f_{s^p} - \frac{1}{2}\frac{\nabla^E(f)}{z + v\nabla^E(f)}v_{s^p} = 0, \quad p = 1,\ldots,n.$$
(7.37)

7.2.7 Surface Grid Equations with Respect to a Monitor Surface

General Equations. If the monitor surface S^{rn} over a surface S^{xn} represented by

$$\boldsymbol{x}(s) : S^n \to R^{n+n_1}, \quad \boldsymbol{x} = (x^1, \ldots, x^{n+n_1}), \quad (7.38)$$

180   7. Grid Equations with Respect to Intermediate Transformations

is formed by a monitor function $\boldsymbol{f}(\boldsymbol{x}) = (f^1(\boldsymbol{x}), \ldots, f^l(\boldsymbol{x}))$ with the parametrization

$$r(s) : S^n \to R^{n+n_1+l}, \quad r(s) = \{\boldsymbol{x}(s), \boldsymbol{f}[\boldsymbol{x}(s)]\}, \tag{7.39}$$

then, in the grid coordinates $\xi^1, \ldots, \xi^n$,

$$\begin{aligned}
g_{ij}^\xi &= g_{ij}^{x\xi} + \boldsymbol{f}_{\xi^i} \cdot \boldsymbol{f}_{\xi^j}, \quad i,j = 1, \ldots, n, \\
g_{ij}^{x\xi} &= \boldsymbol{x}_{\xi^i} \cdot \boldsymbol{x}_{\xi^j}, \quad i,j = 1, \ldots, n, \\
\boldsymbol{r}_{\xi^m \xi^j} &= (\boldsymbol{x}_{\xi^m \xi^j}, \boldsymbol{f}_{\xi^m \xi^j}), \quad j, m = 1, \ldots, n, \\
\boldsymbol{r}_{\xi^i} &= (\boldsymbol{x}_{\xi^i}, \boldsymbol{f}_{\xi^i}), \quad i = 1, \ldots, n,
\end{aligned} \tag{7.40}$$

where g_{ij}^ξ and $g_{ij}^{x\xi}$ are the covariant metric tensors of S^{rn} and S^{xn}, respectively. Consequently the grid equations (7.5) are as follows:

$$g_\xi^{mj} (\boldsymbol{x}_{\xi^m \xi^j} \cdot \boldsymbol{x}_{\xi^k} + \boldsymbol{f}_{\xi^m \xi^j} \cdot \boldsymbol{f}_{\xi^k}) = 0, \quad j, k, m = 1, \ldots, n. \tag{7.41}$$

Since

$$\begin{aligned}
\boldsymbol{x}_{\xi^m \xi^j} \cdot \boldsymbol{x}_{\xi^k} &= \frac{\partial}{\partial \xi^m}\left(\frac{\partial \boldsymbol{x}}{\partial s^p} \frac{\partial s^p}{\partial \xi^j}\right) \cdot \frac{\partial \boldsymbol{x}}{\partial s^a} \frac{\partial s^a}{\partial \xi^k} \\
&= \left(\frac{\partial^2 s^p}{\partial \xi^m \partial \xi^j} g_{ap}^{xs} + \frac{\partial^2 \boldsymbol{x}}{\partial s^p \partial s^b} \cdot \frac{\partial \boldsymbol{x}}{\partial s^a} \frac{\partial s^p}{\partial \xi^j} \frac{\partial s^b}{\partial \xi^m}\right) \frac{\partial s^a}{\partial \xi^k},
\end{aligned}$$

$a, b, j, k, m, p = 1, \ldots, n$,

we obtain, after multiplying the system (7.41) by $(\partial \xi^k / \partial s^b) g_{sx}^{bi}$,

$$\begin{aligned}
&g_\xi^{mj}(\boldsymbol{x}_{\xi^m \xi^j} \cdot \boldsymbol{x}_{\xi^k} + \boldsymbol{f}_{\xi^m \xi^j} \cdot \boldsymbol{f}_{\xi^k}) \frac{\partial \xi^k}{\partial s^b} g_{sx}^{bi} \\
&= g_\xi^{mj}\left(\frac{\partial^2 s^i}{\partial \xi^m \partial \xi^j} + \boldsymbol{f}_{\xi^m \xi^j} \cdot \boldsymbol{f}_{s^b} g_{sx}^{bi}\right) + g_s^{pj}(\boldsymbol{x}_{s^p s^j} \cdot \boldsymbol{x}_{s^b}) g_{sx}^{bi} = 0,
\end{aligned} \tag{7.42}$$

$b, i, j, k, m = 1, \ldots, n$.

Taking into account that

$$(\boldsymbol{x}_{s^p s^j} \cdot \boldsymbol{x}_{s^b}) g_{sx}^{bi} =^x \Upsilon_{pj}^i, \quad i, j, p = 1, \ldots, n,$$

where ${}^x\Upsilon_{pj}^i$ is the Christoffel symbol of the second kind of the surface S^{xn} in the coordinates $s^1, \ldots, s^n$ we obtain from (7.42) that the grid system (7.41) resolved with respect to $s^i(\boldsymbol{\xi})$, $i = 1, \ldots, n$, has the following form

$$g_\xi^{mj}\left(\frac{\partial^2 s^i}{\partial \xi^m \partial \xi^j} + \boldsymbol{f}_{\xi^m \xi^j} \cdot \boldsymbol{f}_{s^b} g_{sx}^{bi}\right) = -g_s^{pj}{}^x\Upsilon_{pj}^i, \quad b, i, j, m, p = 1, \ldots, n, \tag{7.43}$$

Analogously to (7.8) we get, in the grid coordinates $\xi^1, \ldots, \xi^n$,

$$g_\xi^{mj} \frac{\partial^2 \boldsymbol{f}}{\partial \xi^m \partial \xi^j} = L^\xi[\boldsymbol{f}] = \Delta_B[\boldsymbol{f}], \quad j, m = 1, \ldots, n,$$

therefore equations (7.43) are also expressed as

$$L^\xi[s^i] + \Delta_B[f^k]\frac{\partial f^k}{\partial s^b}g^{bi}_{sx} = -g^{pj\,x}_s \Upsilon^i_{pj},$$
(7.44)

$$b, i, j, p = 1, \ldots, n, \quad k = 1, \ldots, l.$$

Simplified Equations. Equations (7.43) and (7.44) are more complicated in comparison with the equations (7.21) prescribed for generating grids in domains. So they are less malleable for the implementation into numerical codes in the case when the process of grid generation on a surface S^{xn} is coupled with the computation of this surface and the monitor function $\boldsymbol{f}(\boldsymbol{x})$ since the quantities $\Delta_B[f^k]$ and $^x\Upsilon^i_{pj}$ in (7.43) and (7.44) include the second derivatives with respect to s^i, $i = 1, \ldots, n$, of the function $\boldsymbol{f}[\boldsymbol{x}(\boldsymbol{s})]$ and the surface parametrization $\boldsymbol{x}(\boldsymbol{s})$. However, we can come to the equations of the form (7.21) for generating grids on the surface S^{xn} if we consider as a monitor function over S^{xn} the following function

$$\boldsymbol{f}_1(\boldsymbol{s}) = \{\boldsymbol{s}, \boldsymbol{f}[\boldsymbol{x}(\boldsymbol{s})]\}.$$
(7.45)

The monitor surface $S^{r_1 n}$ over S^{xn} with this monitor function $\boldsymbol{f}_1(\boldsymbol{s})$ is represented by the parametrization

$$\boldsymbol{r}_1(\boldsymbol{s}) : S^n \to S^{r_1 n}, \quad \boldsymbol{r}_1(\boldsymbol{s}) = \{\boldsymbol{x}(\boldsymbol{s}), \boldsymbol{s}, \boldsymbol{f}[\boldsymbol{x}(\boldsymbol{s})]\}.$$
(7.46)

Note the monitor surface $S^{r_2 n}$ over S^n with the monitor function $\boldsymbol{f}_2(\boldsymbol{s}) = \{\boldsymbol{x}(\boldsymbol{s}), \boldsymbol{f}[\boldsymbol{x}(\boldsymbol{s})]\}$, and represented correspondingly by

$$\boldsymbol{r}_2(\boldsymbol{s}) : S^n \to S^{r_2 n}, \quad \boldsymbol{r}_2(\boldsymbol{s}) = \{\boldsymbol{s}, \boldsymbol{x}(\boldsymbol{s}), \boldsymbol{f}[\boldsymbol{x}(\boldsymbol{s})]\},$$
(7.47)

has the same metric tensor as the surface $S^{r_1 n}$. Hence the equations for finding the intermediate transformation $\boldsymbol{s}(\boldsymbol{\xi}) : \Xi^n \to S^n$ with these monitor surfaces are identical and have, in accordance with (7.21), the following form

$$g^{mj}_\xi \left(\frac{\partial^2 s^i}{\partial \xi^m \partial \xi^j} + \boldsymbol{x}_{\xi^m \xi^j} \cdot \boldsymbol{x}_{s^i} + \boldsymbol{f}_{\xi^m \xi^j} \cdot \boldsymbol{f}_{s^i} \right) = 0, \quad i, j, m = 1, \ldots, m,$$
(7.48)

where g^{mj}_ξ is the contravariant metric tensor of the monitor surface $S^{r_2 n}$ in the grid coordinates $\xi^1, \ldots, \xi^n$. Note, for the covariant metric tensor of $S^{r_2 n}$ in these coordinates we have

$$g^\xi_{ij} = \frac{\partial \boldsymbol{s}(\boldsymbol{\xi})}{\partial \xi^i} \cdot \frac{\partial \boldsymbol{s}(\boldsymbol{\xi})}{\partial \xi^j} + \frac{\partial \boldsymbol{x}[\boldsymbol{s}(\boldsymbol{\xi})]}{\partial \xi^i} \cdot \frac{\partial \boldsymbol{x}[\boldsymbol{s}(\boldsymbol{\xi})]}{\partial \xi^j} + \frac{\partial \boldsymbol{f}\{\boldsymbol{x}[\boldsymbol{s}(\boldsymbol{\xi})]\}}{\partial \xi^i} \cdot \frac{\partial \boldsymbol{f}\{\boldsymbol{x}[\boldsymbol{s}(\boldsymbol{\xi})]\}}{\partial \xi^j},$$

$$i, j = 1, \ldots, n.$$

Using the notion of the basic elliptic operator (7.7) the equations (7.48) also have the form

$$L^\xi[s^i] + x^p_{s^i} L^\xi[x^p] + f^k_{s^i} L^\xi[f^k] = 0,$$
(7.49)

$$i = 1, \ldots, n, \quad p = 1, \ldots, n + n_1, \quad k = 1, \ldots, l.$$

Equations (7.48) or (7.49) with respect to the components $s^i(\boldsymbol{\xi})$, $i = 1,\ldots,n$, of the intermediate transformation $\boldsymbol{s}(\boldsymbol{\xi})$ include only the first derivatives in s^i, $i = 1,\ldots,n$, of the functions $\boldsymbol{x}(\boldsymbol{s})$ and $\boldsymbol{f}[\boldsymbol{x}(\boldsymbol{s})]$, therefore they are more convenient for the implementation into numerical codes in comparison with the equations (7.43) and (7.44). Remind, the grid in S^{xn} is obtained by mapping with $\boldsymbol{x}(\boldsymbol{s})$ a grid in S^n generated through $\boldsymbol{s}(\boldsymbol{\xi})$.

Instead of (7.46) we can also use the parametrization $\boldsymbol{r}_1(\boldsymbol{s})$ in the form

$$\boldsymbol{r}_1(\boldsymbol{s}) = \{N\boldsymbol{x}(\boldsymbol{s}), \boldsymbol{s}, N\boldsymbol{f}[\boldsymbol{x}(\boldsymbol{s})]\}$$

where N is some positive constant. With respect to this parametrization the equations (7.48) are as follows:

$$g_\xi^{mj}\left(\frac{\partial^2 s^i}{\partial \xi^m \partial \xi^j} + N^2 \boldsymbol{x}_{\xi^m \xi^j} \cdot \boldsymbol{x}_{s^i} + N^2 \boldsymbol{f}_{\xi^m \xi^j} \cdot \boldsymbol{f}_{s^i}\right) = 0 ,$$

(7.50)

$i, j, m = 1,\ldots,n$.

It is obvious that the difference between the solutions of (7.43) and (7.50) is lessend when N is increased.

Note that, similarly to (7.23), the equations (7.50) also have the following form

$$L^\xi[s^i] + N^2 L^\xi[x^j]\frac{\partial x^j}{\partial s^i} + N^2 L^\xi[f^k]\frac{\partial f^k}{\partial s^i} = 0 ,$$

$$\Delta_B[s^i] + N^2 \Delta_B[x^j]\frac{\partial x^j}{\partial s^i} + N^2 \Delta_B[f^k]\frac{\partial f^k}{\partial s^i} = 0 ,$$

(7.51)

$i = 1,\ldots,n$, $j = 1,\ldots,n+n_1$, $k = 1,\ldots,l$.

7.2.8 Surface Grid Equations with Respect to a Monitor Metric

A better control of the grid behavior on the surface S^{xn} represented by (7.38) can be realized with the introduction of a more general monitor metric with weight and monitor functions in the form (6.3), for example, as

$$g_{ij}^s = \delta_j^i + w(\boldsymbol{s})g_{ij}^{xs} + v^k(\boldsymbol{s})f_{s^i}^k(\boldsymbol{s})f_{s^j}^k(\boldsymbol{s}) ,$$

$$= \boldsymbol{s}_{s^i} \cdot \boldsymbol{s}_{s^j} + w(\boldsymbol{s})\boldsymbol{x}_{s^i} \cdot \boldsymbol{x}_{s^j} + v^k(\boldsymbol{s})f_{s^i}^k(\boldsymbol{s})f_{s^j}^k(\boldsymbol{s}) ,$$

(7.52)

$i, j = 1,\ldots,n$, $k = 1,\ldots,l$,

where g_{ij}^{xs} is the covariant metric tensor of S^{xn} in the parametric coordinates $s^1,\ldots,s^n$. This metric is similar to (7.25) therefore the surface grid equations with respect to the metric (7.52) are readily obtained from (7.29) assuming $z(\boldsymbol{s}) = 1$. Thus they have the following form

$$L^\xi[s^p] + w(s)x^i_{s^p}L^\xi[x^i] + v^k f^k_{s^p}L^\xi[f^k]$$
$$+x^m_{s^p}\nabla(w,x^m) - \frac{1}{2}w_{s^p}\nabla(x^j,x^j)$$
$$+f^k_{s^p}\nabla(v^k,f^k) - \frac{1}{2}v^k_{s^p}\nabla(f^k) = 0 ,$$
(7.53)

$i,j,m = 1,\ldots,n_1$, $p = 1,\ldots,n$, $k = 1,\ldots,l$.

In particular, if $w(s) \equiv const = c$ then the system (7.53) is as

$$L^\xi[s^p] + cx^i_{s^p}L^\xi[x^i] + v^k f^k_{s^p}L^\xi[f^k]$$
$$+f^k_{s^p}\nabla(v^k,f^k) - \frac{1}{2}v^k_{s^p}\nabla(f^k) = 0 ,$$
(7.54)

$i = 1,\ldots,n_1$, $p = 1,\ldots,n$, $k = 1,\ldots,l$.

7.3 Role of the Mean Curvature in the Grid Equations

In the case of a regular surface $S^{rn} \subset R^{n+1}$ defined by the parametrization

$$\boldsymbol{r}(s): S^n \to R^{n+1}, \quad \boldsymbol{r}(s) = [r^1(s),\ldots,r^{n+1}(s)] ,$$

with its natural metric tensor (g^s_{ij}) specified in the coordinates $s^1,\ldots,s^n$ as

$$g^s_{ij} = \boldsymbol{r}_{s^i} \cdot \boldsymbol{r}_{s^j} , \quad i,j = 1,\ldots,n ,$$

we have, from (7.2), (4.22), and (4.24),

$$\Delta_B[s^i] = -g^{kj}_s \Upsilon^i_{kj} = -g^{kj}_s g^{im}_s (\boldsymbol{r}_{s^k s^j} \cdot \boldsymbol{r}_{s^m}) , \quad i,j,k,m = 1,\ldots,n . \quad (7.55)$$

Now remind that the quantity

$$K_m = \frac{1}{n}g^{kj}_s \boldsymbol{r}_{s^k s^j} \cdot \boldsymbol{n} , \quad j,k = 1,\ldots,n , \quad (7.56)$$

where $\boldsymbol{n}$ is an $(n+1)$-dimensional unit normal vector to S^{rn} in R^{n+1}, is the mean curvature of the monitor surface S^{rn} with respect to the normal $\boldsymbol{n}$.

Note the mean curvature is an invariant of parametrizations of S^{rn}.

It appears that the mean curvature of S^{rn} is connected with the parametrization $\boldsymbol{r}(s): S^n \to S^{rn}$ by the following relation

$$\Delta_B[\boldsymbol{r}] = nK_m\boldsymbol{n} . \quad (7.57)$$

Indeed we have

$$\Delta_B[\boldsymbol{r}] = \frac{1}{\sqrt{g^s}}\frac{\partial}{\partial s^j}(\sqrt{g^s}g^{gi}_s \boldsymbol{r}_{s^i}) = g^{ji}_{sr}\boldsymbol{r}_{s^j s^i} + \Delta_B[s^k]\boldsymbol{r}_{s^k} ,$$
(7.58)

$i,j,k = 1,\cdots,n$,

and applying the relation (7.55) to the last item of (7.58) we obtain

$$\Delta_B[\boldsymbol{r}] = g^{ji}_{sr}[\boldsymbol{r}_{s^j s^i} - g^{mk}_{sr}(\boldsymbol{r}_{s^j s^i} \cdot \boldsymbol{r}_{s^m})\boldsymbol{r}_{s^k}] , \quad i,j,k,m = 1,\cdots,n . \quad (7.59)$$

Now taking into account equation (2.6), yielding that the expansion of any vector $\boldsymbol{b}$ in the basis $(\boldsymbol{r}_{s^1}, \cdots, \boldsymbol{r}_{s^n}, \boldsymbol{n})$ is expressed as

$$\boldsymbol{b} = g_{sr}^{mk}(\boldsymbol{b} \cdot \boldsymbol{r}_{s^m})\boldsymbol{r}_{s^k} + (\boldsymbol{b} \cdot \boldsymbol{n})\boldsymbol{n} ,$$

we find, applying this expansion to (7.59) and using (7.56),

$$\Delta_B[\boldsymbol{r}] = g_{sr}^{ji}(\boldsymbol{r}_{s^j s^i} \cdot \boldsymbol{n})\boldsymbol{n} = nK_m \boldsymbol{n} , \quad i,j = 1, \cdots, n ,$$

i.e. equation (7.57) is valid. Consequently from the relation (7.58) we obtain

$$g_{sr}^{ji}\boldsymbol{r}_{s^j s^i} + \Delta_B[s^k]\boldsymbol{r}_{s^k} = nK_m\boldsymbol{n} , \quad i,j,k = 1, \cdots, n . \tag{7.60}$$

Let the parametric transformation $\boldsymbol{r}(\boldsymbol{s})$ for S^{rn} be specified as $\boldsymbol{r}(\boldsymbol{s}) = [\boldsymbol{s}, f(\boldsymbol{s})]$ with a scalar-valued monitor function $f(\boldsymbol{s})$. Then $S^{rn} \subset R^{n+1}$ is a monitor surface over S^n, whose covariant metric tensor in the coordinates $s^1, \ldots, s^n$ is computed as

$$g_{ij}^s = \boldsymbol{r}_{s^i} \cdot \boldsymbol{r}_{s^j} = \delta_j^i + f_{s^i} f_{s^j} , \quad i,j = 1, \cdots, n . \tag{7.61}$$

Availing us of (6.6) and (6.7), we find

$$g_s^{im} f_{s^m} = \left(\delta_m^i - \frac{1}{g^s} f_{s^i} f_{s^m}\right) f_{s^m} = f_{s^i}\left(1 - \frac{1}{g^s} f_{s^m} f_{s^m}\right) = \frac{1}{g^s} f_{s^i} , \tag{7.62}$$

$$i, m = 1, \ldots, n .$$

So, taking advantage of (7.2), we conclude that

$$\Delta_B[s^i] = -g_s^{kj} g_s^{im} f_{s^k s^j} f_{s^m} = -\frac{1}{g^s} g_s^{kj} f_{s^k s^j} f_{s^i} , \quad i,j,k,m = 1, \ldots, n . \tag{7.63}$$

For the parametrization $\boldsymbol{r}(\boldsymbol{s}) = [\boldsymbol{s}, f(\boldsymbol{s})]$ we find

$$\boldsymbol{r}_{s^i} = (\underbrace{0, \ldots, 0}_{i-1}, 1, \underbrace{0, \ldots, 0}_{n-i}, f_{s^i}) , \quad i = 1, \ldots, n ,$$

so it is obvious that for the unit normal $\boldsymbol{n}$ we can offer

$$\boldsymbol{n} = \frac{1}{\sqrt{g^s}}(-f_{s^1}, \ldots, -f_{s^n}, 1) . \tag{7.64}$$

Therefore

$$\boldsymbol{r}_{s^k s^j} \cdot \boldsymbol{n} = \frac{1}{\sqrt{g^s}} f_{s^k s^j} , \quad j,k = 1, \ldots, n ,$$

and, in accordance with (7.56), the mean curvature of S^{rn} with respect to the normal (7.64) is computed by the following formula

$$K_m = \frac{1}{n\sqrt{g^s}} g_s^{kj} f_{s^k s^j} , \quad j,k = 1, \ldots, n . \tag{7.65}$$

Thus in the metric (7.61) equations (7.63), as well as (7.2), have the following form

$$\Delta_B[s^l] = -\frac{n}{\sqrt{g^s}} K_m f_{s^l} , \quad l = 1, \ldots, n . \tag{7.66}$$

Consequently the grid equations (7.9) in application to a domain $X^n = S^n$ with a scalar monitor function $f(\boldsymbol{x})$ are expressed through the mean curvature of S^{rn} as follows:

$$g_\xi^{ij}\frac{\partial^2 s^k}{\partial \xi^i \partial \xi^j} = -\frac{n}{\sqrt{g^s}}K_m f_{s^k}, \quad i,j,k = 1,\ldots,n. \tag{7.67}$$

Another form of the expression (7.65) for the mean curvature can be computed using the $(n+1)$th component of (7.57). Thus we have

$$K_m = \frac{1}{n}\sqrt{g^s}\Delta_B[f]. \tag{7.68}$$

Availing us of (6.7) and (7.62), in the case of the parametrization $\boldsymbol{r}(\boldsymbol{s}) = [\boldsymbol{s}, f(\boldsymbol{s})]$, we find

$$\Delta_B[f] = \frac{1}{\sqrt{g^s}}\frac{\partial}{\partial s^j}\left(\sqrt{g^s}g_s^{jk}\frac{\partial f}{\partial s^k}\right) = \frac{1}{\sqrt{g^s}}\frac{\partial}{\partial s^j}\left(\frac{1}{\sqrt{g^s}}\frac{\partial f}{\partial s^j}\right), \quad j,k = 1,\ldots,n.$$

Thus from (7.68)

$$K_m = \frac{1}{n}\frac{\partial}{\partial s^j}\left(\frac{1}{\sqrt{g^s}}\frac{\partial f}{\partial s^j}\right), \quad j = 1,\ldots,n. \tag{7.69}$$

Since the points of the monitor surface $S^{rn} \subset R^{n+1}$ in the coordinates $s^1, \ldots, s^n, s^{n+1}$ can be found from the solution of the equation

$$s^{n+1} - f(s^1, \ldots, s^n) = 0,$$

the formula (7.69) is a particular case of (4.111) for

$$\varphi(s^1, \ldots, s^n, s^{n+1}) \equiv s^{n+1} - f(s^1, \ldots, s^n)$$

and

$$g_{ij}^{rs} = \delta_j^i, \quad i,j = 1,\ldots,n+1.$$

7.4 Practical Grid Equations

This section discusses in detail one-, two-, and three-dimensional grid equations adjusted to generating grids on the boundary curves and surfaces of three-dimensional blocks and in the interior of the blocks. The equations are considered with respect to both monitor surfaces and monitor metrics.

7.4.1 Equations for Generating Grids on Curves

Equations with Respect to a Monitor Surface.

General Equations. Let a curve in the k-dimensional space R^k, designated by S^{x1}, be represented as

$$\boldsymbol{x}(s) : [0,1] \to R^k \,, \quad \boldsymbol{x} = (x^1, \ldots, x^k) \,. \tag{7.70}$$

Let also $\boldsymbol{f}(\boldsymbol{x}) = [f^1(\boldsymbol{x}), \ldots, f^l(\boldsymbol{x})]$ be a monitor function, determining the monitor curve S^{r1} over S^{x1}. Then S^{r1} is parametrized by the following transformation

$$\boldsymbol{r}(s) : [0,1] \to R^{l+k} \,, \quad \boldsymbol{r}(s) = \{\boldsymbol{x}(s), \boldsymbol{f}[\boldsymbol{x}(s)]\} \,, \tag{7.71}$$

consequently the covariant g^s and contravariant g_s metric tensor of S^{r1} in the coordinate s is

$$g^s = \boldsymbol{r}_s \cdot \boldsymbol{r}_s = g^{xs} + \boldsymbol{f}_s \cdot \boldsymbol{f}_s \,, \quad g_s = 1/g^s \,, \tag{7.72}$$

respectively, where

$$g^{xs} = \boldsymbol{x}_s \cdot \boldsymbol{x}_s \tag{7.73}$$

is the covariant metric tensor of S^{x1}, while $\boldsymbol{f}_s = \partial \boldsymbol{f}[\boldsymbol{x}(s)]/\partial s$. So the intermediate transformation $s(\xi)$ specified on the interval [0,1] for generating a grid on S^{x1} is subject, in accordance with (7.9), to the following equation

$$g_\xi \frac{d^2 s}{d\xi^2} = \frac{1}{\sqrt{g^s}} \frac{d}{ds}(\sqrt{g^s} g_s) \,, \tag{7.74}$$

where g_ξ is the contravariant tensor of S^{r1} in the grid coordinate ξ, i.e.

$$g_\xi = 1/g^\xi \,, \quad g^\xi = g^s (ds/d\xi)^2 \,.$$

Equation (7.74) is readily converted to

$$\frac{d^2 s}{d\xi^2} = \sqrt{g^\xi} \frac{d}{d\xi}\left(\frac{1}{\sqrt{g^s}}\right) = -\frac{1}{2g^s} \frac{ds}{d\xi} \frac{d}{d\xi} g^s \,, \tag{7.75}$$

and consequently to the following divergent form

$$\frac{d}{d\xi}\left(\frac{ds}{d\xi} \sqrt{g^s}\right) = 0 \,, \quad \xi \in [0,1] \,. \tag{7.76}$$

Simplified Equation. Note if $k = 1$, i.e. S^{x1} is an interval $[a,b]$ then $\boldsymbol{x}(s)$ is a scalar function $x(s)$ mapping the interval [0,1] onto $[a,b]$ and consequently the metric tensor g^s in the grid equations (7.74 – 7.76) has the form

$$g^s = (x_s)^2 + \boldsymbol{f}_s \cdot \boldsymbol{f}_s \,.$$

In particular, when $x(s) \equiv s$, then

$$g^s = 1 + \boldsymbol{f}_s \cdot \boldsymbol{f}_s \,,$$

while the grid equation in this particular case also is, in accordance with (7.5),

$$\frac{d^2 s}{d\xi^2} + \frac{d^2 \boldsymbol{f}[s(\xi)]}{d\xi^2} \cdot \frac{d\boldsymbol{f}(s)}{ds} = 0 . \tag{7.77}$$

If the monitor function $\boldsymbol{f}(s)$ is a scalar-valued function $f(s)$ then applying (3.10) with the identification $x = s$ and $u = f$ we find

$$\frac{d^2 f}{d\xi^2} = \frac{d^2 f}{ds^2}\left(\frac{ds}{d\xi}\right)^2 + f_s \frac{d^2 s}{d\xi^2} = k[1 + (f_s)^2]^{3/2}\left(\frac{ds}{d\xi}\right)^2 + f_s \frac{d^2 s}{d\xi^2} ,$$

where k is the curvature of the monitor surface S^{r1}. Substituting this expression in (7.77) gives the following one-dimension grid equation

$$\frac{d^2 s}{d\xi^2} + k\sqrt{1 + (f_s)^2} f_s \left(\frac{ds}{d\xi}\right)^2 = 0 . \tag{7.78}$$

A form of a curve grid equation similar to (7.77) can be obtained for an orbitrary curve S^{x1} parametrized by (7.70) if we, analogously to (7.45), specify a monitor function $\boldsymbol{f}_1(s)$ over S^{x1} as

$$\boldsymbol{f}_1(s) = \{s, \boldsymbol{f}[\boldsymbol{x}(s)]\} .$$

Then, similarly to (7.48), the curve grid equation is as follows:

$$\frac{d^2 s}{d\xi^2} + \frac{d^2 \boldsymbol{x}[s(\xi)]}{d\xi^2} \cdot \frac{d\boldsymbol{x}(s)}{ds} + \frac{d^2 \boldsymbol{f}\{\boldsymbol{x}[s(\xi)]\}}{d\xi^2} \cdot \frac{d\boldsymbol{f}(s)}{ds} = 0 . \tag{7.79}$$

Equation with Respect to a Monitor Metric. Let the grid behavior on the curve S^{x1} parametrized by (7.70) be controlled by a monitor metric in the form (7.52) for $n = 1$, i.e. the covariant metric tensor g^s in the parametric coordinates s of a one-dimensional monitor manifold M^1 over S^{x1} is defined by

$$g^s = 1 + w(s)g^{xs} + v^m(s)(f_s^m)^2 , \quad m = 1, \ldots, l . \tag{7.80}$$

One curve grid equation with respect to the monitor metric (7.80) is expressed by (7.76).

Another form of the equation can be found from (7.53) for $n = 1$. To write down this equation we note that in one-dimension

$$L^\xi[y] = g_\xi \frac{d^2 y}{d\xi^2} ,$$

$$g_\xi = \frac{1}{g^\xi} , \quad g^\xi = g^s \left(\frac{ds}{d\xi}\right)^2 , \tag{7.81}$$

$$w_s \nabla(x^j, x^j) = x_s^j \nabla(w, x^j) = w_s g^{xs} g_s , \quad j = 1, \ldots, k ,$$

$$f_s^m \nabla(v^m, f^m) = v_s^m \nabla(f^m) = v_s^m g_s (f_s^m)^2 , \quad m = 1, \ldots, l .$$

188   7. Grid Equations with Respect to Intermediate Transformations

Substituting these relations into (7.53) for $n = 1$ yields the following grid equation with respect to the monitor metric (7.80)

$$\frac{d^2 s}{d\xi^2} + wx_s^i \frac{d^2 x^i}{d\xi^2} + v^m f_s^m \frac{d^2 f^m}{d\xi^2} ,$$
$$+ \frac{1}{2} \left(\frac{ds}{d\xi}\right)^2 [w_s g^{xs} + v_s^m (f_s^m)^2] = 0 , \quad i = 1, \ldots, k , \quad m = 1, \ldots, l .$$
(7.82)

7.4.2 Equations for Generating Grids on Two-Dimensional Surfaces

Equations with Respect to Monitor Surfaces. Let us consider here a monitor surface S^{r2} formed by the values of a vector-valued function $\boldsymbol{f}(\boldsymbol{x}) = [f^1(\boldsymbol{x}), \ldots, f^l(\boldsymbol{x})]$ over a two-dimensional surface S^{x2} represented by a parametrization

$$\boldsymbol{x}(s) : S^2 \to R^k , \quad \boldsymbol{x} = (x^1, \ldots, x^k) . \qquad (7.83)$$

The monitor surface S^{r2} is represented in the coordinates s^1, s^2 by the following parametrization

$$\boldsymbol{r}(s) : S^2 \to R^{l+k} , \quad \boldsymbol{r}(s) = \{\boldsymbol{x}(s), \boldsymbol{f}[\boldsymbol{x}(s)]\} , \quad s = (s^1, s^2) . \qquad (7.84)$$

Consequently for the elements g_{ij}^{xs} and g_{ij}^s in the coordinates s^1, s^2 of the covariant metric tensor of S^{x2} and S^{r2}, respectively, we find

$$g_{ij}^{xs} = \boldsymbol{x}_{s^i} \cdot \boldsymbol{x}_{s^j} , \quad i,j = 1,2 ,$$
$$g_{ij}^s = \boldsymbol{r}_{s^i} \cdot \boldsymbol{r}_{s^j} = g_{ij}^{xs} + \boldsymbol{f}_{s^i} \cdot \boldsymbol{f}_{s^j} , \quad i,j = 1,2 .$$
(7.85)

The grid on S^{x2} generated through the equations (7.9) or (7.44) with $n = 2$ (each of these systems is equivalent to the generalized Laplace system in (5.4) with $n = 2$) is obtained by mapping a reference grid in Ξ^2 on S^{x2} by the function $\boldsymbol{x}[s(\boldsymbol{\xi})]$ where $s(\boldsymbol{\xi}) : \Xi^2 \to S^2$ is the intermediate transformation whose components $s^i(\boldsymbol{\xi})$, $i = 1, 2$, satisfy these equations. In the same way, the grid on S^{x2} is obtained by mapping with $\boldsymbol{x}(s)$ the grid in S^2 found by the numerical solution of (7.9) or (7.44) on the reference grid in Ξ^2.

Note that, similarly to (2.21), in the two-dimensional case the elements g_ξ^{ij} of the contravariant metric tensor of the monitor surface S^{r2} in the grid coordinates ξ^1, ξ^2 are computed through the elements g_{ij}^ξ of the covariant metric tensor of S^{r2} in the same coordinates ξ^1, ξ^2 by the formula

$$g_\xi^{ij} = \frac{(-1)^{i+j}}{g^\xi} g_{3-i 3-j}^\xi , \quad i,j = 1,2 ; \quad i,j - \text{fixed} , \qquad (7.86)$$

where

7.4 Practical Grid Equations

$$g_{ij}^{\xi} = g_{ij}^{x\xi} + \frac{\partial f\{x[s(\xi)]\}}{\partial \xi^i} \cdot \frac{\partial f\{x[s(\xi)]\}}{\partial \xi^j}, \quad i,j = 1,2,$$

$$g_{ij}^{x\xi} = \frac{\partial x[s(\xi)]}{\partial \xi^i} \cdot \frac{\partial x[s(\xi)]}{\partial \xi^j}, \quad i,j = 1,2, \quad (7.87)$$

$$g^{\xi} = \det(g_{ij}^{\xi}) = g^s J^2, \quad J = \det\left(\frac{\partial s^i}{\partial \xi^j}\right).$$

Now introducing the operator L^2 by the formula

$$L^2[v] = g_{22}^{\xi} \frac{\partial^2 v}{\partial \xi^1 \partial \xi^1} - 2g_{12}^{\xi} \frac{\partial^2 v}{\partial \xi^1 \partial \xi^2} + g_{11}^{\xi} \frac{\partial^2 v}{\partial \xi^2 \partial \xi^2} \quad (7.88)$$

and taking into account that

$$L^2[v] = g^{\xi} L^{\xi}[v], \quad (7.89)$$

where L^{ξ} is the basic operator defined by (7.7) for $n = 2$, we find from the grid systems (7.9) and (7.44) with $n = 2$ suitable to generate grids on two-dimensional surfaces the following systems

$$L^2[s^i] = g^{\xi} \Delta_B[s^i], \quad i = 1,2, \quad (7.90)$$

and

$$L^2[s^i] + g^{\xi}\left[\Delta_B[f^a]\frac{\partial f^a}{\partial s^b}g_{sx}^{bi} + g_s^{pjx}\Upsilon_{pj}^i\right] = 0,$$
$$b, i, j, p = 1,2, \quad a = 1,\ldots,l, \quad (7.91)$$

respectively.

Simplified Equations. The equations (7.90) and (7.91) are simplified if the monitor function over S^{x2} is chosen in the form

$$f[x(s)] = [s, v(s)], \quad v(s) = [v^1(s), \ldots, v^l(s)]. \quad (7.92)$$

The metric tensor of this monitor surface S^{r2} over S^{x2}, formed by the monitor function (7.92), coincides with the metric tensor of the monitor surface over S^2 formed by the monitor function $f_1[s] = [x(s), v(s)]$ so, in accordance with (7.49) and (7.89), the grid equations with respect to $s^i(\xi)$ are in this case as follows:

$$L^2[s^i] + \frac{\partial x^m}{\partial s^i} L^2[x^m] + \frac{\partial v^p}{\partial s^i} L^2[v^p] = 0,$$
$$i = 1,2, \quad m = 1,\ldots,k, \quad p = 1,\ldots,l. \quad (7.93)$$

Note the metric elements g_{ij}^{ξ} in the formula (7.89) for the operator L^2 are computed with respect to the monitor function (7.92), i.e.

$$g_{ij}^{\xi} = \frac{\partial s^a(\xi)}{\partial \xi^i} \frac{\partial s^a(\xi)}{\partial \xi^j} + \frac{\partial x^m[s(\xi)]}{\partial \xi^i} \frac{\partial x^m[s(\xi)]}{\partial \xi^j} + \frac{\partial v^p[s(\xi)]}{\partial \xi^i} \frac{\partial v^p[s(\xi)]}{\partial \xi^j},$$
$$i,j,a = 1,2, \quad m = 1,\ldots,k, \quad p = 1,\ldots,l.$$

7. Grid Equations with Respect to Intermediate Transformations

If the surface S^{x2} lies in R^3 and is represented by a parametrization

$$\boldsymbol{x}(\boldsymbol{s}) = (\boldsymbol{s}, x^3(\boldsymbol{s})) = (s^1, s^2, x^3(s^1, s^2)) \qquad (7.94)$$

then the simplified equations can be obtained for an arbitrary monitor function $\boldsymbol{f}(\boldsymbol{x})$ since the parametrization (7.84) of the monitor surface S^{r2} has the following form

$$\boldsymbol{r}(\boldsymbol{s}) = \{s^1, s^2, x^3(s^1, s^2), \boldsymbol{f}[\boldsymbol{x}(s^1, s^2)]\}, \qquad (7.95)$$

i.e. S^{r2} is also the monitor surface over the domain S^2, formed by the values of the vector-valued function

$$\boldsymbol{g}(\boldsymbol{s}) = \{x^3(\boldsymbol{s}), \boldsymbol{f}[\boldsymbol{x}(\boldsymbol{s})]\}.$$

Note a nondegenerate, smooth surface $S^{x2} \subset R^3$ is always locally represented as the graph of a bivariate function, i.e. in the form similar to (7.94). In the case of the parametrization (7.95) of the monitor surface S^{r2} over S^2 we can also generate grids on S^{x2} with the use of equations (7.22), for $n = 2$, which analogously to (7.93) are converted to

$$L^2[s^i] + \frac{\partial x^3}{\partial s^i} L^2[x^3] + \frac{\partial f^p}{\partial s^i} L^2[f^p] = 0, \qquad (7.96)$$

$$i = 1, 2, \quad p = 1, \ldots, l,$$

where

$$\frac{\partial x^3}{\partial s^i} = \frac{\partial x^3[\boldsymbol{s}]}{\partial s^i}, \quad \frac{\partial f^p}{\partial s^i} = \frac{\partial f^p[\boldsymbol{x}(\boldsymbol{s})]}{\partial s^i}, \quad i = 1, 2,$$

while the metric elements g_{ij}^ξ in the description (7.88) of the operator L^2 are as follows:

$$g_{ij}^\xi = \frac{\partial \boldsymbol{s}(\boldsymbol{\xi})}{\partial \xi^i} \cdot \frac{\partial \boldsymbol{s}(\boldsymbol{\xi})}{\partial \xi^j} + \frac{\partial x^3[\boldsymbol{s}(\boldsymbol{\xi})]}{\partial \xi^i} \frac{\partial x^3[\boldsymbol{s}(\boldsymbol{\xi})]}{\partial \xi^j} + \frac{\partial \boldsymbol{f}[\boldsymbol{x}(\boldsymbol{s}(\boldsymbol{\xi}))]}{\partial \xi^i} \cdot \frac{\partial \boldsymbol{f}[\boldsymbol{x}(\boldsymbol{s}(\boldsymbol{\xi}))]}{\partial \xi^j},$$

$$i, j = 1, 2.$$

The form (7.94) of the parametrization of S^{x2} allows one to include the mean curvature K_2 of this surface into the grid equations with respect to its metric of this surface, i.e. without a monitor function. Indeed the parametrizaton (7.94) also represents the parametrization of the monitor surface S^{r2} over S^2 with a scalar-valued monitor function $f(\boldsymbol{s}) = x^3(\boldsymbol{s})$. Therefore, availing us of (7.67) for $n = 2$ and (7.85–7.89), we obtain the following grid equation for generating a fixed grid on S^{x2}

$$L^2[s^i] + 2\sqrt{g^s} J^2 K_2 x^3_{s^i} = 0, \quad i = 1, 2, \qquad (7.97)$$

where $J = \det(\partial s^i/\partial \xi^j)$.

7.4 Practical Grid Equations

Equations with Respect to a Monitor Metric. Introducing a monitor metric g^s_{ij} over S^{x2} in the form (7.52) for $n = 2$ produces the following surface grid equations

$$\begin{aligned}&L^2[s^p] + w(s)x^i_{s^p}L^2[x^i] + v^a f^a_{s^p}L^2[f^a] \\ &+ g^{\xi}[x^m_{s^p}\nabla(w, x^m) - \frac{1}{2}w_{s^p}\nabla(x^j, x^j) \\ &+ f^b_{s^p}\nabla(v^b, f^b) - \frac{1}{2}v^c_{s^p}\nabla(f^c)] = 0 \,, \\ &i, j, m = 1, \ldots, k \,, \quad p = 1, 2 \,, \quad a, b, c = 1, \ldots, l \,,\end{aligned} \quad (7.98)$$

obtained from (7.53) and (7.89).

7.4.3 Equations for Generating Grids in Domains

Two-Dimensional Domains. There are two approaches for controlling the grid behavior in a two-dimensional domain $X^2 \subset R^2$ with use of monitor surfaces and monitor manifolds.

Equations with Respect to a Monitor Surface. Let

$$\boldsymbol{f}(\boldsymbol{x}) : X^2 \to R^l \,, \quad \boldsymbol{x} = (x^1, x^2) \,, \quad \boldsymbol{f} = (f^1, \ldots, f^l) \,,$$

be a monitor function over X^2. Then the monitor surface S^{r2} over X^2 is parametrized as follows:

$$\boldsymbol{r}(\boldsymbol{x}) : X^2 \to R^{2+l} \,, \quad \boldsymbol{r}(\boldsymbol{x}) = [x^1, x^2, \boldsymbol{f}(\boldsymbol{x})] \,.$$

The grid equations in this situation are obtained from (7.22) or (7.23) with $n = 2$ by the assumption $x^i = s^i$, $i = 1, 2$. Since (7.86–7.89) the equations are as follows:

$$L^2[x^i] + \frac{\partial f^k}{\partial x^i}L^2[f^k] = 0 \,, \quad i = 1, 2 \,, \quad k = 1, \ldots, l \,. \quad (7.99)$$

Here the elements g^{ξ}_{ij} of the metric tensor of S^{r2} in the definition (7.88) of the operator L^2 are computed by the formula

$$g^{\xi}_{ij} = \frac{\partial x^m}{\partial \xi^i}\frac{\partial x^m}{\partial \xi^j} + \frac{\partial f^k[\boldsymbol{x}(\boldsymbol{\xi})]}{\partial \xi^i}\frac{\partial f^k[\boldsymbol{x}(\boldsymbol{\xi})]}{\partial \xi^j} \,, \quad i, j, m = 1, 2 \,, \quad k = 1, \ldots, l \,.$$

In the case $\boldsymbol{f}(\boldsymbol{x})$ is a scalar-valued monitor function $f(\boldsymbol{x})$ over X^2, the grid equations are also represented by (7.67) with $n = 2$, $s^i = x^i$, $i = 1, 2$. Since (7.87–7.89) the equations (7.67) have one more equivalent form

$$L^2[x^i] + 2g^{\xi}\sqrt{g^x}J^2 K_m f_{x^i} = 0 \,, \quad i = 1, 2 \,, \quad (7.100)$$

where

$$g^\xi_{ij} = \frac{\partial x^m}{\partial \xi^i}\frac{\partial x^m}{\partial \xi^j} + \frac{\partial f[\boldsymbol{x}(\boldsymbol{\xi})]}{\partial \xi^i}\frac{\partial f[\boldsymbol{x}(\boldsymbol{\xi})]}{\partial \xi^j}, \quad i,j,m = 1,2,$$

$$g^\xi = \det(g^\xi_{ij}), \quad g^x = 1 + (f_{x^1})^2 + (f_{x^2})^2,$$

$$K_m = \frac{1}{2\sqrt{g^x}} g^{kj}_x f_{x^k x^j}, \quad j,k = 1,2,$$

g^{kj}_x, $j,k = 1,2$, are the elements of the contravariant metric tensor of S^{r2} in the coordinates x^1, x^2. Since the elements g^x_{ij}, $i,j = 1,2$, of the covariant metric tensor of S^{r2} in the coordinates x^1, x^2, are expressed by

$$g^x_{ij} = \delta^i_j + f_{x^i}f_{x^j}, \quad i,j = 1,2,$$

and analogously to (7.86),

$$g^{ij}_x = (-1)^{i+j} g^x_{3-i 3-j}/g^x, \quad i,j = 1,2, \quad i,j \text{ fixed},$$

we find

$$K_m = \frac{1}{2(g^x)^{3/2}} B, \tag{7.101}$$

where

$$B = (1 + f_{x^2}f_{x^2})f_{x^1 x^1} - 2f_{x^1}f_{x^2}f_{x^1 x^2} + (1 + f_{x^1}f_{x^1})f_{x^2 x^2}.$$

Thus equations (7.100) can be written in the following form

$$L^2[x^i] + \frac{J^2 B}{1 + (f_{x^1})^2 + (f_{x^2})^2} f_{x^i} = 0, \quad i = 1,2. \tag{7.102}$$

Equations with Respect to a Monitor Metric. The grid behavior in the domain X^2 can also be controlled by the monitor metric g^x_{ij} over X^2 in the form

$$g^x_{ij} = z(\boldsymbol{x})\delta^i_j + v^k(\boldsymbol{x})f^k_{x^i}(\boldsymbol{x})f^k_{x^j}(\boldsymbol{x}), \quad i,j = 1,2, \quad k = 1,\ldots,l. \tag{7.103}$$

The grid equations with respect to thic monitor metric a computed from (7.29) for $n = 2$ and with the identification $x^i = s^i$. Namely, taking into account (7.89), these equations are as follows:

$$L^2[x^p] + \frac{1}{z} v^k f^k_{x^p} L^2[f^k] + \frac{g^\xi}{z}\left[f^k_{x^p}\nabla(v^k, f^k) - \frac{1}{2} v^k_{x^p}\nabla(f^k) \right] = 0, \tag{7.104}$$
$i,p = 1,2, \quad k = 1,\ldots,l.$

In particular these equations can be applied to the metric of the form (5.28).

Three-Dimensional Domains. Analogously to the two-dimensional case discussed above we consider two types of three-domensional grid equations for generating grids in a three-dimensional domain $X^3 \subset R^3$.

7.4 Practical Grid Equations

Equations with Respect to a Monitor Surface. Let a monitor function

$$\boldsymbol{f}(\boldsymbol{x}) : X^3 \to R^l \,, \quad \boldsymbol{x} = (x^1, x^2, x^3)\,, \quad \boldsymbol{f} = (f^1, \ldots, f^l)\,,$$

over X^3 has been specified. Then the monitor surface S^{r3} over X^3 has the following parametrization

$$\boldsymbol{r}(\boldsymbol{x}) : X^3 \to R^{3+l}\,, \quad \boldsymbol{r}(\boldsymbol{x}) = [x^1, x^2, x^3, \boldsymbol{f}(\boldsymbol{x})]\,.$$

The grid equations in X^3 are, for example, the equations (7.21) in which $n = 3$, $x^i = s^i$, $i = 1,2,3$, while the contravariant metric tensor (g_ξ^{ij}), $i,j = 1,2,3$, of the surface S^{r3} in the grid coordinates ξ^1, ξ^2, ξ^3, is the matrix inverse to the covariant metric tensor (g_{ij}^ξ), $i,j = 1,2,3$, where

$$g_{ij}^\xi = \frac{\partial x^m(\boldsymbol{\xi})}{\partial \xi^i} \frac{\partial x^m(\boldsymbol{\xi})}{\partial \xi^j} + \frac{\partial f^k[\boldsymbol{x}(\boldsymbol{\xi})]}{\partial \xi^i} \frac{\partial f^k[\boldsymbol{x}(\boldsymbol{\xi})]}{\partial \xi^j}\,, \tag{7.105}$$

$$i,j,m = 1,2,3\,, \quad k = 1,\ldots,l\,.$$

The elements of the inverse matrix (g_ξ^{ij}), $i,j = 1,2,3$, can be found from the elements of (g_{ij}^ξ) by the following general formula

$$g_\xi^{ij} = \frac{1}{g^\xi}(g_{i+1j+1}^\xi g_{i+2j+2}^\xi - g_{i+1j+2}^\xi g_{i+2j+1}^\xi)\,, \quad i,j = 1,\ldots,n\,, \tag{7.106}$$

$$g^\xi = \det(g_{ij}^\xi)\,,$$

in which any index, say i, is identified with $i \pm 3$, so, for instance, $g_{45}^\xi = g_{12}^\xi$. Thus the equations for gridding the domain X^3 with the monitor function $\boldsymbol{f}(\boldsymbol{x}) : X^3 \to R^l$ are transformed from (7.22) to

$$L^3[x^i] + L^3[f^k]\frac{\partial f^k}{\partial x^i} = 0\,, \quad i = 1,2,3\,, \quad k = 1,\ldots,l\,, \tag{7.107}$$

where

$$L^3[v] = [g_{22}^\xi g_{33}^\xi - (g_{23}^\xi)^2]\frac{\partial^2 v}{\partial \xi^1 \partial \xi^1} + 2[g_{23}^\xi g_{13}^\xi - g_{12}^\xi g_{33}^\xi]\frac{\partial^2 v}{\partial \xi^1 \partial \xi^2}$$

$$+ 2[g_{12}^\xi g_{23}^\xi - g_{22}^\xi g_{13}^\xi]\frac{\partial^2 v}{\partial \xi^1 \partial \xi^3} + [g_{11}^\xi g_{33}^\xi - (g_{13}^\xi)^2]\frac{\partial^2 v}{\partial \xi^2 \partial \xi^2}$$

$$+ 2[g_{12}^\xi g_{13}^\xi - g_{11}^\xi g_{23}^\xi]\frac{\partial^2 v}{\partial \xi^2 \partial \xi^3} + [g_{11}^\xi g_{22}^\xi - (g_{12}^\xi)^2]\frac{\partial^2 v}{\partial \xi^3 \partial \xi^3}\,.$$

If $\boldsymbol{f}(\boldsymbol{x})$ is a scalar-valued monitor function then, analogously to the two-dimensional case considered above, the grid equations are represented by (7.67) with $n = 3$, $s^i = x^i$, $i = 1,2,3$. Similarly to (7.100) these equations also have the form

$$L^3[x^i] + 3g^\xi \sqrt{g^x} J^2 K_m f_{x^i} = 0\,, \quad i = 1,2,3\,, \tag{7.108}$$

where, in accordance with (6.6) and (7.65),

194   7. Grid Equations with Respect to Intermediate Transformations

$$g^x = 1 + (f_{x^1})^2 + (f_{x^2})^2 + (f_{x^3})^2 \, ,$$

$$K_m = \frac{1}{3\sqrt{g^x}} g_x^{kj} f_{x^k x^j} \, , \quad j, k = 1, 2, 3 \, ,$$

$$J = \det\left(\frac{\partial x^i}{\partial \xi^j}\right) .$$

Here g_x^{kj}, $j, k = 1, 2, 3$, are the elements of the contravariant metric tensor of S^{r3} in the coordinates x^1, x^2, x^3. These elements are computed, for example, by the formula (6.8) with $n = 3$ in which s is identified with x, i.e.

$$g_x^{ij} = \delta_k^j - \frac{1}{g\mathbf{x}} f_{x^j} f_{x^k} \, , \quad j, k = 1, 2, 3 \, .$$

Equations with Respect to a Monitor Metric. The grid behavior in the domain X^3 can also be controlled by the monitor metric g_{ij}^x over X^3 in the form

$$g_{ij}^x = z(\boldsymbol{x})\delta_j^i + v^k(\boldsymbol{x}) f_{x^i}^k(\boldsymbol{x}) f_{x^j}^k(\boldsymbol{x}) \, , \quad i, j = 1, 2, 3 \, , \quad k = 1, \ldots, l \, . (7.109)$$

The grid equations with respect to this monitor metric are computed from (7.29) for $n = 3$ and with identification $x^i = s^i$, namely, taking into account (7.89) these equations are as follows:

$$L^3[x^p] + \frac{1}{z} v^k f_{x^p}^k L^3[f^k] + \frac{g^\xi}{z}\bigg[\nabla(z, x^p)$$

$$- \frac{1}{2} z_{x^p} \nabla(x^i, x^i) + f_{x^p}^k \nabla(v^k, f^k) - \frac{1}{2} v_{x^p}^k \nabla(f^k)\bigg] = 0 \, , \quad (7.110)$$

$$i, p = 1, 2, 3 \, , \quad k = 1, \ldots, l \, .$$

8. Control of Grid Clustering

This chapter establishes some relations between the qualitative properties of grid lines and/or surfaces generated by popular comprehensive grid systems, the forms of monitor functions, and the geometric characteristics of the physical domains or surfaces undergoing a gridding process.

We consider n-dimensional regular surfaces S^{xn}, $n \geq 1$, locally represented by parametrizations of the kind

$$\boldsymbol{x}(\boldsymbol{s}) : S^n \to R^{n+k} , \quad \boldsymbol{x} = (x^1, \ldots, x^{n+k}) , \quad \boldsymbol{s} = (s^1, \ldots, s^n) , \quad n > 0 , \quad (8.1)$$

and whose covariant metric tensor designated as (g_{ij}^{xs}) in the coordinates s^i, $i = 1, \ldots, n$, is defined through such parametrizations as follows:

$$g_{ij}^{xs} = \boldsymbol{x}_{s^i} \cdot \boldsymbol{x}_{s^j} , \quad i, j = 1, \ldots, n . \tag{8.2}$$

Typically, in practical applications, S^{xn} is a physical geometry associated for $l = 0$ with an n-dimensional domain (interval when $n = 1$), for $n = 1$ and $l = 1$ ($l = 2$) with a boundary curve of a two-dimensional (three-dimensional) domain, and for $n = 2$ and $l = 1$ with a boundary surface of a three-dimensional domain.

For the purpose of controlling the qualitative properties of the grids obtained by comprehensive grid generators we will use both monitor regular surfaces and monitor manifolds over S^{xn}.

A monitor surface over S^{xn}, designated by S^{rn}, is represented in the coordinates $s^1, \ldots, s^n$ by the parametrization of the form

$$\boldsymbol{r}(\boldsymbol{s}) : S^n \to R^{n+k+l} , \quad \boldsymbol{r}(\boldsymbol{s}) = [\boldsymbol{x}(\boldsymbol{s}), \boldsymbol{f}(\boldsymbol{s})] , \quad \boldsymbol{f} = (f^1, \ldots, f^l) , \tag{8.3}$$

where $\boldsymbol{f}(\boldsymbol{s})$ is some vector-valued monitor function specified by the user or found in the process of the numerical computation. Thus the formulation (8.3) is similar to (8.1). The covariant metric tensor of the monitor surface S^{rn} in the coordinates $s^1, \ldots, s^n$ is also defined analogously to (8.2). In order to distinguish this tensor we designate it in the coordinates $s^1, \ldots, s^n$ by (g_{ij}^s). So

$$g_{ij}^s = \boldsymbol{r}_{s^i} \cdot \boldsymbol{r}_{s^j} = g_{ij}^{xs} + \boldsymbol{f}_{s^i} \cdot \boldsymbol{f}_{s^j} , \quad i, j = 1, \ldots, n . \tag{8.4}$$

It is obvious that the monitor surface S^{rn} is, in fact, a regular surface. On the other hand the regular surface S^{xn} can be identified with the monitor surface over it if $\boldsymbol{f}$ is a constant vector function.

A monitor manifold is a generalization of the monitor surface S^{rn}. The points and parametrizations of monitor manifolds over S^{xn} are represented by (8.1), however the elements of the covariant metric tensor designated in the coordinate $s^1, \ldots, s^n$ by (g^{ms}_{ij}) are defined, for example, by the form (5.23)

$$g^{ms}_{ij} = z(\boldsymbol{s})g^{xs}_{ij} + v^m(\boldsymbol{s})f^m_{s^i}f^m_{s^j}, \quad i,j = 1,\ldots,n, \quad m = 1,\ldots,l, \qquad (8.5)$$

where g^{xs}_{ij} is the (ij)th element of the covariant metric tensor of S^{xn}, determined by (8.2), $z(\boldsymbol{s}) > 0$ and $v^m(\boldsymbol{s}) > 0$ are weight functions, $f^m(\boldsymbol{s})$, $m = 1,\ldots,l$, are monitor functions. Remind repeated indices mean the summation over them, i.e. in (8.5)

$$v^m(\boldsymbol{s})f^m_{s^i}f^m_{s^j} = \sum_{m=1}^{l} v^m(\boldsymbol{s})f^m_{s^i}f^m_{s^j} .$$

A more general formulation of monitor metrics for monitor manifolds over S^{xn} was presented by (5.24).

8.1 Fundamental Formula for Grid Clustering

It is well-known that when S^{xn} is a domain with the Eucledian metric then the operator of Beltrami in this metric is the Laplace operator and the spacing between $(n-1)$-dimensional grid surfaces $\xi^i = $ const in S^{xn} related to the solution of the Laplace equations, for both $n = 2$ and $n = 3$, increases near a boundary convex segment and, conversely, the spacing decreases when the boundary segment in concave (see Fig. 8.1). It is shown further that similar facts are also valid for the grid hypersurfaces related to the solution of the generalized Laplace equations with respect to the metric of arbitrary n-dimensional regular surfaces.

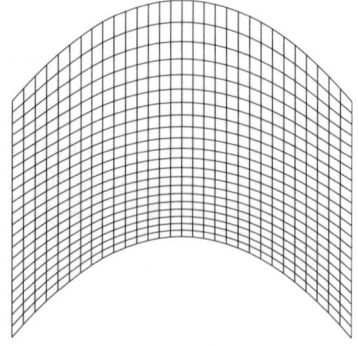

Fig. 8.1. Grid rarefaction (clustering) near a convex (concave) boundary segment

8.1.1 Relative Spacing Between Coordinate Surfaces

Throughout this subsection we assume i_0, $1 \leq i_0 \leq n$ as a fixed index, i.e. the summation in the formulas with i_0 repeated is not carried out over this index.

Let $s^1, \ldots, s^n$ be an arbitrary local coordinate system of an n-dimensional regular surface S^{xn} represented by (8.1). We first consider on the surface S^{xn} a family of the coordinate $(n-1)$-dimensional hypersurfaces $s^{i_0} = \text{const}$. Analogously to the specification by (4.12) we can readily find that the vector $\boldsymbol{n}_{i_0}$ lying in the tangent n-dimensional plane to S^{xn} and which is expressed in the form

$$\boldsymbol{n}_{i_0} = \frac{1}{\sqrt{g_{sx}^{i_0 i_0}}} g_{sx}^{i_0 j} \boldsymbol{x}_{s^j} , \quad j = 1, \ldots, n , \tag{8.6}$$

is a unit normal to the coordinate surface $s^{i_0} = c_0$. Here g_{sx}^{ij} is the (ij)th element of the contravariant metric tensor of S^{xn} in the coordinates $s^1, \ldots, s^n$ and the values of g_{sx}^{ij} and $\boldsymbol{x}_{s^j}$ are considered at the points of S^n for which $s^{i_0} = c_0$. Indeed

$$\boldsymbol{n}_{i_0} \cdot \boldsymbol{x}_{s^k} = \frac{1}{\sqrt{g_{sx}^{i_0 i_0}}} g_{sx}^{i_0 j} \boldsymbol{x}_{s^j} \cdot \boldsymbol{x}_{s^k} = \frac{1}{\sqrt{g_{sx}^{i_0 i_0}}} g_{sx}^{i_0 j} g_{jk}^{xs} = \frac{1}{\sqrt{g_{sx}^{i_0 i_0}}} \delta_k^{i_0} ,$$
$$j, k = 1, \ldots, n ,$$

i.e. $\boldsymbol{n}_{i_0}$ is orthogonal to the coordinate surface $s^{i_0} = c_0$ on S^{xn}. Further

$$\boldsymbol{n}_{i_0} \cdot \boldsymbol{n}_{i_0} = \frac{1}{\sqrt{g_{sx}^{i_0 i_0}}} g_{sx}^{i_0 j} \boldsymbol{x}_{s^j} \cdot \frac{1}{\sqrt{g_{sx}^{i_0 i_0}}} g_{sx}^{i_0 k} \boldsymbol{x}_{s^k}$$
$$= \frac{1}{g_{sx}^{i_0 i_0}} g_{sx}^{i_0 j} g_{sx}^{i_0 k} g_{jk}^{xs} = 1 , \quad j, k = 1, \ldots, n ,$$

i.e. $\boldsymbol{n}_{i_0}$ is a unit vector.

Let us denote by l_h the distance between a point on the coordinate surface $s^{i_0} = c_0$ and the nearest point on the surface $s^{i_0} = c_0 + h$ in S^{xn}. We have

$$l_h = (\boldsymbol{n}_{i_0} \cdot \boldsymbol{x}_{s^{i_0}}) h + O(h^2) = h \frac{1}{\sqrt{g_{sx}^{i_0 i_0}}} g_{sx}^{i_0 j} \boldsymbol{x}_{s^j} \cdot \boldsymbol{x}_{s^{i_0}} + O(h^2)$$
$$= h \frac{1}{\sqrt{g_{sx}^{i_0 i_0}}} g_{sx}^{i_0 j} g_{ji_0}^{xs} + O(h^2) = h \frac{1}{\sqrt{g_{sx}^{i_0 i_0}}} + O(h^2) , \quad j = 1, \ldots, n .$$

So the quantity $1/\sqrt{g_{sx}^{i_0 i_0}}$ with i_0 fixed reflects the relative spacing between the corresponding points on the coordinate hypersurfaces $s^{i_0} = c_0 + h$ and $s^i = c_0$ on S^{xn} (see Fig. 8.2 for $n = 2$).

8.1.2 Rate of Change of the Relative Spacing

The vector $\boldsymbol{n}_{i_0}$ is orthogonal to the coordinate hypersurface $s^{i_0} = c_0$ on S^{xn} and therefore the derivative of $1/\sqrt{g_{sx}^{i_0 i_0}}$ in the $\boldsymbol{n}_{i_0}$ direction is the rate of

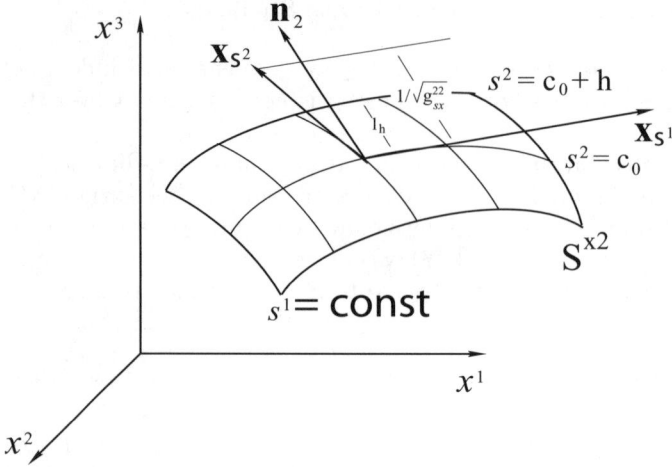

Fig. 8.2. Spacing of the coordinate lines $s^2 = const$ on the regular surface

change of the relative spacing of the coordinate hypersurfaces $s^{i_0} = const$ near this surface. Using (8.6) we obtain

$$\frac{d}{dn_{i_0}}\left(\frac{1}{\sqrt{g_{sx}^{i_0 i_0}}}\right) = \frac{1}{\sqrt{g_{sx}^{i_0 i_0}}} g^{i_0 j} \frac{\partial}{\partial s^j}\left(\frac{1}{\sqrt{g_{sx}^{i_0 i_0}}}\right)$$
$$= \frac{1}{\sqrt{g_{sx}^{i_0 i_0}}} \nabla\left(s^{i_0}, \frac{1}{\sqrt{g_{sx}^{i_0 i_0}}}\right), \quad j = 1, \ldots, n, \quad (8.7)$$

where $\nabla(\,,\,)$ is the Beltrami's mixed differential parameter. On the other hand

$$\frac{d}{dn_{i_0}}\left(\frac{1}{\sqrt{g_{sx}^{i_0 i_0}}}\right) = -\frac{1}{2\sqrt{(g_{sx}^{i_0 i_0})^3}} \frac{d}{dn_{i_0}} g_{sx}^{i_0 i_0}$$
$$= -\frac{1}{2(g_{sx}^{i_0 i_0})^2} g_{sx}^{i_0 j} \frac{\partial}{\partial s^j} g_{sx}^{i_0 i_0}, \quad j = 1, \ldots, n. \quad (8.8)$$

Note that

$$\frac{\partial}{\partial s^j} g_{sx}^{i_0 i_0} = g_{sx}^{i_0 l} g_{lk}^{xs} \frac{\partial}{\partial s^j} g_{sx}^{k i_0} = -g_{sx}^{i_0 l} g_{sx}^{k i_0} \frac{\partial}{\partial s^j} g_{lk}^{xs}, \quad j, k, l = 1, \ldots, n.$$

The application of (4.25) to this equation yields

$$\frac{\partial}{\partial s^j} g_{sx}^{i_0 i_0} = -g_{sx}^{i_0 l} g_{sx}^{k i_0}([lj, k]^s + [kj, l]^s) = -2g_{sx}^{i_0 l} \Upsilon_{lj}^{i_0}, \quad j, k, l = 1, \ldots, n.$$

therefore (8.8) is

$$\frac{d}{dn_{i_0}}\left(\frac{1}{\sqrt{g_{sx}^{i_0 i_0}}}\right) = \frac{1}{(g_{sx}^{i_0 i_0})^2} g_{sx}^{i_0 l} g_{sx}^{i_0 j} \Upsilon_{lj}^{i_0}, \quad j, l = 1, \ldots, n. \quad (8.9)$$

Note this equation is valid for an arbitrary coordinate system $s^1, \ldots, s^n$.

8.1.3 Relations to Geometry Characteristics

In order to connect the rate of change of the relative spacing of the coordinate hypersurfaces with geometrical characteristics we need to consider the following general situation in the theory of matrices. Let $\{a_{ij}\}$, $i,j = 1,\ldots,n$, $n \geq 2$, be a symmetric matrix of rank n and (a^{ij}), $i,j = 1,\ldots,n$, be its inverse matrix. Let $a^{i_0 i_0} \neq 0$ for some fixed index i_0, $1 \leq i_0 \leq n$. Define a matrix (b^{ij}) where

$$b^{ij} = \frac{1}{a^{i_0 i_0}}(a^{i_0 i_0} a^{ij} - a^{i_0 i} a^{i_0 j}), \quad i,j = 1,\ldots,n. \tag{8.10}$$

Let $(a_{ij}^{i_0})$ and $(b_{i_0}^{ij})$ be the $(n-1) \times (n-1)$ matrices obtained by deleting the i_0th row and i_0th column of the matrices (a_{ij}) and (b^{ij}), respectively.

Lemma 8.1.1. *The matrix $(b_{i_0}^{ij})$ is the inverse of $(a_{ij}^{i_0})$.*

Proof. It is sufficient to show that

$$a_{ij}^{i_0} b_{i_0}^{jl} = \delta_l^i, \quad i,j,l = 1,\ldots,n, \text{ and } i,j,l \neq i_0. \tag{8.11}$$

Expound, here and further the entry $i = 1,\ldots,n$, and $i \neq k$ means $i = 1,\ldots,k-1,k+1,\ldots,n$.

From (8.10) we readily see

$$b^{i_0 k} = b^{k i_0} = 0, \quad k = 1,\ldots,n,$$

therefore

$$a_{ij}^{i_0} b_{i_0}^{jl} = a_{im} b^{ml}, \quad i,j,l,m = 1,\ldots,n, \text{ and } i,j,l \neq i_0.$$

Since (8.10)

$$a_{im} b^{ml} = a_{im} a^{ml} - \frac{a_{im}}{a^{i_0 i_0}} a^{i_0 m} a^{i_0 l}$$

$$= \delta_l^i - \frac{\delta_{i_0}^i}{a^{i_0 i_0}} a^{i_0 l} = \delta_l^i, \quad i,l,m = 1,\ldots,n, \text{ and } i,l \neq i_0,$$

i.e. (8.11) is held. This proves the lemma.

Now we apply the matrix consideration given above to the matrix (g_{ij}^{xs}) which is the covariant metric tensor of the regular surface S^{xn} in the coordinates $s^1, \ldots, s^n$.

Multidimensional Case. Designate by $(g_{ij}^{i_0})$ the matrix obtained by deleting the i_0th row and i_0th column of (g_{ij}^{xs}). From (8.1) and (8.2) it is clear that $(g_{ij}^{i_0})$ is the covariant metric tensor of the coordinate hypersurface $s^{i_0} = c_0$ in the coordinates $s^1, \ldots, s^{i_0-1}, s^{i_0+1}, \ldots, s^n$. The matrix which is inverse to $(g_{ij}^{i_0})$ is the contravariant metric tensor of this coordinate hypersurface $s^{i_0} = c_0$ in the same coordinates $s^1, \ldots, s^{i_0-1}, s^{i_0+1}, \ldots, s^n$. Let this matrix be designated by $(g_{i_0}^{ij})$. Since

8. Control of Grid Clustering

$$g_{sx}^{i_0 i_0} = \det(g_{ij}^{i_0})/g^{xs} \neq 0 \;,$$

where $g^{xs} = \det(g_{ij}^{xs})$, we find, availing us of lemma 8.1.1,

$$g_{i_0}^{ij} = \frac{1}{g_{sx}^{i_0 i_0}}(g_{sx}^{i_0 i_0} g_{sx}^{ij} - g_{sx}^{i_0 i} g_{sx}^{i_0 j}) \;, \quad i,j = 1,\ldots,n\;, \text{ and } i,j \neq i_0 \;.$$

Therefore

$$g_{sx}^{i_0 i} g_{sx}^{i_0 j} = g_{sx}^{i_0 i_0}(g_{sx}^{ij} - g_{i_0}^{ij}) \;, \quad i,j = 1,\ldots,n\;, \text{ and } i,j \neq i_0 \;. \tag{8.12}$$

Since

$$g_{sx}^{i_0 l} g_{sx}^{i_0 j} \Upsilon_{lj}^{i_0} = g_{sx}^{i_0 k} g_{sx}^{i_0 p} \Upsilon_{kp}^{i_0} + 2 g_{sx}^{i_0 i_0} g_{sx}^{i_0 j} \Upsilon_{i_0 j}^{i_0} - g_{sx}^{i_0 i_0} g_{sx}^{i_0 i_0} \Upsilon_{i_0 i_0}^{i_0} \;,$$

$$j,k,l,p = 1,\ldots,n \;, \text{ and } k,p \neq i_0 \;,$$

we obtain, using (8.12),

$$\begin{aligned}
g_{sx}^{i_0 l} g_{sx}^{i_0 j} \Upsilon_{lj}^{i_0} &= g_{sx}^{i_0 i_0}(g_{sx}^{kp} - g_{i_0}^{kp})\Upsilon_{kp}^{i_0} + 2 g_{sx}^{i_0 i_0} g_{sx}^{i_0 j} \Upsilon_{i_0 j}^{i_0} - g_{sx}^{i_0 i_0} g_{sx}^{i_0 i_0} \Upsilon_{i_0 i_0}^{i_0} \\
&= g_{sx}^{i_0 i_0} g_{sx}^{lj} \Upsilon_{lj}^{i_0} - g_{sx}^{i_0 i_0} g_{i_0}^{kp} \Upsilon_{kp}^{i_0} \;,
\end{aligned} \tag{8.13}$$

$$j,k,l,p = 1,\ldots,n \;, \text{ and } k,p \neq i_0 \;.$$

Now we note that from (4.28)

$$\Upsilon_{kp}^{i_0} = g_{sx}^{i_0 l}[\boldsymbol{x}_{s^k s^p} \cdot \boldsymbol{x}_{s^l}] \;, \quad k,l,p = 1,\ldots,n \;,$$

so, since (8.6),

$$\Upsilon_{kp}^{i_0} = \sqrt{g_{sx}^{i_0 i_0}}\, \boldsymbol{x}_{s^k s^p} \cdot \boldsymbol{n}_{i_0} \;, \quad k,p = 1,\ldots,n \;,$$

where $\boldsymbol{n}_{i_0}$ is the unit normal to the coordinate hypersurface $s^{i_0} = c_0$ on S^{xn}. Thus, in the designation (4.79),

$$\Upsilon_{kp}^{i_0} = \sqrt{g_{sx}^{i_0 i_0}}\, b_{kp} \;, \quad k,p = 1,\ldots,n \;, \text{ and } k,p \neq i_0 \;, \tag{8.14}$$

where

$$b_{kp} = \boldsymbol{x}_{s^k s^p} \cdot \boldsymbol{n}_{i_0} \;, \quad k,p = 1,\ldots,n \;, \text{ and } k,p \neq i_0 \;,$$

is the (kp)th element of the second fundamental form of the coordinate hypersurface $s^{i_0} = c_0$ on S^{xn}. Therefore, using (4.81) and (8.14), we find

$$g_{i_0}^{kp} \Upsilon_{kp}^{i_0} = \sqrt{g_{sx}^{i_0 i_0}}\, g_{i_0}^{kp} b_{kp} = (n-1)\sqrt{g_{sx}^{i_0 i_0}}\, K_m(s^{i_0}) \;,$$

$$k,p = 1,\ldots,n \;, \text{ and } k,p \neq i_0 \;, \tag{8.15}$$

where $K_m(s^{i_0})$ is the mean curvature of the coordinate hypersurface $s^{i_0} = c_0$ on S^{xn} with respect to the normal $\boldsymbol{n}_{i_0}$. Further, from (7.2), we obtain

$$g_{sx}^{lj} \Upsilon_{lj}^{i_0} = -\Delta_B[s^{i_0}] \;, \quad j,l = 1,\ldots,n \;.$$

Substituting this equation and (8.15) in (8.13) gives

8.1 Fundamental Formula for Grid Clustering

$$g_{sx}^{iol}g_{sx}^{ioj}\Upsilon_{lj}^{io} = -g_{sx}^{ioio}\Delta_B[s^{io}] - (n-1)(g_{sx}^{ioio})^{3/2}K_m(s^{io}), \quad j,l = 1,\ldots,n.$$

Therefore, using (8.9), we conclude that the rate of change of the relative spacing of the coordinate hypersurfaces $s^{io} = const$ is expressed through the mean curvature and the Beltrami's second differential parameter as follows:

$$\frac{d}{dn_{io}}\left(\frac{1}{\sqrt{g_{sx}^{ioio}}}\right) = -\frac{1}{g_{sx}^{ioio}}\Delta_B[s^{io}] - \frac{n-1}{\sqrt{g_{sx}^{ioio}}}K_m(s^{io}). \tag{8.16}$$

The application of (8.7) to this equation gives the following relation between the mean curvature of the coordinate hypersurface $s^{io} = c_0$ and the Beltrami's mixed and second differential parameters:

$$(n-1)K_m(s^{io}) + \nabla\left(s^{io}, \frac{1}{\sqrt{g_{sx}^{ioio}}}\right) + \frac{\Delta_B[s^{io}]}{\sqrt{g_{sx}^{ioio}}} = 0. \tag{8.17}$$

Note the equations (8.16) and (8.17) are readily rewritten for the case of the monitor surface S^{rn} reperesented by (8.3), namely, it suffices to substitute s for sx in these equations. In particular, when S^{rn} is the monitor surface over S^n with a scalar-valued monitor function $f(s)$, i.e. S^{rn} is represented by

$$\boldsymbol{r}(s) : S^n \to R^{n+1}, \quad \boldsymbol{r}(s) = [s, f(s)],$$

then, availing us of (7.66), we obtain that the equation (8.16) has the following specific form

$$\frac{d}{dn_{io}}\left(\frac{1}{\sqrt{g_s^{ioio}}}\right) = \frac{n}{\sqrt{g^s g_s^{ioio}}}K_m(S^{rn})f_{s^{io}} - \frac{n-1}{\sqrt{g_s^{ioio}}}K_m(s^{io}) \tag{8.18}$$

where $K_m(S^{rn})$ is the mean curvature of S^{rn} in R^{n+1} with respect to the unit normal (7.64).

Two-Dimensional Case. Let s^1, s^2 be an arbitrary local coordinate system of a two-dimensional regular surface S^{x2}. Taking into account (8.9) and (8.13), we obtain that the rate of change of the relative spacing of the family of the coordinate lines $s^1 = const$ on S^{x2} is expressed as follows:

$$\frac{d}{d\boldsymbol{n}_1}\left(\frac{1}{\sqrt{g_{sx}^{11}}}\right) = \frac{1}{g_{sx}^{11}}\left(g_{sx}^{kl}\Upsilon_{kl}^1 - \frac{1}{g_{22}^{xs}}\Upsilon_{22}^1\right), \quad k,l = 1,2, \tag{8.19}$$

where

$$\boldsymbol{n}_1 = \frac{1}{\sqrt{g_{sx}^{11}}}g_{sx}^{1i}\boldsymbol{x}_{s^i}, \quad i = 1,2,$$

is the unit normal to the coordinate line $s^1 = c_0$ on S^{x2}.

Similarly, for the family of the coordinate lines $s^2 = const$, the rate of change of the relative spacing on S^{x2} has the form

$$\frac{d}{d\boldsymbol{n}_2}\left(\frac{1}{\sqrt{g_{sx}^{22}}}\right) = \frac{1}{g_{sx}^{22}}\left(g_{sx}^{kl}\Upsilon_{kl}^2 - \frac{1}{g_{11}^{xs}}\Upsilon_{11}^2\right), \quad k,l = 1,2, \tag{8.20}$$

where

$$n_2 = \frac{1}{\sqrt{g_{sx}^{22}}} g_{sx}^{2i} x_{s^i}, \quad i = 1, 2,$$

is the unit vector orthogonal to the line $s^2 = c_0$ on S^{x2}. Since for the coordinate line $s^{i_0} = c_0$ on S^{x2}

$$K_m(s^{i_0}) = \sigma_{i_0}, \tag{8.21}$$

where σ_{i_0} is the geodesic curvature of the curve $s^{i_0} = c_0$, the equations derived from (8.16) in the two-dimensional case have the form

$$\frac{d}{dn_1}\left(\frac{1}{\sqrt{g_{sx}^{11}}}\right) = -\frac{1}{g_{sx}^{11}} \Delta_B[s^1] - \frac{\sigma_1}{\sqrt{g_{sx}^{11}}},$$

$$\frac{d}{dn_2}\left(\frac{1}{\sqrt{g_{sx}^{22}}}\right) = -\frac{1}{g_{sx}^{22}} \Delta_B[s^2] - \frac{\sigma_2}{\sqrt{g_{sx}^{22}}}. \tag{8.22}$$

While equation (8.17) in the two-dimensional case yields

$$\sigma_1 + \nabla(s^1, 1/\sqrt{g_{sx}^{11}}) + \Delta_B[s^1]/\sqrt{g_{sx}^{11}} = 0,$$

$$\sigma_2 + \nabla(s^2, 1/\sqrt{g_{sx}^{22}}) + \Delta_B[s^2]/\sqrt{g_{sx}^{22}} = 0. \tag{8.23}$$

Two-Dimensional Domain. Let S^{x2} be a two-dimensional domain X^2 with a curvilinear coordinate system s^1, s^2 specified by a one-to-one smooth transformation

$$\boldsymbol{x}(s) : S^2 \to X^2, \quad \boldsymbol{x} = (x^1, x^2), \quad s = (s^1, s^2). \tag{8.24}$$

In this case there is the inverse of $\boldsymbol{x}(s)$:

$$s(\boldsymbol{x}) : X^2 \to S^2, \quad s(\boldsymbol{x}) = (s^1(\boldsymbol{x}), s^2(\boldsymbol{x}))$$

while the contravariant metric tensor (g_{sx}^{ij}) of X^2 in the coordinates s^1, s^2 is determined as

$$g_{sx}^{ij} = \frac{\partial s^i}{\partial x^l} \frac{\partial s^j}{\partial x^l}, \quad i, j, l = 1, 2. \tag{8.25}$$

As for the Beltramian operator, we have

$$\Delta_B[s^i] = \frac{1}{\sqrt{g^{xs}}} \frac{\partial}{\partial s^j}(\sqrt{g^{xs}} g_{sx}^{ij}) = \frac{1}{\sqrt{g^{xs}}} \frac{\partial}{\partial s^j}\left(\sqrt{g^{xs}} \frac{\partial s^i}{\partial x^l} \frac{\partial s^j}{\partial x^l}\right)$$

$$= \frac{1}{\sqrt{g^{xs}}} \left[\frac{\partial}{\partial s^j}\left(\sqrt{g^{xs}} \frac{\partial s^j}{\partial x^l}\right) \frac{\partial s^i}{\partial x^l} + \sqrt{g^{xs}} \frac{\partial}{\partial s^j}\left(\frac{\partial s^i}{\partial x^l}\right) \frac{\partial s^j}{\partial x^l}\right] \tag{8.26}$$

$$= \frac{\partial}{\partial x^l} \frac{\partial s^i}{\partial x^l} = \nabla^2[s^i], \quad i, j, l = 1, 2,$$

since (2.48), as $\sqrt{g^{xs}}$ can be computed by the folowing formula

$$\sqrt{g^{xs}} = \det\left(\frac{\partial x^i}{\partial s^j}\right) = J, \quad i, j = 1, 2.$$

Thus the equations (8.22) have the form

$$\frac{d}{dn_i}\left(\frac{1}{\sqrt{g_{sx}^{ii}}}\right) = -\frac{1}{g_{sx}^{ii}}\nabla^2(s^i) - \frac{\sigma_i}{\sqrt{g_{sx}^{ii}}}, \quad i,j = 1,2, \ i \text{ fixed}. \quad (8.27)$$

Remind the declaration "i fixed" in any formula means that the summation over the repeated index i is not carried out.

Two-Dimensional Monitor Surface over a Domain. When S^{r2} is a monitor surface over S^2 with a scalar-valued monitor function $f(s)$ then, since (8.21) and (8.18),

$$\frac{d}{dn_i}\left(\frac{1}{\sqrt{g_s^{ii}}}\right) = \frac{2}{\sqrt{g^s g_s^{ii}}} K_m(S^{x2}) f_{s^i} - \frac{1}{\sqrt{g_s^{ii}}} \sigma_i, \quad i=1,2, \ i \text{ fixed}.$$

Three-Dimensional Case. We demonstrate here another inference of the general formula connecting the mean curvature of coordinate two-dimensional surfaces and the rate of change of their relative spacing on S^{x3}.

In the case of a three-dimensional regular surface S^{x3} specified by (8.1) the unit normal $\boldsymbol{n}_3$ to the surface $\xi^3 = \text{const}$ in S^{r3} is defined, analogously to (8.6), as follows:

$$\boldsymbol{n}_3 = \frac{1}{\sqrt{g_{sx}^{33}}} g_{sx}^{3i} \boldsymbol{x}_{s^i}, \quad i=1,2,3. \quad (8.28)$$

Further

$$g_{sx}^{kl}\Upsilon_{kl}^3 = \frac{1}{g_{sx}^{33}}[g_{sx}^{33}g_{sx}^{kl}\Upsilon_{kl}^3]$$

$$= \frac{1}{g_{sx}^{33}}[g_{sx}^{3k}g_{sx}^{3l}\Upsilon_{kl}^3 + (g_{sx}^{33}g_{sx}^{1l} - g_{sx}^{31}g_{sx}^{3l})\Upsilon_{1l}^3 + (g_{sx}^{33}g_{sx}^{2l} - g_{sx}^{32}g_{sx}^{3l})\Upsilon_{2l}^3]$$

$$= \frac{1}{g_{sx}^{33}}[g_{sx}^{3k}g_{sx}^{3l}\Upsilon_{kl}^3 + (g_{sx}^{33}g_{sx}^{11} - g_{sx}^{31}g_{sx}^{31})\Upsilon_{11}^3 + 2(g_{sx}^{33}g_{sx}^{12} - g_{sx}^{31}g_{sx}^{32})\Upsilon_{12}^3$$

$$+ (g_{sx}^{33}g_{sx}^{22} - g_{sx}^{32}g_{sx}^{32})\Upsilon_{22}^3], \quad k,l=1,2,3,$$

and applying (4.9) yields

$$g_{sx}^{kl}\Upsilon_{kl}^3 = \frac{1}{g_{sx}^{33}}\left[g_{sx}^{3k}g_{sx}^{3l}\Upsilon_{kl}^3 + \frac{1}{g^{xs}}(g_{22}^{xs}\Upsilon_{11}^3 - 2g_{12}^{xs}\Upsilon_{12}^3 + g_{11}^{xs}\Upsilon_{22}^3)\right], \quad (8.29)$$
$$k,l=1,2,3.$$

The two-dimensional surface $s^3 = \text{const}$ lying on S^{r3} has a natural covariant metric tensor (g_{ij}^3) in the coordinates s^1, s^2 such that

$$g_{ij}^3 = g_{ij}^{xs}, \quad i,j=1,2,$$

and, consequently, the elements of its contravariant metric tensor (g_3^{ij}) are defined as follows:

8. Control of Grid Clustering

$$g_3^{ij} = (-1)^{i+j} g_{3-i3-j}^3 / \det(g_{ij}^3), \quad i, j = 1, 2.$$

It is apparent that

$$g^{xs} = \det(g_{ij}^3)/g_{sx}^{33}, \quad i, j = 1, 2,$$

where $g^{xs} = \det(g_{ij}^{xs})$, therefore

$$g_3^{ij} = (-1)^{i+j} g_{3-i3-j}^3 / (g^{xs} g_{sx}^{33}), \quad i, j = 1, 2. \tag{8.30}$$

Note the equation

$$g_3^{ij} = g_{sx}^{ij}, \quad i, j = 1, 2,$$

is not valid, in general. From (8.29) and (8.30) we find

$$g_{sx}^{kl} \Upsilon_{kl}^3 = \frac{1}{g_{sx}^{33}} [g_{sx}^{3k} g_{sx}^{3l} \Upsilon_{kl}^3 + g_{sx}^{33} g_3^{ij} \Upsilon_{ij}^3], \quad i, j = 1, 2, \quad k, l = 1, 2, 3. \tag{8.31}$$

Since (8.9), for $n = 3$, $i_0 = 3$,

$$\frac{d}{dn_3}\left(\frac{1}{\sqrt{g_{sx}^{33}}}\right) = \frac{1}{(g_{sx}^{33})^2} g_{sx}^{3i} g_{sx}^{3l} \Upsilon_{il}^3, \quad i, l = 1, 2, 3,$$

therefore equation (8.31) is equivalent to

$$\frac{d}{dn_3}\left(\frac{1}{\sqrt{g_{sx}^{33}}}\right) = \frac{1}{g_{sx}^{33}} (g_{sx}^{kl} \Upsilon_{kl}^3 - g_3^{ij} \Upsilon_{ij}^3), \quad i, j = 1, 2, \quad k, l = 1, 2, 3. \tag{8.32}$$

Note that the quantity

$$K_m(s^3) = \frac{1}{2} \frac{1}{\sqrt{g_{sx}^{33}}} g_3^{ij} \Upsilon_{ij}^3, \quad i, j = 1, 2,$$

is the mean curvature of the surface $s^3 = \text{const}$ on S^{x3}. So (8.32) yields

$$\frac{d}{dn_3}\left(\frac{1}{\sqrt{g_{sx}^{33}}}\right) = \frac{1}{\sqrt{g_{sx}^{33}}} [g_{sx}^{kl} \Upsilon_{kl}^3 - 2\sqrt{g_{sx}^{33}} K_m(s^3)].$$

Analogously, for the coordinate surface $s^i = c_0$, we obtain

$$\frac{d}{dn_i}\left(\frac{1}{\sqrt{g_{sx}^{ii}}}\right) = \frac{1}{g_{sx}^{ii}} [g_{sx}^{kl} \Upsilon_{kl}^i - 2\sqrt{g_{sx}^{ii}} K_m(s^i)], \tag{8.33}$$

$$i, k, l = 1, 2, 3, \quad i \text{ fixed},$$

where $K_m(s^i)$ is the mean curvature of the coordinate surface $s^i = c_0$.

If S^{x3} is a three-dimensional domain X^3 then, similar to (8.27), the equations (8.33) have the following form

$$\frac{d}{dn_i}\left(\frac{1}{\sqrt{g_{sx}^{ii}}}\right) = -\frac{1}{g_{sx}^{ii}} \nabla^2(s^i) - \frac{2}{\sqrt{g_{sx}^{ii}}} K_m(s^i), \quad i = 1, 2, 3, \quad i \text{ fixed}. \tag{8.34}$$

These equations also have the form (8.18) when S^{r3} is a monitor surface over S^3 with a scalar-valued monitor function $f(s)$:

$$\frac{d}{dn_i}\left(\frac{1}{\sqrt{g_s^{ii}}}\right) = \frac{3}{\sqrt{g^s g_s^{ii}}} K_m(S^{rn}) f_{s^i} - \frac{2}{\sqrt{g_s^{ii}}} K_m(s^i) ,$$

$$i = 1, 2, 3 , \quad i \text{ fixed} .$$

8.1.4 Basic Relation to Grid Coordinates

Let us apply formulas (8.16) and (8.22) to the grid coordinates $\xi^1, \ldots, \xi^n$ in S^{xn} obtained by the composition of the parametrization (5.1) and intermediate transformation (5.2).

We designate by v_p the rate of change of the relative spacing between the grid hypersurfaces $\xi^p = const$ in S^{xn}, i.e., analogously to (4.63),

$$v_p = \frac{d}{dn_p}\left(\frac{1}{\sqrt{g_{\xi x}^{pp}}}\right) = \frac{1}{(g_{\xi x}^{pp})^2} g_{\xi x}^{pl} g_{\xi x}^{pj} \Upsilon_{lj}^p , \quad j, l, p = 1, \ldots, n , \; p \text{ fixed} , \quad (8.35)$$

here Υ_{lj}^p is the Christoffel symbol of the second rank of S^{xn} in the grid coordinates $\xi^1, \ldots, \xi^n$, $\boldsymbol{n}_p$ is the normal to the grid hypersurface $\xi^p = const$, namely, similar to (8.6),

$$\boldsymbol{n}_p = \frac{1}{\sqrt{g_{\xi x}^{pp}}} g_{\xi x}^{pj} \frac{\partial}{\partial \xi^j} \boldsymbol{x}[s(\boldsymbol{\xi})] , \quad j, p = 1, \ldots, n , \; p \text{ fixed} . \quad (8.36)$$

We see that if $v_p < 0 (v_p > 0)$ then the nodes of the coordinate grid cluster (rarefy) in the $\boldsymbol{n}_p$ direction, i.e. v_p is a measure of change of grid spacing. We also call it a measure of grid clustering. The formulas (8.16) and (8.22), rewritten in the grid coordinates, result in the following:

Theorem 8.1.1. *Let $\boldsymbol{x}(s)$ in (5.1) and $\boldsymbol{s}(\boldsymbol{\xi})$ in (5.2) be nondegenerate transformations of the class $C^2[S^n]$ and $C^2[\Xi^n]$, respectively. Then*

$$v_p = -\frac{1}{g_{\xi x}^{pp}} \Delta_B[\xi^p] - \frac{n-1}{\sqrt{g_{\xi x}^{pp}}} K_m(\xi^p) , \quad p = 1, \ldots, n , \; p \text{ fixed} , \quad (8.37)$$

where Δ_B is the operator of Beltrami defined by (5.3); the function $\xi^p(\boldsymbol{s})$ is the pth component of the transformation $\boldsymbol{\xi}(\boldsymbol{s}) : S^n \to \Xi^n$ inverse to (5.2); $K_m(\xi^p)$ is the geodesic curvature of the curve $\xi^p = c_0$ in S^{x2} when $n = 2$, while $K_m(\xi^p)$, when $n > 2$, is the mean curvature of the grid hypersurface $\xi^p = c_0$ in S^{xn}.

8.1.5 Remarks

We assume here that the logical domain Ξ^n in the boundary value problem (5.4) formulated for generating boundary conforming grids is a rectangular

n-dimensional parallelepiped $0 \leq \xi^i \leq l_i$, $i = 1, \ldots, n$, and the $(n-1)$-dimensional boundary plane $\xi^i = 0$ or $\xi^i = l_i$ is mapped onto some $(n-1)$-dimensional segment of the boundary of S^{xn}.

Formula (8.37) demonstrates how the measure v_p of grid clustering near a boundary grid hypersurface of S^{xn} depends on its mean curvature. In particular, when the grid coordinate function $\xi^p(s)$, $p = 1, \ldots, n$, is subject to the corresponding pth equation of the grid system

$$\Delta_B[\xi^i] = 0, \quad i = 1, \ldots, n, \tag{8.38}$$

in the original metric (8.2) of the physical geometry S^{xn} then

$$v_p = -\frac{n-1}{\sqrt{g_{\xi x}^{pp}}} K_m(\xi^p), \quad p = 1, \ldots, n, \quad p \text{ fixed}. \tag{8.39}$$

So the sign of v_p is determined by the sign of $K_m(\xi^p)$. Note the equations (8.38) proposed for two-dimensional domains by Winslow (1967) and for surfaces by Warsi (1981) are the most popular for the generation of fixed grids in domains and on two-dimensional surfaces.

If S^{xn} is a domain S^n with the Eucledian metric then the equations (8.38) are the Laplace equations

$$\Delta[\xi^i] \equiv \frac{\partial}{\partial s^j}\left(\frac{\partial \xi^i}{\partial s^j}\right) = 0, \quad i, j = 1 \ldots, n. \tag{8.40}$$

In the monographs of Thompson J.F., Warsi Z.U.A., and Mastin C.W. (1985) and Liseikin V.D. (1999) there was proved that the nodes of the coordinate grid obtained in S^n by the solution of (8.40) for a rectangular computational domain $\Xi^n : 0 \leq \xi^i \leq l_i$, $i = 1, \ldots, n$, cluster near concave boundary segments of S^n and rarefy near its convex boundary segments (see Fig. 8.1). However the formula (8.39) yields more strong conclusions for n-dimensional domains when $n \geq 3$. In order to formulate these results we first note that the unit normal $\boldsymbol{n}_p$, specified by (8.36), in this case is as follows

$$\boldsymbol{n}_p = \frac{1}{\sqrt{g_{\xi s}^{pp}}} g_{\xi s}^{pj} \boldsymbol{s}_{\xi^j}, \quad j, p = 1, \ldots, n, \quad p \text{ fixed},$$

where

$$g_{\xi s}^{pj} = \frac{\partial \boldsymbol{\xi}}{\partial s^p} \cdot \frac{\partial \boldsymbol{\xi}}{\partial s^j}, \quad j, p = 1, \ldots, n.$$

Since

$$\boldsymbol{n}_p \cdot \boldsymbol{s}_{\xi^p} = \frac{1}{\sqrt{g_{\xi s}^{pp}}} > 0, \quad p = 1, \ldots, n, \quad p \text{ fixed},$$

the unit normal $\boldsymbol{n}_p$ is directed to the interior of S^n at the points of the boundary hypersurface $\xi^p = 0$. Contrary, at the points of the hypersurface $\xi^p = l_p$ it is directed to the exterior of S^n. Therefore the inequality $v_p >$

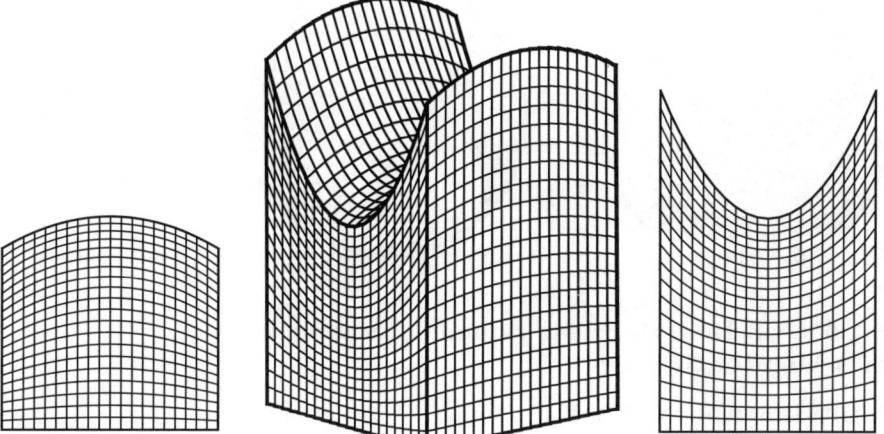

Fig. 8.3. Grid clustering near the upper boundary segment having the negative mean curvature

0 ($v_p < 0$) at the points of the grid hypersurface $\xi^p = 0$ means that the grid nodes cluster (rarefy) near it. Contrary for the hypersurface $\xi^p = l_p$ in S^n the inequality $v_p > 0$ ($v_p < 0$) means rarefaction (clustering) of grid nodes near it. Thus the nodes of the grid produced by (8.40) may also cluster (rarefy) near the boundary $\xi^p = l_p$ if this segment is not concave (convex), for example, it is a saddle surface but having the negative (positive) mean curvature (see Figs. 8.3 and 8.4).

Formula (8.39) also yields a new result in the theory of surface grid generation. Namely, the nodes of the coordinate grid on the surface S^{x2} generated through (8.38) cluster (rarefy) near concave (convex) segments of the boundary of S^{x2}. Figure 8.5 of the grid generated through the solution of (8.38) for $n = 2$ on the part of a sphere with a concave boundary viewed under different angles demonstrates node clustering near its boundary.

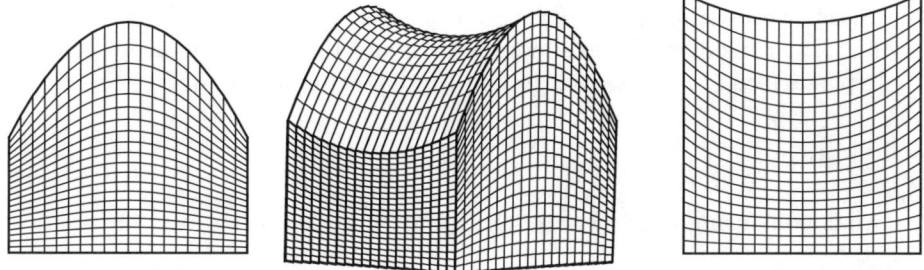

Fig. 8.4. Grid rarefaction near a boundary segment having the positive mean curvature

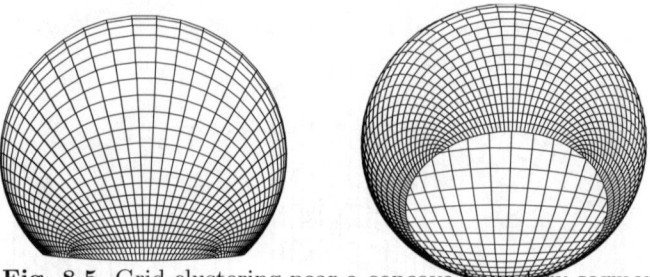

Fig. 8.5. Grid clustering near a concave boundary segment of a cut sphere

8.1.6 Grid Behavior near Boundary Segments of a Monitor Surface

Multidimensional Case. Let now S^{rn} be a monitor surface represented by (8.3) and $\xi^1, \ldots, \xi^n$ be the grid coordinate system of S^{rn} obtained by the solution of the boundary value problem (5.4). Then the elements of the contravariant metric tensor and the Christoffel symbols of S^{rn} in the grid coordinates $\xi^1, \ldots, \xi^n$ satisfy the grid equations (7.3), i.e.

$$\Delta_B[\xi^i] \equiv -g_\xi^{kj}\Upsilon_{kj}^i = 0, \quad i, k, j = 1, \ldots, n. \tag{8.41}$$

On the other hand, there are valid the identities (8.37) which, in application to the monitor surface S^{rn} and the coordinates $\xi^1, \ldots, \xi^n$, are of the form

$$\frac{\mathrm{d}}{\mathrm{d}\boldsymbol{n}_i}\left(\frac{1}{\sqrt{g_\xi^{ii}}}\right) = -\frac{1}{g_\xi^{ii}}\Delta_B[\xi^i] - \frac{n-1}{\sqrt{g_\xi^{ii}}}K_m(\xi^i), \quad i = 1, \ldots, n, \; i \text{ fixed}, \tag{8.42}$$

where

$$\boldsymbol{n}_i = \frac{1}{\sqrt{g_\xi^{ii}}}g_\xi^{ij}\boldsymbol{r}_{\xi^j}, \quad i, j = 1, \ldots, n, \quad i \text{ fixed},$$

is the unit normal vector to the grid hypersurface $\xi^i = c_0$ on S^{rn}, while $K_m(\xi^i)$ is the mean curvature of this hypersurface. So, taking into account (8.41), we obtain that the diagonal contravariant metric elements g_ξ^{ii}, $i = 1, \ldots, n$, i fixed, in the grid coordinates and the mean curvature of the grid hypersurface $\xi^i = c_0$ are subject to the following restriction

$$\frac{\mathrm{d}}{\mathrm{d}\boldsymbol{n}_i}\left(\frac{1}{\sqrt{g_\xi^{ii}}}\right) = -\frac{n-1}{\sqrt{g_\xi^{ii}}}K_m(\xi^i), \quad i = 1, \ldots, n, \quad i \text{ fixed}. \tag{8.43}$$

Analogous using of (8.41) and (8.17) gives the following form of (8.43)

$$(n-1)K_m(\xi^i) + \nabla\left(\xi^i, \frac{1}{\sqrt{g_\xi^{ii}}}\right) = 0, \quad i = 1, \ldots, n, \quad i \text{ fixed}, \tag{8.44}$$

with respect to the grid coordinates $\xi^1, \ldots, \xi^n$ on S^{rn}.

8.1 Fundamental Formula for Grid Clustering

Two-Dimensional Case. In the two-dimensional case

$$\sigma_i = K_m(\xi^i), \quad i = 1, 2,$$

where σ_i is the geodesic curvature of the coordinate line $\xi^i = c_0$ on the two-dimensional monitor surface S^{r2}. Therefore the equations (8.43) are identical to

$$\frac{d}{dn_1}\left(\frac{1}{\sqrt{g_\xi^{11}}}\right) = -\frac{1}{\sqrt{g_\xi^{11}}}\sigma_1,$$

$$\frac{d}{dn_2}\left(\frac{1}{\sqrt{g_\xi^{22}}}\right) = -\frac{1}{\sqrt{g_\xi^{22}}}\sigma_2. \tag{8.45}$$

Thus if the geodesic curvature σ_i of the boundary curve $\xi^i = 0$ on S^{r2} is negative (positive) then

$$\frac{d}{dn_i}\left(\frac{1}{\sqrt{g_\xi^{ii}}}\right) > 0(< 0), \quad i = 1, 2, \; i \text{ fixed}.$$

This means that the coordinate curves $\xi^i = const$ on S^{r2} obtained through (5.4) cluster near the curve $\xi^l = 0$ when it is a boundary curve if $\sigma_i < 0$ and vice versa the coordinate lines become sparser when approaching the boundary curve $\xi^i = 0$ if $\sigma_i > 0$. Note the sign of the geodesic curvature indicates the convexity ($\sigma_i > 0$) or concavity ($\sigma_i < 0$) of the boundary curve $\xi^i = 0$ on S^{r2}, while the condition $\sigma_i = 0$ means that the coordinate line $\xi^i = 0$ is a geodesic line on S^{r2}. Analogous results are valid for the lines $\xi^i = 1, i = 1, 2$.

So we conclude, finally, that the grid lines obtained from (5.4) are repelled from the convex segments of the boundary lines of S^{r2} and attracted to their concave segments. Figure 8.6 illustrates this conclusion in the case of a concave boundary segment.

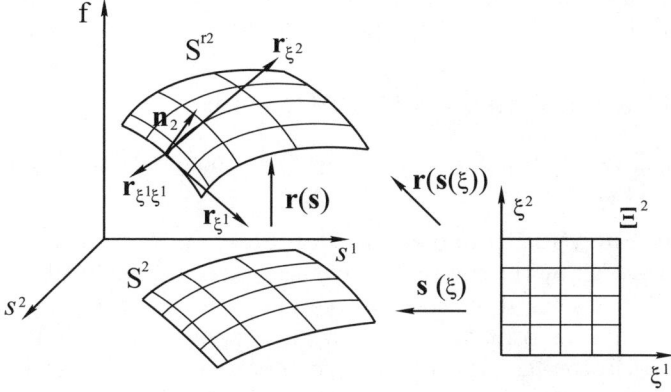

Fig. 8.6. Illustration of grid clustering near the concave boundary segment of the monitor surface

Three-Dimensional Case. If the transformation $\boldsymbol{\xi}(\boldsymbol{s})$ is defined through (5.4) with $n = 3$ then, availing us of (8.43), we have in the three-dimensional case for the coordinate surface $\xi^3 = c_0$ on S^{r3}

$$\frac{\mathrm{d}}{\mathrm{d}n_3}\left(\frac{1}{\sqrt{g_\xi^{33}}}\right) = -\frac{2}{\sqrt{g_\xi^{33}}} K_m(\xi^3) \,. \tag{8.46}$$

Similar equations, obtained by substituting i for 3 in (8.46), are valid for the grid hypersurfaces $\xi^i = const$, $i = 1, 2$.

Thus we conclude, analogously to the two-dimensional case considered above, that the grid hypersurfaces $\xi^i = c$, $i = 1, 2, 3$, on S^{r3} obtained from (5.4) are repelled from the coordinate boundary segment of S^{r3} whose mean curvature in S^{r3} with respect to the corresponding normal $\boldsymbol{n}_i$ is positive and attracted to it if its mean curvature in S^{r3} is negative.

8.2 Application of Theorem to Popular Elliptic Models

Formula (8.37) gives one an opportunity for finding explicitly the necessary values of control functions in elliptic equations to provide the generation of grids whose nodes cluster or rarefy, if it is required, near specified boundary segments of an arbitrary physical geometry S^{xn}.

We assume that the logical domain $\varXi^n$ in the boundary value problems (5.4) formulated for generating grids is the standard unit cube, i.e. $0 \leq \xi^i \leq 1$, $i = 1, \ldots, n$. Besides this, let the $(n-1)$-dimensional boundary plane $\xi^i = 0$ or $\xi^i = 1$ of $\varXi^n$ for some i, $1 \leq i \leq n$, is mapped onto some segment of the boundary of S^{rn}, i.e. the segment is an $(n-1)$-dimensional coordinate surface. For the sake of definitiveness we consider further the boundary segments $\xi^p = 0$ whose normal (8.36) is directed to the interior of the physical geometry S^{xn}. So the condition $v_p > 0 (v_p < 0)$ at the points of such a boundary segment means grid clustering (rarefaction) near it. For the segments with opposite direction of the normal (8.36) all results are readily reformulated.

8.2.1 Control of Grids near Boundaries of Domains

Here we study the behavior of grid lines and surfaces near boundary segments of parametric and physical domains.

General Relations for Multidimensional Domains. In the process of gridding a physical area X^n by the comprehensive methods outlined in Chap. 5 the area is typically considered as a logical parametric domain S^n. Consequently, while applying the formulas of the current section to the physical domain X^n, it suffices to replace the parametric coordinates $s^1, \ldots, s^n$ by physical coordinates $x^1, \ldots, x^n$.

8.2 Application of Theorem to Popular Elliptic Models

The logical domain S^n, parametrized by the identity

$$\boldsymbol{I}(\boldsymbol{s}): S^n \to S^n, \quad \boldsymbol{I}(\boldsymbol{s}) = \boldsymbol{s}, \quad \boldsymbol{s} = (s^1, \ldots, s^n),$$

has a natural Euclidean metric

$$g_{ij}^{ss} = \delta_j^i, \quad i, j = 1, \ldots, n,$$

in the original parametric coordinates $s^1, \ldots, s^n$. In the process of grid generation this domain S^n is parametrized by an intermediate one-to-one smooth transformation

$$\boldsymbol{s}(\boldsymbol{\xi}) : \Xi^n \to S^n, \quad \boldsymbol{\xi} = (\xi^1, \ldots, \xi^n), \tag{8.47}$$

having the inverse

$$\boldsymbol{\xi}(\boldsymbol{s}) : S^n \to \Xi^n, \quad \boldsymbol{\xi}(\boldsymbol{s}) = [\xi^1(\boldsymbol{s}), \ldots, \xi^n(\boldsymbol{s})].$$

The elements of the Euclidean covariant and contravariant metric tensors of S^n in arbitrary curvilinear coordinates $\xi^1, \ldots, \xi^n$ are designated, in accordance with (8.2), as $g_{ij}^{s\xi}$ and $g_{\xi s}^{ij}$, $i = 1, \ldots, n$, respectively. Availing us of (8.2), (2.15), (2.18), and (2.38) we have

$$g_{ij}^{s\xi} = \boldsymbol{s}_{\xi^i} \cdot \boldsymbol{s}_{\xi^j} = \frac{\partial s^k}{\partial \xi^i} \frac{\partial s^k}{\partial \xi^j}, \quad i,j,k = 1, \ldots, n,$$

$$g_{\xi s}^{ij} = \boldsymbol{\nabla}\xi^i \cdot \boldsymbol{\nabla}\xi^j = \frac{\partial \xi^i}{\partial s^k} \frac{\partial \xi^j}{\partial s^k}, \quad i,j,k = 1, \ldots, n, \tag{8.48}$$

$$\Gamma_{ij}^m = \boldsymbol{s}_{\xi^i \xi^j} \cdot \boldsymbol{\nabla}\xi^m = \frac{\partial^2 s^l}{\partial \xi^i \partial \xi^j} \frac{\partial \xi^m}{\partial s^l}, \quad i,j,l,m = 1, \ldots, n,$$

here Γ_{ij}^m, $i,j,m = 1, \ldots, n$, are the Christoffel symbols of S^n in the coordinates $\xi^1, \ldots, \xi^n$,

$$\boldsymbol{\nabla}\xi^i = \left(\frac{\partial \xi^i}{\partial s^1}, \ldots, \frac{\partial \xi^i}{\partial s^n}\right), \quad i = 1, \ldots, n.$$

Let us consider a family of $(n-1)$-dimensional grid hypersurfaces $\xi^n = const$ in S^n (curves when $n = 2$). In analogy with (8.35), we conclude that the rate of change of the relative spacing of these grid hypersurfaces in S^n is as follows:

$$v_n = \frac{d}{d n_n^{s\xi}}\left(\frac{1}{\sqrt{g_{\xi s}^{nn}}}\right) = \frac{1}{(g_{\xi s}^{nn})^2} g_{\xi s}^{ni} g_{\xi s}^{nl} \Gamma_{li}^n, \quad i,l = 1, \ldots, n, \tag{8.49}$$

where

$$\boldsymbol{n}_n^{s\xi} = \frac{1}{\sqrt{g_{\xi s}^{nn}}} g_{\xi s}^{ni} \boldsymbol{s}_{\xi^i}, \quad i = 1, \ldots, n,$$

is a unit normal vector to the coordinate surface $\xi^n = c_0$ in S^n. We also find, similar to (8.13) and taking into account that $g_{\xi s}^{kl} \Gamma_{lk}^2$ may not be equal zero, that

8. Control of Grid Clustering

$$g^{ni}_{\xi s} g^{nl}_{\xi s} \Gamma^n_{li} = g^{nn}_{\xi s}(g^{kl}_{\xi s}\Gamma^n_{lk} - g^{pj}_n \Gamma^n_{pj}) , \tag{8.50}$$

$$i,j,k,l = 1,\ldots,n , \quad j,p = 1,\ldots,n-1 ,$$

where, in accordance with (8.12) the quantities

$$g^{pj}_n = \frac{1}{g^{nn}_{\xi s}}(g^{nn}_{\xi s} g^{pj}_{\xi s} - g^{pn}_{\xi s} g^{jn}_{\xi s}) , \quad j,p = 1,\ldots,n-1 , \tag{8.51}$$

are the elements of the contravariant metric tensor in the coordinates $\xi^1,\ldots,\xi^{n-1}$ of the hypersurface $\xi^n = c_0$. Note, since (8.48),

$$g^{ij}_{\xi s}\Gamma^l_{ij} = \frac{\partial \xi^i}{\partial s^k}\frac{\partial \xi^j}{\partial s^k}\frac{\partial^2 s^p \xi^i}{\partial \xi^i \partial \xi^j}\frac{\partial \xi^l}{\partial s^p}$$

$$= \frac{\partial^2 \xi^i}{\partial s^k \partial s^m}\frac{\partial s^m}{\partial \xi^j}\frac{\partial \xi^j}{\partial s^k}\frac{\partial s^p}{\partial \xi^i}\frac{\partial \xi^l}{\partial s^p} \tag{8.52}$$

$$= \frac{\partial}{\partial s^k}\frac{\partial \xi^l}{\partial s^k} = -\nabla^2[\xi^l] , \quad i,j,k,l,m,p = 1,\ldots,n ,$$

therefore, similarly to (8.16) and (8.17), equations (8.49) and (8.50) result in the following formulas for the rate of change of the relative spacing of the grid hypersurfaces $\xi^n = const$ in S^n:

$$\frac{d}{dn^{s\xi}_n}\left(\frac{1}{\sqrt{g^{nn}_{\xi s}}}\right) = -\frac{1}{g^{nn}_{\xi s}}\nabla^2[\xi^n] - \frac{n-1}{\sqrt{g^{nn}_{\xi s}}}K^{s\xi}_m(\xi^n) , \tag{8.53}$$

and for the connection between the mean curvature and Beltrami's mixed and second differential parameters:

$$(n-1)K^{s\xi}_m(\xi^n) + \nabla\left(\xi^n, \frac{1}{\sqrt{g^{nn}_{\xi s}}}\right) + \frac{1}{\sqrt{g^{nn}_{\xi s}}}\nabla^2[\xi^n] = 0 , \tag{8.54}$$

respectively, where $K^{s\xi}_m(\xi^n)$ is the mean curvature of the coordinate hypersurface $\xi^n = c_0$ in S^n computed, in accordance with (4.81), by the equation

$$K^{s\xi}_m(\xi^n) = \frac{1}{n-1}g^{jp}_n s_{\xi^j \xi^p} \cdot \mathbf{n}^{s\xi}_n = \frac{1}{(n-1)\sqrt{g^{nn}_{\xi s}}}g^{jp}_n \Gamma^n_{jp} , \tag{8.55}$$

$$j,p = 1,\ldots,n-1 .$$

Note this equation is a particular case of (4.81).

The same relations are valid for the grid hypersurface $\xi^i = const$ in S^n; they are obtained by substituting the fixed index i for n in (8.53 – 8.55).

Two-Dimensional Case. Formula (8.51), in the two-dimensional case, is of the form

$$g^{11}_2 = \frac{1}{g^{22}_{\xi s}}[g^{22}_{\xi s}g^{11}_{\xi s} - (g^{12}_{\xi s})^2] = \frac{1}{g^{22}_{\xi s}g^{s\xi}} = \frac{1}{g^{s\xi}_{11}} ,$$

where

$$g^{s\xi} = \det(g^{s\xi}_{ij}) .$$

Therefore, from (8.55),

$$K_m^{s\xi}(\xi^2) = \frac{1}{g_{11}^{s\xi}} s_{\xi^1\xi^1} \cdot n_2^{s\xi} = k_2,$$

where, according to (3.5) and (8.21) k_2 is the curvature of the coordinate curve $\xi^2 = c_0$ with respect to the normal $n_2^{s\xi}$. Note, in the case of the Euclidian metric in S^2 the curvature k coincides with the geodesic curvature σ.

Thus the equation (8.53), in the two-dimensional case, is as follows:

$$\frac{d}{dn_2^{s\xi}}\left(\frac{1}{\sqrt{g_{\xi s}^{22}}}\right) = -\frac{1}{g_{\xi s}^{22}}\nabla^2[\xi^2] - \frac{k_2}{\sqrt{g_{\xi s}^{22}}}. \tag{8.56}$$

Analogously we have at the points of the coordinate curve $\xi^1 = c_0$

$$\frac{d}{dn_1^{s\xi}}\left(\frac{1}{\sqrt{g_{\xi s}^{11}}}\right) = -\frac{1}{g_{\xi s}^{11}}\nabla^2[\xi^1] - \frac{k_1}{\sqrt{g_{\xi s}^{11}}}, \tag{8.57}$$

where k_1 is the curvature of the curve $\xi^1 = c_0$ in S^2.

Control by a Scalar-Valued Monitor Function. We establish in this section a relation between the Beltrami's second differential parameter $\Delta_B[\xi^i]$ with respect to the monitor surface S^{rn} over S^n and the magnitude $\nabla^2[\xi^i]$ which, in fact, is also the Beltrami's second differential parameter of $\xi^i(s)$ with respect to the Euclidean metric of S^n.

General Multidimensional Case. Let a grid on S^n be controlled by a monitor surface S^{rn} over S^n represented by the parametrization of the type (8.3) with a scalar-valued monitor function $f(s)$, i.e. by

$$r(s): S^n \to R^{n+1}, \quad r(s^1,\ldots,s^n) = [s^1,\ldots,s^n, f(s^1,\ldots,s^n)], \tag{8.58}$$

and whose metric is borrowed from R^{n+1}, i.e.

$$g_{ij}^s = r_{s^i} \cdot r_{s^j} = \delta_j^i + f_{s^i} f_{s^j}, \quad i,j = 1,\ldots,n. \tag{8.59}$$

This situation occurs when adaptive grids are built by the comprehensive grid generation method in an n-dimensional domain X^n which is identified with S^n.

Let the functions $\xi^i(s)$, $i = 1,\ldots,n$, comprise a new coordinate system. We establish now a relation between $\nabla^2[\xi^i]$ and $\Delta_B[\xi^i]$, where Δ_B is the operator of Beltrami with respect to the monitor surface S^{rn} represented by (5.3). Expanding the differentiation in $\Delta_B[\xi^i]$ gives

$$\Delta_B[\xi^i] = \frac{1}{\sqrt{g^s}}\frac{\partial}{\partial s^j}\left(\sqrt{g^s} g_s^{jk}\frac{\partial \xi^i}{\partial s^k}\right)$$

$$= g_s^{jk}\frac{\partial^2 \xi^i}{\partial s^j \partial s^k} + \Delta_B[s^k]\frac{\partial \xi^i}{\partial s^k}, \quad i,j,k = 1,\ldots,n. \tag{8.60}$$

214   8. Control of Grid Clustering

As the covariant metric elements g^s_{ij}, $i,j = 1,\ldots,n$, are specified by (8.59) so, in accordance with (6.6) and (6.7) for $v(s) \equiv 1$,

$$g^s = \det(g^s_{ij}) = 1 + \sum_{j=1}^{n} \left(\frac{\partial f}{\partial s^j}\right)^2,$$
$$g^{ik}_s = \delta^i_k - \frac{1}{g^s}\frac{\partial f}{\partial s^j}\frac{\partial f}{\partial s^k}, \quad j,k = 1,\ldots,n.$$
(8.61)

Therefore equations (8.60) also have the following form

$$\Delta_B[\xi^i] = \left(\delta^j_k - \frac{1}{g^s}\frac{\partial f}{\partial s^j}\frac{\partial f}{\partial s^k}\right)\frac{\partial^2 \xi^i}{\partial s^j \partial s^k} + \Delta_B[s^k]\frac{\partial \xi^i}{\partial s^k}$$

$$= \nabla^2[\xi^i] - \frac{1}{g^s}\frac{\partial f}{\partial s^j}\frac{\partial f}{\partial s^k}\frac{\partial^2 \xi^i}{\partial s^j \partial s^k} + \Delta_B[s^k]\frac{\partial \xi^i}{\partial s^k}, \quad i,j,k=1,\ldots,n,$$

where $\nabla^2 = \dfrac{\partial}{\partial s^j}\dfrac{\partial}{\partial s^j}$, $j = 1,\ldots,n$. These equations give the following connection between $\nabla^2[\xi^i]$ and $\Delta_B[\xi^i]$:

$$\nabla^2[\xi^i] = \Delta_B[\xi^i] - \Delta_B[s^k]\frac{\partial \xi^i}{\partial s^k} + \frac{1}{g^s}\frac{\partial f}{\partial s^j}\frac{\partial f}{\partial s^k}\frac{\partial^2 \xi^i}{\partial s^j \partial s^k},$$
(8.62)
$$i,j,k = 1,\ldots,n.$$

Note the Laplace operator ∇^2 is the Beltrami operator with respect to the Eucledian metric in S^n therefore, from (7.3), as well as from (8.52),

$$\nabla^2[\xi^n] = -g^{kl}_{\xi s}\Gamma^n_{lk}, \quad l,k = 1,\ldots,n,$$

where the metric elements and Christoffel symbols are in the coordinates $\xi^1,\ldots,\xi^n$ and defined by the formulas in (8.48). Now, applying (7.3) to $\Delta_B[\xi^i]$ and $\Delta_B[s^k]$, we find from (8.62)

$$g^{lj}_\xi {}^\xi\Upsilon^i_{lj} - g^{lj}_s {}^s\Upsilon^k_{lj}\frac{\partial \xi^i}{\partial s^k} - g^{lj}_{\xi s}\Gamma^i_{lj} - \frac{1}{g^s}\frac{\partial f}{\partial s^j}\frac{\partial f}{\partial s^k}\frac{\partial^2 \xi^i}{\partial s^j \partial s^k} = 0,$$
(8.63)
$$i,j,k,l = 1,\ldots,n,$$

where ${}^\xi\Upsilon^i_{lj}$, ${}^s\Upsilon^k_{lj}$ are the Christoffel symbols of the second kind of the monitor surface S^{rn} in the coordinates $\xi^1,\ldots,\xi^n$ and $s^1,\ldots,s^n$, respectively, while Γ^i_{lj} are the Christoffel symbols of the second kind of the parametric domain S^n in the coordinates $\xi^1,\ldots,\xi^n$. Note the equations (8.60), (8.62), and (8.63) are valid for an arbitrary coordinate system $\xi^1,\ldots,\xi^n$.

Let now the coordinates $\xi^1,\ldots,\xi^n$ be obtained by the solution of (5.4) with respect to the metric (8.59). Then applying (8.62) with $i = n$ and the condition $\Delta_B[\xi^n] = 0$ to (8.53) we obtain

8.2 Application of Theorem to Popular Elliptic Models

$$\frac{d}{dn_n^{s\xi}}\left(\frac{1}{\sqrt{g_{\xi s}^{nn}}}\right) = -\frac{1}{g_{\xi s}^{nn}}\nabla^2 \xi^n - \frac{n-1}{\sqrt{g_{\xi s}^{nn}}}K_m^{s\xi}(\xi^n)$$

$$= -\frac{1}{g_{\xi s}^{nn} g^s}\frac{\partial f}{\partial s^j}\frac{\partial f}{\partial s^k}\frac{\partial^2 \xi^n}{\partial s^j \partial s^k} \quad (8.64)$$

$$+ \frac{1}{g_{\xi s}^{nn}}\Delta_B[s^k]\frac{\partial \xi^n}{\partial s^k} - \frac{n-1}{\sqrt{g_{\xi s}^{nn}}}K_m^{s\xi}(\xi^n),$$

$$j, k = 1, \ldots, n.$$

This formula establishes a connection between the rate of change of the relative spacing of the grid hypersurface $\xi^n = const$ in S^n and the monitor function and the mean curvature of the hypersurface $\xi^n = c_0$.

Analogous formula, for the coordinate surfaces $\xi^i = const$, is obtained from (8.64) by substituting the fixed index i for n.

Now we establish a relation between the Christoffel symbols of the second kind of the monitor surface S^{rn} and domain S^n in the coordinates $\xi^1, \ldots, \xi^n$ designated as ${}^\xi \Upsilon^i_{jk}$ and Γ^i_{jk}, respectively. Using (6.16) with the identification $\xi^i = s^i$ $v(s) \equiv 1$ we find

$${}^\xi \Upsilon^i_{jk} = \Gamma^i_{jk} + \frac{1}{g^s}\frac{\partial f}{\partial s^p}\frac{\partial \xi^i}{\partial s^p}\nabla^\xi_{jk}(f), \quad i, j, k, l, m = 1, \ldots, n, \quad (8.65)$$

where

$$\nabla^\xi_{jk}(f) = (f_{\xi^j \xi^k} - f_{\xi^l}\Gamma^l_{jk})$$

$$= \left(f_{\xi^j \xi^k} - f_{\xi^l}\frac{\partial^2 s^m}{\partial \xi^j \partial \xi^k}\frac{\partial \xi^l}{\partial s^m}\right) \quad (8.66)$$

$$= \left(f_{\xi^j \xi^k} - \frac{\partial f}{\partial s^l}\frac{\partial^2 s^l}{\partial \xi^j \partial \xi^k}\right), \quad i, k, l, m = 1, \ldots, n,$$

is the mixed derivative of $f[s(\boldsymbol{\xi})]$ in the metric $(g_{ij}^{s\xi})$. Since $\nabla^\xi_{jk}(f)$ is a covariant tensor of the second rank we obtain

$$\nabla^\xi_{jk}(f) = f_{s^l}f_{s^p}\frac{\partial s^l}{\partial \xi^j}\frac{\partial s^p}{\partial \xi^k}, \quad j, k, l, p = 1, \ldots, n.$$

Further for g_ξ^{jk} we have from (8.61) and (6.10)

$$g_\xi^{jk} = g_{\xi s}^{jk} - \frac{1}{g^s}\frac{\partial f}{\partial s^m}\frac{\partial \xi^j}{\partial s^m}\frac{\partial f}{\partial s^l}\frac{\partial \xi^k}{\partial s^l}, \quad j, k, l, m = 1, \ldots, n,$$

therefore

216   8. Control of Grid Clustering

$$g_\xi^{jk\xi}\Upsilon_{jk}^i = \left(g_{\xi s}^{jk} - \frac{1}{g^s}\frac{\partial f}{\partial s^m}\frac{\partial \xi^j}{\partial s^m}\frac{\partial f}{\partial s^l}\frac{\partial \xi^k}{\partial s^l}\right)$$
$$\times \left(\Gamma_{jk}^i + \frac{1}{g^s}\frac{\partial f}{\partial s^p}\frac{\partial \xi^i}{\partial s^p}\nabla_{jk}^\xi(f)\right)$$
$$= g_{\xi s}^{jk}\Gamma_{jk}^i - \frac{1}{g^s}\frac{\partial f}{\partial s^m}\frac{\partial \xi^j}{\partial s^m}\frac{\partial f}{\partial s^l}\frac{\partial \xi^k}{\partial s^l}\Gamma_{jk}^i \qquad (8.67)$$
$$+ \frac{1}{g^s}\frac{\partial f}{\partial s^p}\frac{\partial \xi^i}{\partial s^p}\frac{\partial^2 f}{\partial s^m \partial s^m}$$
$$- \frac{1}{(g^s)^2}\frac{\partial f}{\partial s^m}\frac{\partial f}{\partial s^l}\frac{\partial^2 f}{\partial s^m \partial s^l}\frac{\partial f}{\partial s^p}\frac{\partial \xi^i}{\partial s^p},$$
$$i,j,k,l,m,p = 1,\ldots,n.$$

In accordance with (6.16)

$$^s\Upsilon_{ij}^k = -\frac{1}{g^s}\frac{\partial^2 f}{\partial s^i \partial s^j}\frac{\partial f}{\partial s^k}, \quad i,j,k = 1,\ldots,n,$$

so using (7.3) and (8.61) gives

$$-\Delta_B[s^p] = \frac{1}{g^s}g_s^{lm}\frac{\partial^2 f}{\partial s^l \partial s^m}\frac{\partial f}{\partial s^p}$$
$$= \frac{1}{g^s}\left(\delta_m^l - \frac{1}{g^s}\frac{\partial f}{\partial s^l}\frac{\partial f}{\partial s^m}\right)\frac{\partial^2 f}{\partial s^l \partial s^m}\frac{\partial f}{\partial s^p}$$
$$= \frac{1}{g^s}\left(\frac{\partial^2 f}{\partial s^m \partial s^m} - \frac{1}{g^s}\frac{\partial f}{\partial s^m}\frac{\partial f}{\partial s^l}\frac{\partial^2 f}{\partial s^m \partial s^l}\right)\frac{\partial f}{\partial s^p},$$
$$l,m,p = 1,\ldots,n,$$

and consequently equations (8.67) also have the form

$$g_\xi^{jk}\Upsilon_{jk}^i = g_{\xi s}^{jk}\Gamma_{jk}^i - \Delta_B[s^p]\frac{\partial \xi^i}{\partial s^p} - \frac{1}{g^s}\frac{\partial f}{\partial s^m}\frac{\partial \xi^j}{\partial s^m}\frac{\partial f}{\partial s^l}\frac{\partial \xi^k}{\partial s^l}\Gamma_{jk}^i,$$
$$i,j,k,l,m,p = 1,\ldots,n,$$

which is equivalent to (8.62) and (8.63). Hence, in the case of the grid coordinates, satisfying (5.4), we find

$$\nabla^2[\xi^i] = -\Delta_B[s^p]\frac{\partial \xi^i}{\partial s^p} - \frac{1}{g^s}\frac{\partial f}{\partial s^m}\frac{\partial \xi^j}{\partial s^m}\frac{\partial f}{\partial s^l}\frac{\partial \xi^k}{\partial s^l}\Gamma_{jk}^i, \qquad (8.68)$$
$$i,j,k,l,m,p = 1,\ldots,n,$$

Availing us of (8.68) we also obtain another form of (8.64), assuming $n = i$,

8.2 Application of Theorem to Popular Elliptic Models

$$\frac{\mathrm{d}}{\mathrm{d}\boldsymbol{n}_i^{s\xi}}\left(\frac{1}{\sqrt{g_{\xi_s}^{ii}}}\right) = \frac{1}{g_{\xi_s}^{ii}g^s}\frac{\partial f}{\partial s^p}\frac{\partial \xi^j}{\partial s^p}\frac{\partial f}{\partial s^l}\frac{\partial \xi^k}{\partial s^l}\Gamma_{jk}^i$$
$$+ \frac{1}{g_{\xi_s}^{ii}}\Delta_B[s^k]\frac{\partial \xi^i}{\partial s^k} - \frac{n-1}{\sqrt{g_{\xi_s}^{ii}}}K_m^{s\xi}(\xi^i)\,, \quad (8.69)$$
$$i,j,k,l,p = 1,\ldots,n\,, \quad i \text{ fixed}\,.$$

Using (7.66) and the relation

$$\mathrm{grad}\xi^i = \sqrt{g_{\xi_s}^{ii}}\boldsymbol{n}_i^{s\xi}\,, \quad i = 1,\ldots,n\,, \quad i \text{ fixed}\,,$$

we conclude that

$$\Delta_B[s^k]\frac{\partial \xi^i}{\partial s^k} = -n\frac{\sqrt{g_{\xi_s}^{ii}}}{\sqrt{g^s}}\mathrm{grad}f \cdot \boldsymbol{n}_i^{s\xi}K_m(S^{rn})\,, \quad i=1,\ldots,n\,, \quad i \text{ fixed}\,,$$

where the quantity

$$K_m(S^{rn}) = \frac{1}{n\sqrt{g^s}}g_s^{kj}f_{s^k s^j}\,, \quad j,k = 1,\ldots,n\,,$$

is, in accordance with (7.65), the mean curvature of the monitor surface S^{rn} in R^{n+1} with respect to the unit normal (7.64). Therefore (8.69) also has the following form

$$\frac{\mathrm{d}}{\mathrm{d}\boldsymbol{n}_i^{s\xi}}\left(\frac{1}{\sqrt{g_{\xi_s}^{ii}}}\right) = \frac{1}{g_{\xi_s}^{ii}g^s}\frac{\partial f}{\partial s^p}\frac{\partial \xi^j}{\partial s^p}\frac{\partial f}{\partial s^l}\frac{\partial \xi^k}{\partial s^l}\Gamma_{jk}^i$$
$$- \frac{1}{\sqrt{g_{\xi_s}^{ii}}}\left[\frac{nK_m(S^{rn})}{\sqrt{g^s}}\mathrm{grad}f \cdot \boldsymbol{n}_i^{s\xi} + (n-1)K_m^{s\xi}(\xi^i)\right], \quad (8.70)$$
$$i,j,k,l,p = 1,\ldots,n\,, \quad i \text{ fixed}\,.$$

Now we establish a relation between the mean curvature of the grid hypersurfaces $\xi^i = c_0$ on S^{rn} and $\xi^i = c_0$ in S^n. Let the mean curvature of this hypersurface on S^{rn} be designated by $K_m^{r\xi}(\xi^i)$. For the sake of simplicity, we assume $i = n$.

We designate by g_{sn}^{ij} and g_{rn}^{ij} the elements of the contravariant metric tensor in the coordinates $\xi^1,\ldots,\xi^{n-1}$ of the grid hypersurface $\xi^n = c_0$ in S^n and on S^{rn}, respectively. Then from (8.51) we obtain

$$g_{sn}^{ij} = \frac{1}{g_{\xi_s}^{nn}}(g_{\xi_s}^{nn}g_{\xi_s}^{ij} - g_{\xi_s}^{in}g_{\xi_s}^{jn})\,, \quad i,j = 1,\ldots,n-1\,,$$

$$g_{rn}^{ij} = \frac{1}{g_\xi^{nn}}(g_\xi^{nn}g_\xi^{ij} - g_\xi^{in}g_\xi^{jn})\,, \quad i,j = 1,\ldots,n-1\,.$$

Since the elements of the covariant metric tensor of the grid hypersurface $\xi^n = c_0$ on S^{rn} in the coordinates $\xi^1,\ldots,\xi^n$ are defined by

$$g_{ij}^\xi = \boldsymbol{r}_{\xi^i} \cdot \boldsymbol{r}_{\xi^j} = \boldsymbol{s}_{\xi^i} \cdot \boldsymbol{s}_{\xi^j} + \frac{\partial f}{\partial \xi^i}\frac{\partial f}{\partial \xi^j} = g_{ij}^{s\xi} + \frac{\partial f}{\partial \xi^i}\frac{\partial f}{\partial \xi^j}\,, \quad i,j = 1,\ldots,n-1\,,$$

we also find from (6.6) and (6.7)

$$g_{rn}^{ij} = g_{sn}^{ij} - \frac{1}{1+\nabla^n(f)} g_{sn}^{il} g_{sn}^{jp} \frac{\partial f}{\partial \xi^l} \frac{\partial f}{\partial \xi^p}, \quad i,j,k,l,m,p = 1,\ldots,n-1,$$

where

$$\nabla^n(f) = g_{sn}^{kl} \frac{\partial f}{\partial \xi^k} \frac{\partial f}{\partial \xi^l}, \quad k,l = 1,\ldots,n-1.$$

Therefore, availing us of (8.15), (8.55), and (8.65), we conclude that

$$\begin{aligned} K_m^{r\xi}(\xi^n) &= \frac{1}{(n-1)\sqrt{g_\xi^{nn}}} g_{rn}^{ij}\,{}^\xi\Upsilon_{ij}^n \\ &= \frac{1}{(n-1)\sqrt{g_\xi^{nn}}} \left(g_{sn}^{ij} - \frac{1}{1+\nabla^n(f)} g_{sn}^{il} g_{sn}^{jp} \frac{\partial f}{\partial \xi^l} \frac{\partial f}{\partial \xi^p} \right) \\ &\quad \times \left(\Gamma_{ij}^n + \frac{1}{g^s} \frac{\partial f}{\partial s^k} \frac{\partial \xi^n}{\partial s^k} \nabla_{\xi^i \xi^j}(f) \right) \\ &= \frac{\sqrt{g_{s\xi}^{nn}}}{\sqrt{g_\xi^{nn}}} K_m^{s\xi}(\xi^n) + \frac{1}{(n-1)\sqrt{g_\xi^{nn}}} \left[\frac{1}{g^s} \frac{\partial f}{\partial s^k} \frac{\partial \xi^n}{\partial s^k} \nabla_{\xi^i \xi^j}(f) g_{rn}^{ij} \right. \\ &\quad \left. - \frac{1}{1+\nabla^n(f)} g_{sn}^{il} g_{sn}^{jp} \frac{\partial f}{\partial \xi^l} \frac{\partial f}{\partial \xi^p} \Gamma_{ij}^n \right], \end{aligned} \quad (8.71)$$

$$i,j,l,p = 1,\ldots,n-1, \quad k = 1,\ldots,n.$$

Control of Grid Clustering near Boundary Segments. Let now the monitor function be specified as

$$f(s) = g[\varphi(s)], \quad s \in S^n, \quad (8.72)$$

where $\varphi(s)$ is such scalar-valued function that the $(n-1)$-dimensional surface defined from the equation $\varphi(s) = 0$ concides with the coordinate surface $\xi^i = c_0$ for some c_0. We designate this $(n-1)$-dimensional surface by $S_{c_0}^{n-1}$. We also assume that all expressions throughout this paragraph are considered at the points of the surface $S_{c_0}^{n-1}$ and

$$\mathrm{grad}\varphi(s) \cdot \nabla \xi^i(s) = \varphi_{s^k}(s) \frac{\partial \xi^i}{\partial s^k}(s) > 0, \quad s \in S_{c_0}^{n-1}, \quad (8.73)$$

$$i,k = 1,\ldots,n, \quad i \text{ fixed}.$$

With these assumptions it is readily found that

$$\boldsymbol{n}_i^{s\xi} = \mathrm{grad}\varphi/|\mathrm{grad}\varphi|,$$

$$\mathrm{grad} f = g'\mathrm{grad}\varphi,$$

and, consequently,

$$\mathrm{grad} f \cdot \boldsymbol{n}_i^{s\xi} = g'|\mathrm{grad}\varphi|.$$

Further, availing us of (4.106), we have

8.2 Application of Theorem to Popular Elliptic Models

$$K_m(S_{c_0}^{n-1}) = \frac{1}{(n-1)|\text{grad}\varphi|^3}(\varphi_{s^i}\varphi_{s^j}\varphi_{s^i s^j} - |\text{grad}\varphi|^2 \varphi_{s^l s^l}),$$

$$i, j, l = 1, \ldots, n,$$
(8.74)

where $K_m(S_{c_0}^{n-1})$ is the mean curvature of the surface $S_{c_0}^{n-1}$ in S^n.

Analogously we can compute the mean curvature $K_m(S^{rn})$ of the monitor surface S^{rn} in R^{n+1} with respect to the normal (4.77), using the formula (4.83). For this purpose we find, taking advantage of (8.72),

$$f_{s^i} = g'\varphi_{s^i}, \quad i = 1, \ldots, n,$$
$$g^s = 1 + \nabla(f) = 1 + (g')^2|\text{grad}\varphi|^2.$$
(8.75)

Similarly

$$f_{s^i s^j} = g'\varphi_{s^i s^j} + g''\varphi_{s^i}\varphi_{s^j}, \quad i, j = 1, \ldots, n,$$

so

$$\nabla^2[f] = g'\nabla^2[\varphi] + g''|\text{grad}\varphi|^2,$$

$$f_{s^i}f_{s^j}f_{s^i s^j} = (g')^2\varphi_{s^i}\varphi_{s^j}(g'\varphi_{s^i s^j} + g''\varphi_{s^i}\varphi_{s^j})$$
$$= (g')^3\varphi_{s^i}\varphi_{s^j}\varphi_{s^i s^j} + (g')^2 g''|\text{grad}\varphi|^4, \quad i, j = 1, \ldots, n.$$

Availing us of these relations and (8.74), we obtain

$$\nabla^2[f] - \frac{1}{g^s}f_{s^i}f_{s^j}f_{s^i s^j}$$
$$= \frac{1}{g^s}\{[1 + (g')^2|\text{grad}\varphi|^2][g'\nabla^2[\varphi] + g''|\text{grad}\varphi|^2]$$
$$- (g')^3\varphi_{s^i}\varphi_{s^j}\varphi_{s^i s^j} - (g')^2 g''|\text{grad}\varphi|^4\}$$
$$= \frac{1}{g^s}\{g'\nabla^2[\varphi] - (n-1)(g')^3|\text{grad}\varphi|^3 K_m(S_{c_0}^{n-1}) + g''|\text{grad}\varphi|^2\},$$
$$i, j = 1, \ldots, n.$$

Hence (4.105) gives

$$K_m(S^{rn}) = \frac{1}{n(g^s)^{3/2}}\{g'\nabla^2[\varphi]$$
$$- (n-1)(g')^3|\text{grad}\varphi|^3 K_m(S_{c_0}^{n-1}) + g''|\text{grad}\varphi|^2\},$$
(8.76)

Now using this equation we conclude that

$$\frac{n}{\sqrt{g^s}}\text{grad}f \cdot \boldsymbol{n}_i^{s\xi}K_m(S^{rn}) + (n-1)K_m(S_{c_0}^{n-1})$$
$$= \frac{ng'|\text{grad}\varphi|}{\sqrt{g^s}}K_m(S^{rn}) + (n-1)K_m(S_{c_0}^{n-1})$$
$$= \frac{g'|\text{grad}\varphi|}{(g^s)^2}\{g'\nabla^2[\varphi] + g''|\text{grad}\varphi|^2\}$$
$$+ (n-1)\left(1 - \frac{(g')^4|\text{grad}\varphi|^4}{(g^s)^2}\right)K_m(S_{c_0}^{n-1}), \quad i = 1, \ldots, n.$$
(8.77)

Further we note that in the case of (8.72) we obtain, availing us of the equation (8.49),

$$\frac{\partial f}{\partial s^p}\frac{\partial \xi^j}{\partial s^p}\frac{\partial f}{\partial s^l}\frac{\partial \xi^k}{\partial s^l}\Gamma^i_{jk} = \frac{\partial f}{\partial \xi^p}g_{\xi s}^{pj}\frac{\partial f}{\partial \xi^l}g_{\xi s}^{lk}\Gamma^i_{jk}$$

$$= \left(\frac{\partial f}{\partial \xi^i}\right)^2 g_{\xi s}^{ij} g_{\xi s}^{ik}\Gamma^i_{jk}$$

$$= (g_{\xi s}^{ii})^2 \left(\frac{\partial f}{\partial \xi^i}\right)^2 \frac{d}{dn_i^{s\xi}}\left(\frac{1}{\sqrt{g_{\xi s}^{ii}}}\right) \quad (8.78)$$

$$= g_{\xi s}^{ii}\nabla(f)\frac{d}{dn_i^{s\xi}}\left(\frac{1}{\sqrt{g_{\xi s}^{ii}}}\right),$$

$$i,j,k,l,p=1,\ldots,n, \; i \text{ fixed},$$

since (8.49) with $n = i$. Now, substituting (8.77) and (8.78) in (8.70), we find

$$\frac{d}{dn_i^{s\xi}}\left(\frac{1}{\sqrt{g_{\xi s}^{ii}}}\right) = -\frac{1}{\sqrt{g_{\xi s}^{ii}g^s}}b[s,g'(0),g''(0)], \quad (8.79)$$

$$i = 1,\ldots,n, \; i \text{ fixed},$$

where

$$b[s,g'(0),g''(0)] = g'(0)|\text{grad}\varphi|\{g'(0)\nabla^2[\varphi] + g''(0)|\text{grad}\varphi|^2\}$$

$$+ (n-1)[1 + 2(g'(0))^2|\text{grad}\varphi|^2]K_m(S_{c_0}^{n-1}).$$

Thus the positive (negative) sign of the expression $b[s,g'(0),g''(0)]$ indicates the grid clustering (rarefaction) near the coordinate surface $\varphi(s) = 0$. If this coordinate surface is a boundary surface of the domain S^n then the quantity $K_m(S_{c_0}^{n-1})$ is known and, by a suitable specification of the constants $a = g'(0)$ and $b = g''(0)$, we can realize the necessary sign of this expression. Consequently the monitor function $f = g(\varphi)$ in (8.72) can be chosen, for example, by

$$g(\varphi) = a\varphi + \frac{1}{2}b\varphi^2. \quad (8.80)$$

Another example of the monitor function gives the following formula

$$g(\varphi) = \frac{a^2}{b}\exp\left(\frac{b}{a}\varphi\right).$$

Let $g'(0)$ be specified such that $g'(0) \neq 0$ then

$$b[s,g'(0),g''(0)] > 0$$

if

$$g''(0) > \max_{s \in S_{c_0}^{n-1}} -\frac{1}{g'(0)|\text{grad}\varphi|^3}D_1, \quad (8.81)$$

where
$$D_1 = [g'(0)]^2|\text{grad}\varphi|\nabla^2[\varphi] + (n-1)[1 + 2[g'(0)]^2|\text{grad}\varphi|^2]K_m(S_{c_0}^{n-1}),$$
while
$$b[s, g'(0), g''(0)] < 0$$
if
$$g''(0) < \min_{s \in S_{c_0}^{n-1}} -\frac{1}{g'(0)|\text{grad}\varphi|^3} D_1. \tag{8.82}$$

For example, assuming $g'(0) = 1$, we readily find that
$$b[s, 1, g''(0)] > 0$$
if
$$g''(0) > \max_{s \in S_{c_0}^{n-1}} -\frac{1}{|\text{grad}\varphi|^3} D_2, \tag{8.83}$$
where
$$D_2 = |\text{grad}\varphi|\nabla^2[\varphi] + (n-1)(1 + 2|\text{grad}\varphi|^2)K_m(S_{c_0}^{n-1}).$$
Analogously
$$b[s, 1, g''(0)] < 0$$
if
$$g''(0) < \min_{s \in S_{c_0}^{n-1}} -\frac{1}{|\text{grad}\varphi|^3} D_2. \tag{8.84}$$

Since the terms in the formula (8.79) are computed locally, we can choose a single function of the form (8.72) to realize the necessary requirements of grid clustering near opposite segments of the boundary of S^n.

For example, let $\xi^i = c_0$ and $\xi^i = c_1$ be two opposite boundary segments of S^n found from the equations $\varphi_0(s) = 0$ and $\varphi_1(s) = 0$, respectively. Then assuming in (8.72) and (8.80)
$$\varphi(s) = \varphi_0(s)\varphi_1(s),$$
we can, analogously to (8.81) and (8.82), find $g'(0)$ and $g''(0)$ to provide the necessary grid behavior near these boundary segments.

Two-Dimensional Case. In the two-dimensional case the mean curvature in (8.69) is replaced by the curvature of the corresponding curvilinear line, in accordance with the relation,
$$K_m^{s\xi}(\xi^i) = k_i, \quad i = 1, 2,$$
where k_i is the curvature of the curve $\xi^i = c_0$ in S^2. We find from (8.71), in the two-dimensional case,

222   8. Control of Grid Clustering

$$\sigma_2 = \frac{\xi \Upsilon_{11}^2}{\sqrt{g_\xi^{22} g_{11}^\xi}} = \frac{\sqrt{g_{\xi s}^{22} g_{11}^{s\xi}}}{\sqrt{g_\xi^{22} g_{11}^\xi}} (k_2 + b_2) , \tag{8.85}$$

where σ_2 is the geodesic curvature of the curve $\xi^2 = c_0$ on S^{r2}, k_2 is the curvature of the curve $\xi^2 = c_0$ in S^2, while

$$b_2 = \frac{1}{g^s \sqrt{g_{\xi s}^{22} g_{11}^{s\xi}}} \frac{\partial f}{\partial s_k} \frac{\partial \xi_2}{\partial s_k} \frac{\partial^2 f}{\partial s_l \partial s_m} \frac{\partial s_m}{\partial \xi_1} \frac{\partial s_l}{\partial \xi_1} , \quad k, l, m = 1, 2 .$$

Thus

$$k_2 = \frac{\sqrt{g_\xi^{22}}}{\sqrt{g_{\xi s}^{22} g_{11}^{s\xi}}} \sigma_2 - b_2 , \tag{8.86}$$

where

$$k_2 = \frac{1}{\sqrt{g_{\xi s}^{22} g_{11}^{s\xi}}} \frac{\partial^2 s^l}{\partial \xi^1 \partial \xi^1} \frac{\partial \xi^2}{\partial s^l} = \frac{1}{\sqrt{g_{\xi s}^{22} g_{11}^{s\xi}}} \Upsilon_{11}^2 , \tag{8.87}$$

is the curvature of the coordinate line $\xi^2 = \text{const}$ in the parametric domain S^2. Substituting this expression in (8.56) gives

$$\frac{d}{dn_2^{s\xi}} \left(\frac{1}{\sqrt{g_\xi^{22}}} \right) = \frac{1}{g_{\xi s}^{22}} \Delta_B[s_k] \frac{\partial \xi^2}{\partial s^k} - \frac{\sqrt{g_\xi^{22}}}{g_{\xi s}^{22} g_{11}^{s\xi}} \sigma_2 + \frac{b_2}{\sqrt{g_{\xi s}^{22}}} \\ - \frac{1}{g_{\xi s}^{22} g^s} \frac{\partial f}{\partial s^j} \frac{\partial f}{\partial s^k} \frac{\partial^2 \xi^2}{\partial s^j \partial s^k} , \quad k = 1, 2 . \tag{8.88}$$

Since

$$\frac{b_2}{\sqrt{g_{\xi s}^{22}}} - \frac{1}{g_{\xi s}^{22} g^s} \frac{\partial f}{\partial s^j} \frac{\partial f}{\partial s^k} \frac{\partial^2 \xi^2}{\partial s^j \partial s^k} = \frac{1}{g^s g_{\xi s}^{22} g_{11}^{s\xi}} \left(\frac{\partial f}{\partial s^k} \frac{\partial \xi^2}{\partial s^k} \frac{\partial^2 f}{\partial s^l \partial s^m} \frac{\partial s^m}{\partial \xi^1} \frac{\partial s^l}{\partial \xi^1} \right. \\ \left. - g_{11}^{s\xi} \frac{\partial f}{\partial s^j} \frac{\partial f}{\partial s^k} \frac{\partial^2 \xi^2}{\partial s^j \partial s^k} \right) ,$$

$$j, k, l, m = 1, 2 ,$$

so in contrast to (8.45) a sign of the change rate of the grid spacing near a boundary segment in S^2 is not determined by the curvature of this segment only; it depends as well on the derivatives of the monitor function $f(\boldsymbol{s})$. Therefore one can control the grid spacing near the boundary of S^2 with a proper choice of this function.

Thus using (8.88) with the choice of the monitor function $f(s^1, s^2)$ and the boundary transformation $\boldsymbol{s}(\xi^1, \text{const})$ we can, in principle, provide grid nodes clustering near the boundary curve $\xi^2 = 0$ on S^{r2} or in S^2, respectively, regardless of its geometry in S^2. Analogous results are obtained for the grid lines near the boundaries $\xi^2 = 1$ and $\xi^1 = 1$ and $\xi^1 = 0$.

Control by a Vector-Valued Monitor Function.

General Formula. Let now the monitor surface over S^n be formed by the vector-valued function $\boldsymbol{f}(\boldsymbol{s}) : \boldsymbol{f} = (f^1, f^2)$, i.e. S^{rn} in parametrized as

$$\boldsymbol{r}(\boldsymbol{s}) : S^n \to R^{n+2}, \quad \boldsymbol{r}(\boldsymbol{s}) = [\boldsymbol{s}, \boldsymbol{f}(\boldsymbol{s})] . \tag{8.89}$$

Then the elements of the covariant metric tensor of S^{rn} in the coordinates $s^1, \ldots, s^n$ are as follows:

$$g^s_{ij} = \delta^i_j + \frac{\partial f^1}{\partial s^i}\frac{\partial f^1}{\partial s^j} + \frac{\partial f^2}{\partial s^i}\frac{\partial f^2}{\partial s^j}, \quad i,j = 1,\ldots,n , \tag{8.90}$$

and, availing us of (6.48), as well as (6.52) for $l = 2$, we have the following expression for the elements of the contravariant metric tensor of S^{rn} in the coordinates $s^1, \ldots, s^n$

$$g^{ij}_s = \delta^i_j - \frac{1}{g^s} v^{ij}, \quad i,j = 1,\ldots,n , \tag{8.91}$$

where

$$g^s = [1 + \nabla(f^1)][1 + \nabla(f^2)] - [\nabla(f^1, f^2)]^2 ,$$

$$\nabla(f) = \frac{\partial f}{\partial s^i}\frac{\partial f}{\partial s^i}, \quad \nabla(f^1, f^2) = \frac{\partial f^1}{\partial s^i}\frac{\partial f^2}{\partial s^i}, \quad i = 1,\ldots,n ,$$

$$v^{ij} = [1 + \nabla(f^1)]\frac{\partial f^2}{\partial s^i}\frac{\partial f^2}{\partial s^j} + [1 + \nabla(f^2)]\frac{\partial f^1}{\partial s^i}\frac{\partial f^1}{\partial s^j}$$

$$- \nabla(f^1, f^2)\left(\frac{\partial f^1}{\partial s^i}\frac{\partial f^2}{\partial s^j} + \frac{\partial f^2}{\partial s^i}\frac{\partial f^1}{\partial s^j}\right), \quad i,j = 1,\ldots,n .$$

Therefore

$$g^{jk}_s \frac{\partial^2 \xi^i}{\partial s^j \partial s^k} = \nabla^2[\xi^i] - \frac{1}{g^s} v^{jk} \frac{\partial^2 \xi^i}{\partial s^j \partial s^k}, \quad i,j,k = 1,\ldots,n ,$$

and applying (2.49)

$$g^{jk}_s \frac{\partial^2 \xi^i}{\partial s^j \partial s^k} = \nabla^2[\xi^i] + \frac{1}{g^s} v^{jk} \frac{\partial \xi^l}{\partial s^j}\frac{\partial \xi^p}{\partial s^k} \Gamma^i_{lp}, \quad i,j,k,l,p = 1,\ldots,n , \tag{8.92}$$

We note now that equations (8.60) are valid for an arbitrary metric of S^{rn}, in particular, for that specified by (8.90). Hence (8.60) and (8.92) yield

$$\Delta_B[\xi^i] = \nabla^2[\xi^i] + \Delta_B[s^k]\frac{\partial \xi^i}{\partial s^k} + \frac{1}{g^s} v^{jk} \frac{\partial \xi^l}{\partial s^j}\frac{\partial \xi^p}{\partial s^k} \Gamma^i_{lp} ,$$

$$i,j,k,l,p = 1,\ldots,n ,$$

where Δ_B is the Beltramian operator with respect to the metric (8.90). So, similarly to (4.63) and (8.69), we find, at the points of the grid hypersurface $\xi^i = c_0$ in S^n,

8. Control of Grid Clustering

$$\frac{d}{dn_i^{s\xi}}\left(\frac{1}{\sqrt{g_{\xi s}^{ii}}}\right) = \frac{1}{g_{\xi s}^{ii} g^s} v^{jk} \frac{\partial \xi^l}{\partial s^j} \frac{\partial \xi^p}{\partial s^k} \Gamma_{lp}^i + \frac{1}{g_{\xi s}^{ii}} \Delta_B[s^k] \frac{\partial \xi^i}{\partial s^k}$$

$$- \frac{n-1}{\sqrt{g_{\xi s}^{ii}}} K_m^{s\xi}(\xi^i) , \qquad (8.93)$$

$$i, j, k, l, p = 1, \ldots, n , \; i \text{ fixed} .$$

From (7.2), (4.28), and (8.91) we obtain

$$\Delta_B[s^k] = -g_s^{jp} \Upsilon_{jp}^k = -g_s^{jp} g_s^{kl} (\boldsymbol{r}_{s^j s^p} \cdot \boldsymbol{r}_{s^l})$$

$$= -g_s^{jp} g_s^{kl} \left(\frac{\partial^2 f^1}{\partial s^j \partial s^p} \frac{\partial f^1}{\partial s^l} + \frac{\partial^2 f^2}{\partial s^j \partial s^p} \frac{\partial f^2}{\partial s^l} \right) , \qquad j, k, l, p = 1, \ldots, n .$$

As

$$g_s^{kl} \frac{\partial f^1}{\partial s^l} = \left(\delta_l^k - \frac{1}{g^s} v^{kl} \right) \frac{\partial f^1}{\partial s^l} = \frac{\partial f^1}{\partial s^k} - \frac{1}{g^s} \left\{ [1 + \nabla(f^1)] \nabla(f^1 f^2) \frac{\partial f^2}{\partial s^k} \right.$$

$$+ [1 + \nabla(f^2)] \nabla(f^1) \frac{\partial f^1}{\partial s^k} - \nabla(f^1 f^2) \left[\frac{\partial f^1}{\partial s^k} \nabla(f^1 f^2) + \frac{\partial f^2}{\partial s^k} \nabla(f^1) \right] \right\}$$

$$= \frac{1}{g^s} \left\{ [1 + \nabla(f^2)] \frac{\partial f^1}{\partial s^k} - \nabla(f^1 f^2) \frac{\partial f^2}{\partial s^k} \right\} , \qquad k, l = 1, \ldots, n ,$$

and analogously

$$g_s^{kl} \frac{\partial f^2}{\partial s^l} = \frac{1}{g^s} \left\{ [1 + \nabla(f^1)] \frac{\partial f^2}{\partial s^k} - \nabla(f^1 f^2) \frac{\partial f^1}{\partial s^k} \right\} , \qquad k, l = 1, \ldots, n .$$

Therefore

$$\Delta_B[s^k] = -\frac{1}{g^s} g_s^{jp} b_{jp}^k , \qquad j, k, p = 1, \ldots, n ,$$

where

$$b_{jp}^k = \frac{\partial^2 f^1}{\partial s^j \partial s^p} \left\{ [1 + \nabla(f^2)] \frac{\partial f^1}{\partial s^k} - \nabla(f^1, f^2) \frac{\partial f^2}{\partial s^k} \right\}$$

$$+ \frac{\partial^2 f^2}{\partial s^j \partial s^p} \left\{ [1 + \nabla(f^1)] \frac{\partial f^2}{\partial s^k} - \nabla(f^1, f^2) \frac{\partial f^1}{\partial s^k} \right\} ,$$

$$j, k, p = 1, \ldots, n .$$

Hence (8.93) also has the following form

$$\frac{d}{dn_i^{s\xi}}\left(\frac{1}{\sqrt{g_{\xi s}^{ii}}}\right) = \frac{1}{g_{\xi s}^{ii} g^s} v^{jk} \frac{\partial \xi^l}{\partial s^j} \frac{\partial \xi^p}{\partial s^k} \Gamma_{lp}^i - \frac{1}{g_{\xi s}^{ii} g^s} g_s^{jp} b_{jp}^k \frac{\partial \xi^i}{\partial s^k}$$

$$- \frac{n-1}{\sqrt{g_{\xi s}^{ii}}} K_m^{s\xi}(\xi^i) , \qquad (8.94)$$

$$j, k, l, p = 1, \ldots, n , \; i \text{ fixed} .$$

8.2 Application of Theorem to Popular Elliptic Models 225

Particular Specification of the Monitor Functions. Let us, analogously to (8.72), suggest that

$$f^1(s) = g_1[\varphi(s)], \quad f^2(s) = g_2[\varphi(s)], \quad s \in S^n, \tag{8.95}$$

where the scalar-valued function $\varphi(s)$ satisfies (8.73) and the grid hypersurface $\xi^i = c_0$ in S^n designated by $S^{n-1}_{c_0}$ coincides with the surface $\varphi(s) = 0$. Then at the points of this hypersurface we have

$$\frac{\partial f^\alpha}{\partial \xi^j} = \frac{\partial f^\alpha[s(\xi)]}{\partial \xi^j} = 0, \quad \alpha = 1,2, \quad j \neq i,$$

$$\frac{\partial \varphi}{\partial \xi^j} = \frac{\partial \varphi[s(\xi)]}{\partial \xi^j} = 0, \quad j \neq i,$$

$$\frac{\partial^2 f^\alpha}{\partial s^j \partial s^k} = g''_\alpha(0)\varphi_{s^j}\varphi_{s^k} + g'_\alpha(0)\varphi_{s^j s^k}, \quad \alpha = 1,2, \quad j,k = 1,\ldots,n,$$

$$g^s = 1 + \{[g'_1(0)]^2 + [g'_2(0)]^2\}\nabla(\varphi),$$

$$\nabla(f^\alpha, f^\beta) = g'_\alpha(0)g'_\beta(0)\nabla(\varphi), \quad \alpha,\beta = 1,2,$$

$$v^{ij} = \{[g'_1(0)]^2 + [g'_2(0)]^2\}\varphi_{s^i}\varphi_{s^j}, \quad i,j = 1,\ldots,n,$$

$$b^k_{jp} = g'_\alpha(0)\varphi_{s^k}[g''_\alpha(0)\varphi_{s^j}\varphi_{s^p} + g'_\alpha(0)\varphi_{s^j s^p}], \quad \alpha = 1,2, \quad j,k,p = 1,\ldots,n.$$

Further using the relation (8.49), as for proving (8.78), we find

$$\varphi_{s^j}\varphi_{s^k}\frac{\partial \xi^l}{\partial s^j}\frac{\partial \xi^p}{\partial s^k}\Gamma^i_{lp} = \frac{\partial \varphi}{\partial \xi^i}\frac{\partial \varphi}{\partial \xi^i}g^{il}_{\xi s}g^{ip}_{\xi s}\Gamma^i_{lp}$$

$$= g^{ii}_{\xi s}\nabla(\varphi)\frac{\mathrm{d}}{\mathrm{d}\boldsymbol{n}^{s\xi}_i}\left(\frac{1}{\sqrt{g^{ii}_{\xi s}}}\right),$$

$$\alpha,\beta = 1,2, \quad j,l,k,p = 1,\ldots,n\,; \quad i \text{ fixed}.$$

Therefore

$$v^{jk}\frac{\partial \xi^l}{\partial s^j}\frac{\partial \xi^p}{\partial s^k}\Gamma^i_{lp} = g^{ii}_{\xi s}\nabla(\varphi)\{[g'_1(0)]^2 + [g'_2(0)]^2\}\frac{\mathrm{d}}{\mathrm{d}\boldsymbol{n}^{s\xi}_i}\left(\frac{1}{\sqrt{g^{ii}_{\xi s}}}\right),$$

$$i,j,k,l,p = 1,\ldots,n, \quad i \text{ fixed},$$

$$g^{jp}_s b^k_{jp}\frac{\partial \xi^i}{\partial s^k} = \sqrt{g^{ii}_{\xi s}}\sqrt{\nabla(\varphi)}g'_\alpha(0)g^{jp}_s[g''_\alpha(0)\varphi_{s^j}\varphi_{s^p} + g'_\alpha(0)\varphi_{s^j s^p}]$$

$$= \sqrt{g^{ii}_{\xi s}}\sqrt{\nabla(\varphi)}g'_\alpha(0)\{g''_\alpha(0)\nabla(\varphi) + g'_\alpha(0)\nabla^2[\varphi]$$

$$- \frac{1}{g^s}[g'_1(0) + g'_2(0)][g''_\alpha(0)[\nabla(\varphi)]^2 + g'_\alpha(0)\varphi_{s^j}\varphi_{s^p}\varphi_{s^j s^p}]\}$$

$$= \frac{\sqrt{g^{ii}_{\xi s}}\sqrt{\nabla(\varphi)}}{g^s} g'_\alpha(0)\{g''_\alpha(0)\nabla(\varphi) + g'_\alpha(0)\nabla^2[\varphi]$$

$$- (n-1)g'_\alpha(0)[\nabla(\varphi)]^{3/2}[g'_1(0) + g'_2(0)]K_m(S^{n-1}_{c_0}),$$

$$i,j,k,l,p = 1,\ldots,n\,;\quad \alpha,\beta = 1,2\,,\quad i\text{ fixed}\,.$$

So, from (8.94) we obtain

$$\frac{\mathrm{d}}{\mathrm{d}\boldsymbol{n}^{s\xi}_i}\left(\frac{1}{\sqrt{g^{ii}_{\xi s}}}\right) = -\frac{1}{\sqrt{g^{ii}_{\xi s}g^s}} b[\boldsymbol{s},g'_1(0),g'_2(0),g''_1(0),g''_2(0)]\,,\ i\text{ fixed}\,, \quad (8.96)$$

where

$$b[\boldsymbol{s},g'_1(0),g'_2(0),g''_1(0),g''_2(0)] = g'_\alpha(0)\sqrt{\nabla(\varphi)}[g'_\alpha(0)\nabla^2(\varphi) + g''_\alpha(0)\nabla(\varphi)]$$

$$+ (n-1)[1 + 2g'_\alpha(0)g'_\alpha(0)\nabla(\varphi)]K_m(S^{n-1}_{c_0})\,,\quad \alpha = 1,2\,.$$

Analogously to the case discussed for the scalar-valued monitor function we can specify the constants $g'_\alpha(0)$ and $g''_\alpha(0)$, $\alpha = 1,2$, in order to maintain a necessary sign of the function $b[\boldsymbol{s},g'_1(0),g'_2(0),g''_1(0),g''_2(0)]$ at the points of the boundary surface $S^{n-1}_{c_0}$ thus realizing grid clustering (rarefaction) near this boundary segment when $b > 0 (b < 0)$.

Control by Weight Functions.

Multidimensional Case. If the grid coordinate functions $\xi^1(\boldsymbol{s}),\ldots,\xi^n(\boldsymbol{s})$ in a domain S^n are found by the solution of the popular Poisson system

$$\nabla^2[\xi^i] = P^i(\boldsymbol{s})\,,\quad i = 1,\ldots,n\,, \tag{8.97}$$

then, availing us of (8.53) with $n = i$, we find

$$\frac{\mathrm{d}}{\mathrm{d}\boldsymbol{n}^{s\xi}_i}\left(\frac{1}{\sqrt{g^{ii}_{\xi s}}}\right) = -\frac{1}{\sqrt{g^{ii}_{\xi s}}}\left(\frac{P^i}{\sqrt{g^{ii}_{\xi s}}} + (n-1)K^{s\xi}_m(\xi^i)\right)\,, \tag{8.98}$$

$$i = 1,\ldots,n\,,\ i\text{ fixed}\,.$$

So, if the weight function P^i in (8.97) is specified at the points of a boundary segment $\xi^i = c_0$ such that

$$P^i + (n-1)\sqrt{g^{ii}_{\xi s}}K^{s\xi}_m(\xi^i) > 0(< 0)\,,\quad i = 1,\ldots,n\,,\ i\text{ fixed}\,,$$

then the grid surfaces $\xi^i = const$ cluster to (repell from) this boundary segment.

For example, assuming $P^i = -c\sqrt{g^{ii}_{\xi s}}$, i fixed, where

$$-c > \max_{\xi|_{\xi^i=0}}(n-1)K^{s\xi}_m(\xi^i)\,,$$

we obtain, by the solution of (8.97), the grid whose nodes cluster near the boundary segment $\xi^i = 0$ of S^n.

Analogously for the boundary segment $\xi^i = l_i$ the grid nodes cluster (rarefy) if $v_i < 0$ ($v_i > 0$). In particular, if $P^i = c\sqrt{g^{ii}_{\xi s}}$, i fixed, where

$$c > \max_{\xi|_{\xi^i=l_i}} (n-1) K^{s\xi}_m(\xi^i) ,$$

then the grid nodes cluster near the boundary segment $\xi^i = l_i$ of S^n.

Two-Dimensional Case. In the two-dimensional case equation (8.98) for the curve $\xi^2 = c_0$ is transformed to

$$\frac{\mathrm{d}}{\mathrm{d}\boldsymbol{n}^{s\xi}_2}\left(\frac{1}{\sqrt{g^{22}_{\xi s}}}\right) = -\frac{1}{\sqrt{g^{22}_{\xi s}}}\left(\frac{P^2}{\sqrt{g^{22}_{\xi s}}} + k_2\right) , \qquad (8.99)$$

where k_2 is the curvature of the coordinate line $\xi^2 = c_0$ in S^2. Analogous equation is obtained for the grid lines $\xi^1 = c_0$:

$$\frac{\mathrm{d}}{\mathrm{d}\boldsymbol{n}^{s\xi}_1}\left(\frac{1}{\sqrt{g^{11}_{\xi s}}}\right) = -\frac{1}{\sqrt{g^{11}_{\xi s}}}\left(\frac{P^1}{\sqrt{g^{11}_{\xi s}}} + k_1\right) , \qquad (8.100)$$

where k_1 is the curvature of the coordinate line $\xi^1 = c_0$ in the parametric domain S^2.

Thus, if the weight function P^i in (8.97) with $n = 2$ is specified at the points of a boundary curve $\xi^i = c_0$ such that

$$P^i + \sqrt{g^{ii}_{\xi s}} k_i > 0 (< 0) , \quad i = 1, 2 , \quad i \text{ fixed} ,$$

then the grid curves cluster to (repell from) this boundary curve.

8.2.2 Diffusion Equations

The third popular approach for generating grids is based on the solution of diffusion equations

$$\frac{\partial}{\partial s^j}\left[w(\boldsymbol{s})\frac{\partial \xi^i}{\partial s^j}\right] = 0 , \quad i, j = 1, \ldots n , \qquad (8.101)$$

where $w(\boldsymbol{s}) > 0$ is a diffusion function specified by the user to provide grid control. Equations (8.101) were proposed for $n = 2$ by Danaev N.T., Liseikin V.D., and Yanenko N.N. (1980) and Winslow A.M. (1981) for the generation of adaptive grids in domains.

The transformed equations, for finding the intermediate map (5.2), in which the dependent and independent variables are mutually altered have, in accordance with (7.15), the following form

$$g^{kj}_{\xi s}\frac{\partial}{\partial \xi^j}\left(\frac{1}{w(\boldsymbol{s})}\frac{\partial s^l}{\partial \xi^k}\right) = 0 , \quad j, k, l = 1, \ldots n , \qquad (8.102)$$

where

$$g^{kj}_{\xi s} = \frac{\partial \xi^k}{\partial s^i}\frac{\partial \xi^j}{\partial s^i}, \quad i,j,k = 1,\ldots n\,.$$

If S^{xn} is a domain S^n with the Eucledian metric then the operator of Beltrami Δ_B coincides with the Laplace operator ∇^2 and besides this

$$\mathrm{grad}\xi^p = \left(\frac{\partial \xi^p}{\partial s^1},\ldots,\frac{\partial \xi^p}{\partial s^n}\right) = \sqrt{g^{pp}_{\xi x}}\boldsymbol{n}_p\,,\quad p = 1,\ldots,n\,,\quad p\text{ fixed}\,.$$

Therefore we obtain, if (8.101) is held,

$$\nabla^2[\xi^p] = -\frac{\sqrt{g^{pp}_{\xi x}}}{w}\mathrm{grad}w\cdot\boldsymbol{n}_p\,,\quad p = 1,\ldots,n\,,\quad p\text{ fixed}\,,$$

and consequently formulas (8.37) and (8.101) yield

$$v_p = \frac{1}{\sqrt{g^{pp}_{\xi x}}}\left(\frac{1}{w}\mathrm{grad}w\cdot\boldsymbol{n}_p - (n-1)K_m(\xi^p)\right)\,, \quad (8.103)$$
$$p = 1,\ldots,n\,,\quad p\text{ fixed}\,.$$

The normal $\boldsymbol{n}_p$ and the mean curvature of the boundary of the domain S^n is known beforehand so the characteristic v_p of the rate of change of the grid spacing near a boundary segment is defined explicitly by the diffusion function $w(\boldsymbol{s})$ in the vicinity of the segment. Thus the grid points cluster (rarefy) near the boundary segment $\xi^p = 0$ if $w(\boldsymbol{s})$ is such that $v_p > 0$ ($v_p < 0$) in (8.103) at its points.

The specification of the diffusion function $w(\boldsymbol{s})$ is facilitated if it is sought in the form

$$w(\boldsymbol{s}) = g[\varphi(\boldsymbol{s})]$$

where $\varphi(\boldsymbol{s})$ is a scalar-valued function such that the equation $\varphi(\boldsymbol{s}) = 0$ determines the boundary hypersurface $\xi^p = 0$. The vector

$$\mathrm{grad}\varphi = (\varphi_{s^1},\ldots,\varphi_{s^n})$$

is orthogonal to this hypersurface and consequently parallel to $\boldsymbol{n}_p$. We assume that $\varphi(\boldsymbol{s})$ is such that the both vectors $\mathrm{grad}\varphi$ and $\boldsymbol{n}_p$ are of the same direction, i.e.

$$\boldsymbol{n}_p = \frac{1}{|\mathrm{grad}\varphi|}\mathrm{grad}\varphi\,.$$

Since

$$\mathrm{grad}w = g'\mathrm{grad}\varphi,$$

we readily obtain from (8.103)

$$v_p = \frac{1}{\sqrt{g^{pp}_{\xi x}}}\left(\frac{g'}{g}|\mathrm{grad}\varphi| - (n-1)K_m(\xi^p)\right)\,,\quad p = 1,\ldots,n\,,\quad p\text{ fixed}\,.$$

Thus $v_p > 0$ (for grid clustering) if $g(\varphi)\neq 0$ and

$$\frac{g'(0)}{g(0)} > \max_{s|_{\varphi(s)=0}} \frac{n-1}{|\operatorname{grad}\varphi|} K_m(\xi^p) ,$$

while $g(\varphi)$ we can, for example, specify as

$$g(\varphi) = g(0) + g'(0)\varphi$$

or

$$g(\varphi) = g(0) \exp\left[\frac{g'(0)}{g(0)}\varphi\right] .$$

Note for the hypersurface $\varphi(s) = 0$ in S^n the invariant $K_m(\xi^p)$ can, in accordance with (4.111), be computed by the following formula

$$K_m(\xi^p) = -\frac{1}{n-1}\frac{\partial}{\partial s^j}\left(\frac{1}{\sqrt{\nabla(\varphi)}}\varphi_{s^j}\right) , \quad j = 1, \ldots, n ,$$

where

$$\nabla(\varphi) = \varphi_{s^k}\varphi_{s^k} , \quad k = 1, \ldots, n .$$

Two-Dimensional Case. In the two-dimensional case we, analogously to (8.103), have for the coordinate grid curve $\xi^2 = c_0$

$$\frac{\mathrm{d}}{\mathrm{d}\boldsymbol{n}_2^{s\xi}}\left(\frac{1}{\sqrt{g_{\xi s}^{22}}}\right) = \frac{1}{\sqrt{g_{\xi s}^{22}}}\left(\frac{1}{w}\operatorname{grad}w \cdot \boldsymbol{n}_2^{s\xi} - k_2\right) . \tag{8.104}$$

Similarly, for the curve $\xi^1 = c_0$, we find

$$\frac{\mathrm{d}}{\mathrm{d}\boldsymbol{n}_1^{s\xi}}\left(\frac{1}{\sqrt{g_{\xi s}^{11}}}\right) = \frac{1}{\sqrt{g_{\xi s}^{11}}}\left(\frac{1}{w}\operatorname{grad}w \cdot \boldsymbol{n}_1^{s\xi} - k_1\right) . \tag{8.105}$$

Thus the grid lines $\xi^i = const$ obtained by the solution of (8.102) with $n = 2$ cluster to (repell from) a boundary curve $\xi^i = c_0$ if the following inequality is valid at the points of this curve:

$$\frac{1}{w}\operatorname{grad}w \cdot \boldsymbol{n}_i^{s\xi} - k_i < 0 (> 0) , \quad i = 1, 2 ,$$

which can be realized by a proper choice of the diffusion function $w(\boldsymbol{s})$.

8.2.3 Control of Grid Spacing near Boundaries of Physical Surfaces

Multidimensional Relations. We consider now a regular surface S^{xn} with a local parametrization

$$\boldsymbol{x}(\boldsymbol{s}) : S^n \to R^{n+k} , \quad k \geq 0 , \tag{8.106}$$

and corresponding metric (g_{ij}^{xs}) in the coordinates $s^1, \ldots, s^n$:

$$g_{ij}^{xs} = \boldsymbol{x}_{s^i} \cdot \boldsymbol{x}_{s^j} , i, j = 1, \ldots, n . \tag{8.107}$$

8. Control of Grid Clustering

Let

$$s(\boldsymbol{\xi}) : \Xi^n \to S^n \tag{8.108}$$

be an intermediate one-to-one smooth transformation specifying the necessary local grid on S^{xn}. As was emphasized, the original grid equations in the popular elliptic grid algorithms are formulated more easily with respect to the inverse transformation

$$\boldsymbol{\xi}(\boldsymbol{s}) : S^n \to \Xi^n \ . \tag{8.109}$$

Then the mapping (8.108) is obtain as a solution of the transformed equations in which the dependent and independent variables are mutually interchanged.

The composition of the parametrization $\boldsymbol{x}(\boldsymbol{s})$ and the intermediate mapping $\boldsymbol{s}(\boldsymbol{\xi})$ defines both a new parametrization of S^{xn}:

$$\boldsymbol{x}[\boldsymbol{s}(\boldsymbol{\xi})] : \Xi^n \to R^{n+k} \tag{8.110}$$

and the elements of the covariant metric temsor $(g_{ij}^{x\xi})$ of S^{xn} in the coordinates $\xi^1, \ldots, \xi^n$:

$$g_{ij}^{x\xi} = \frac{\partial \boldsymbol{x}[\boldsymbol{s}(\boldsymbol{\xi})]}{\partial \xi^i} \cdot \frac{\partial \boldsymbol{x}[\boldsymbol{s}(\boldsymbol{\xi})]}{\partial \xi^j} \ , \quad i,j = 1 \ldots, n \ . \tag{8.111}$$

Formula (8.16) for the rate of change of the relative spacing of the coordinate hypersurfaces $\xi^i = const$ in S^{xn} is readily rewritten with respect to arbitrary coordinates $\xi^1, \ldots, \xi^n$, namely, as

$$\frac{\mathrm{d}}{\mathrm{d} n_i}\left(\frac{1}{\sqrt{g_{\xi x}^{ii}}}\right) = -\frac{1}{g_{\xi x}^{ii}} \Delta_B[\xi^i] - \frac{n-1}{\sqrt{g_{\xi x}^{ii}}} K_m^{x\xi}(\xi^i) \ , \tag{8.112}$$

$$i = 1, \ldots, n \ , \quad i \text{ fixed} \ ,$$

where $g_{\xi x}^{ij}$ is the (ij)th element of the contravariant metric tensor of S^{xn} in the coordinates $\xi^1, \ldots, \xi^n$,

$$\boldsymbol{n}_i = \frac{1}{\sqrt{g_{\xi x}^{ii}}} g_{\xi x}^{ij} \frac{\partial}{\partial \xi^j} \boldsymbol{x}[\boldsymbol{s}(\boldsymbol{\xi})] \ , \quad i,j = 1, \ldots, n \ , \quad i \text{ fixed} \ , \tag{8.113}$$

$$\Delta_B[\xi^i] = -g_{\xi x}^{lj} \Upsilon_{lj}^i \ , \quad i,j,l = 1, \ldots, n \ ,$$

$K_m^{x\xi}(\xi^i)$ is the mean curvature of the $(n-1)$-dimensional coordinate hypersurface $\xi^i = c_0$ on S^{xn}, Δ_B is the Beltramian operator with respect to the metric (8.107). If this coordinate surface is a boundary surface of S^{xn} then the quantity $K_m^{x\xi}(\xi^i)$ is known, therefore, the rate of change of the relative spacing of the coordinate hypersurfaces $\xi^i = const$ near this boundary surface is dependent on the quantity $\Delta_B[\xi^i]$.

8.2 Application of Theorem to Popular Elliptic Models

Two-Dimensional Case. In the two-dimensional case

$$K_m^{x\xi}(\xi^i) = \sigma_i, \quad i = 1, 2,$$

where σ_i is the geodesic curvature of the coordinate curve $\xi^i = c_0$ on S^{x2}. Therefore the system (8.112) in the two-dimensional case is transformed to

$$\frac{d}{dn_1}\left(\frac{1}{\sqrt{g_{\xi x}^{11}}}\right) = -\frac{1}{g_{\xi x}^{11}}\Delta_B[\xi^1] - \frac{1}{\sqrt{g_{\xi x}^{11}}}\sigma_1,$$

$$\frac{d}{dn_2}\left(\frac{1}{\sqrt{g_{\xi x}^{22}}}\right) = -\frac{1}{g_{\xi x}^{22}}\Delta_B[\xi^2] - \frac{1}{\sqrt{g_{\xi x}^{22}}}\sigma_2.$$
(8.114)

Control by Scalar-Valued Monitor Functions. Now we discuss a situation when the coordinate functions $\xi^i(s)$ on S^{xn} are defined by a solution of the problem (5.4) with respect to the metric of a monitor surface over S^{xn}.

Multidimensional Case for Arbitrary Coordinates. Let the monitor surface S^{rn} over S^{xn} be formed by a scalar-valued monitor function $f(s)$, i.e. S^{rn} is parametrized in the coordinates $s^1, \ldots, s^n$ as

$$\boldsymbol{r}(s) : S^n \to R^{n+k+1}, \quad \boldsymbol{r}(s) = (\boldsymbol{x}(s), f(s)),$$
(8.115)

where $\boldsymbol{x}(s)$ is the mapping (8.106). Consequently the elements of the covariant metric tensor (g_{ij}^s) of S^{rn} in the coordinates $s^1, \ldots, s^n$ are connected with the elements g_{ij}^{xs} as

$$g_{ij}^s = \boldsymbol{r}_{s^i} \cdot \boldsymbol{r}_{s^j} = g_{ij}^{xs} + f_{s^i} f_{s^j}, \quad i, j = 1, \ldots, n.$$
(8.116)

The composition of $\boldsymbol{r}(s)$ and the intermediate transformation $\boldsymbol{s}(\boldsymbol{\xi})$ specified by (8.108) for determining grid coordinates defines a new parametrization of S^{rn} in the coordinates $\xi^1, \ldots, \xi^n$ which produces the elements of the covariant metric tensor (g_{ij}^ξ) of S^{rn} in these coordinates:

$$g_{ij}^\xi = g_{ij}^{x\xi} + \frac{\partial f[s(\boldsymbol{\xi})]}{\partial \xi^i}\frac{\partial f[s(\boldsymbol{\xi})]}{\partial \xi^j}, \quad i, j = 1\ldots, n.$$
(8.117)

If we establish a relation between the Beltramian operators with respect to the metrics (8.107) and (8.116) then we can use formula (8.112) in order to control with the monitor function $f(s)$ the rate of change of the relative spacing of the grid surfaces on S^{xn}. In order to find this relation we designate by $^r\varUpsilon^i_{jk}$ and $^x\varUpsilon^i_{jk}$ the Christoffel symbols in the coordinates $\xi^1, \ldots, \xi^n$ of S^{rn} and S^{xn}, respectively. These symbols are defined by the formula (4.28) with the identification $\boldsymbol{s} = \boldsymbol{\xi}$. Using (6.17) in the coordinates $\xi^1, \ldots, \xi^n$ we obtain

$$^r\varUpsilon^i_{jk} = {}^x\varUpsilon^i_{jk} + \frac{1}{1+\nabla(f)}g_{\xi x}^{il}f_{\xi^l}\nabla^x_{\xi^j\xi^k}(f), \quad i,j,k,l,m,p = 1,\ldots,n,$$
(8.118)

where

$$\nabla(f) = g_{sx}^{ij}f_{s^i}f_{s^j}, \quad i, j = 1, \ldots, n,$$

while

$$f_{s^i} = \frac{\partial f[\mathbf{s}]}{\partial s^i}, \quad f_{\xi^i} = \frac{\partial f[\mathbf{s}(\boldsymbol{\xi})]}{\partial \xi^i}, \quad i = 1, \ldots, n,$$

$$f_{\xi^i \xi^j} = \frac{\partial^2 f[\mathbf{s}(\boldsymbol{\xi})]}{\partial \xi^i \xi^j}, \quad i, j = 1, \ldots, n,$$

and

$$\nabla^x_{\xi^j \xi^k}(f) = f_{\xi^j \xi^k} - f_{\xi^p}\,{}^x\Upsilon^p_{jk}, \quad j, k, p = 1, \ldots, n,$$

is the mixed covariant derivative of $f[\mathbf{s}(\boldsymbol{\xi})]$ with respect to ξ^j and ξ^k in the metric (8.107). Note formula (8.118) generalizes the formula (8.65).

Remind the mixed covariant derivatives of f in the coordinates $\xi^1, \ldots, \xi^n$ and $s^1, \ldots, s^n$ are connected by the tensor relations such as the elements of a covariant metric tensor:

$$\nabla^x_{\xi^j \xi^k}(f) = \nabla^x_{s^l s^m}(f) \frac{\partial s^l}{\partial \xi^j} \frac{\partial s^m}{\partial \xi^k}, \quad i, j, l, m = 1, \ldots, n. \tag{8.119}$$

Now, from (4.56) with $s^i = \xi^i$, (8.118), and (6.10), we find similarly to (6.20)

$$\Delta^r_B[\xi^i] = -g^{jk}_{\xi}{}^r\Upsilon^i_{jk}$$

$$= -\left(g^{jk}_{\xi x} - \frac{1}{1+\nabla(f)} g^{jm}_{\xi x} f_{\xi^m} g^{kp}_{\xi x} f_{\xi^p}\right)$$

$$\times \left({}^x\Upsilon^i_{jk} + \frac{1}{1+\nabla(f)} g^{il}_{\xi x} f_{\xi^l} \nabla^x_{\xi^j \xi^k}(f)\right) \tag{8.120}$$

$$= \Delta^x_B[\xi^i] - \frac{1}{1+\nabla(f)} g^{il}_{\xi x} f_{\xi^l} g^{jk}_{\xi} \nabla^x_{\xi^j \xi^k}(f)$$

$$+ \frac{1}{1+\nabla(f)} g^{jm}_{\xi x} f_{\xi^m} g^{kp}_{\xi x} f_{\xi^p}\,{}^x\Upsilon^i_{jk}, \quad i, j, k, l, m, p = 1, \ldots, n.$$

where Δ^r_B and Δ^x_B is the operator of Beltrami with respect to the metric (8.116) and (8.107), respectively. Note

$$g^{ij}_{\xi x} f_{\xi^j} = g^{kl}_{sx} \frac{\partial \xi^i}{\partial s^k} \frac{\partial \xi^j}{\partial s^l} f_{s^m} \frac{\partial s^m}{\partial \xi^j} = g^{km}_{sx} f_{s^m} \frac{\partial \xi^i}{\partial s^k}, \quad i, j, k, l, m = 1, \ldots, n,$$

$$g^{jk}_{\xi} \nabla^x_{\xi^j \xi^k}(f) = g^{lm}_s \frac{\partial \xi^j}{\partial s^l} \frac{\partial \xi^k}{\partial s^m} \nabla^x_{s^i s^p}(f) \frac{\partial s^i}{\partial \xi^j} \frac{\partial s^p}{\partial \xi^k} = g^{lm}_s \nabla^x_{s^l s^m}(f),$$

$$i, j, k, l, m, p = 1, \ldots, n.$$

Therefore (8.120) is as follows

8.2 Application of Theorem to Popular Elliptic Models

$$\Delta_B^r[\xi^i] = \Delta_B^x[\xi^i] - \frac{1}{1+\nabla(f)} g_{sx}^{kp} f_{s^p} \frac{\partial \xi^i}{\partial s^k} g_s^{lm} \nabla_{s^l s^m}^x(f)$$

$$+ \frac{1}{1+\nabla(f)} g_{\xi x}^{jm} f_{\xi^m} g_{\xi x}^{kp} f_{\xi^p} {}^x\Upsilon_{jk}^i , \qquad (8.121)$$

$$i,j,k,l,m,p = 1,\ldots,n ,$$

where the second Christoffel symbol ${}^x\Upsilon_{jk}^i$ is in the coordinates $\xi^1,\ldots,\xi^n$. Assuming $\xi^i = s^i$, $i=1,\ldots,n$, in (8.118) gives

$${}^x\Upsilon_{jk}^i = {}^r\Upsilon_{jk}^i - \frac{1}{1+\nabla(f)} g_{sx}^{il} f_{s^l} \nabla_{s^j s^k}^x(f) , \qquad i,j,k,l = 1,\ldots,n .$$

Here the Christoffel symbols ${}^x\Upsilon_{jk}^i$ and ${}^r\Upsilon_{jk}^i$ are in the coordinates $s^1,\ldots,s^n$. Using these relations yields

$$\nabla_{s^l s^m}^x(f) = f_{s^l s^m} - f_{s^k} {}^x\Upsilon_{lm}^k$$

$$= f_{s^l s^m} - f_{s^k} {}^r\Upsilon_{lm}^k + \frac{1}{1+\nabla(f)} f_{s^k} g_{sx}^{kp} f_{s^p} \nabla_{s^l s^m}^x(f)$$

$$= \nabla_{s^l s^m}^r(f) + \frac{\nabla(f)}{1+\nabla(f)} \nabla_{s^l s^m}^x(f) , \qquad k,l,m,p = 1,\ldots,n ,$$

where $\nabla_{s^l s^m}^r$ is the mixed derivative with respect to the metric (8.116). Thus the mixed derivatives in the metrics (g_{ij}^{xs}) and (g_{ij}^s) are connected by the following relation

$$\nabla_{s^l s^m}^x(f) = [1+\nabla(f)]\nabla_{s^l s^m}^r(f) , \qquad l,m = 1,\ldots,n , \qquad (8.122)$$

hence (8.120) is also expressed as follows:

$$\Delta_B^r[\xi^i] = \Delta_B^x[\xi^i] - g_{sx}^{kp} f_{s^p} \frac{\partial \xi^i}{\partial s^k} g_s^{lm} \nabla_{s^l s^m}^r(f)$$

$$+ \frac{1}{1+\nabla(f)} g_{\xi x}^{jm} f_{\xi^m} g_{\xi x}^{kp} f_{\xi^p} {}^x\Upsilon_{jk}^i , \qquad (8.123)$$

$$i,j,k,l,m,p = 1,\ldots,n .$$

Multidimensional Case for Grid Coordinates. Let the grid coordinates $\xi^1,\ldots,\xi^n$ be obtained by a solution of the problem (5.4) with respect to the metric (8.116). Then $\Delta_B^r[\xi^i] = 0$, $i=1,\ldots,n$, and from (8.123)

$$\Delta_B^x[\xi^i] = g_{sx}^{kp} f_{s^p} \frac{\partial \xi^i}{\partial s^k} g_s^{lm} \nabla_{s^l s^m}^r(f)$$

$$- \frac{1}{1+\nabla(f)} g_{\xi x}^{jm} f_{\xi^m} g_{\xi x}^{kp} f_{\xi^p} {}^x\Upsilon_{jk}^i , \qquad (8.124)$$

$$i,j,k,l,m,p = 1,\ldots,n .$$

Substituting this expression of $\Delta_B^x[\xi^i]$ for $\Delta_B[\xi^i]$ in (8.112) gives the following formula

234   8. Control of Grid Clustering

$$\frac{d}{d\boldsymbol{n}_i}\left(\frac{1}{\sqrt{g_{\xi x}^{ii}}}\right) = -\frac{1}{g_{\xi x}^{ii}}\left[g_{sx}^{kp}f_{sp}\frac{\partial\xi^i}{\partial s^k}g_s^{lm}\nabla^r_{s^l s^m}(f)\right.$$
$$\left.-\frac{1}{1+\nabla(f)}g_{\xi x}^{jm}f_{\xi^m}g_{\xi x}^{kp}f_{\xi^p}{}^x\varUpsilon^i_{jk}\right] - \frac{n-1}{\sqrt{g_{\xi x}^{ii}}}K_m^{x\xi}(\xi^i)\,, \quad (8.125)$$

$$i,j,k,l,m,p = 1,\ldots,n\,, \quad i \text{ fixed}\,,$$

for the rate of change of the relative spacing of the grid surfaces $\xi^i = const$ on S^{xn}. This formula is an extension of (8.64) to regular surfaces.

Taking into account the following relation

$$g_s^{lm}\nabla^r_{s^l s^m}(f) = g_\xi^{lm}\nabla^r_{\xi^l \xi^m}(f) = g_\xi^{lm}(f_{\xi^l \xi^m} - f_{\xi^i}{}^r\varUpsilon^i_{lm})$$
$$= g_\xi^{lm} f_{\xi^l \xi^m}\,, \quad i,l,m = 1,\ldots,n\,,\ i \text{ fixed}\,, \quad (8.126)$$

since the grid coordinates satisfy the grid equations

$$\Delta_B^r[\xi^j] = -g_\xi^{lm r}\varUpsilon^j_{lm} = 0\,, \quad j,l,m = 1,\ldots,n\,,$$

we conclude that these grid equations also lead to the following formula

$$\Delta_B^r[f] = \frac{1}{\sqrt{g^\xi}}\frac{\partial}{\partial\xi^j}\left(\sqrt{g^\xi}g_\xi^{jk}f_{\xi^k}\right) = \Delta_B^r[\xi^k]f_{\xi^k} + g_\xi^{jk}f_{\xi^j\xi^k}$$
$$= g_\xi^{jk}f_{\xi^j\xi^k} = g_s^{lm}\nabla^r_{s^l s^m}(f)\,, \quad j,k = 1,\ldots,n\,, \quad (8.127)$$

where $g^\xi = \det(g^\xi_{jk})$. Thus, using (8.126) and (8.127) in (8.125), we come to the following relation

$$\frac{d}{d\boldsymbol{n}_i}\left(\frac{1}{\sqrt{g_{\xi x}^{ii}}}\right) = \frac{1}{g_{\xi x}^{ii}}\left[\frac{1}{1+\nabla(f)}g_{\xi x}^{jm}f_{\xi^m}g_{\xi x}^{kp}f_{\xi^p}{}^x\varUpsilon^i_{jk}\right.$$
$$\left.-g_{sx}^{kp}f_{sp}\frac{\partial\xi^i}{\partial s^k}\Delta_B^r[f]\right] - \frac{n-1}{\sqrt{g_{\xi x}^{ii}}}K_m^{x\xi}(\xi^i)\,, \quad (8.128)$$

$$i,j,k,l,m,p = 1,\ldots,n\,, \quad i \text{ fixed}\,,$$

Specification of the Monitor Function. Let us specify, similarly to (8.72), the monitor function $f(s)$ as

$$f(\boldsymbol{s}) = g[\varphi(\boldsymbol{s})]\,, \quad \boldsymbol{s} \in S^n\,, \quad (8.129)$$

i.e. $f(\boldsymbol{s}) = const$ at the points of the $(n-1)$-dimensional surface S^{n-1} in S^n defined from the equation $\varphi(\boldsymbol{s}) = 0$. Let this surface be a coordinate surface $\xi^i = c_0$ in S^n for some c_0 and $\varphi(\boldsymbol{s})$ be such that

$$\frac{d}{d\xi^i}\varphi[\boldsymbol{s}(\boldsymbol{\xi})] > 0\,, \quad \xi^i = c_0\,, \text{ for } \boldsymbol{\xi} = (\xi^1,\ldots,\xi^{i-1},c_0,\xi^{i+1},\ldots,\xi^n)\,. \quad (8.130)$$

We designate this surface by $S^{n-1}_{c_0}$, while the coordinate surface $\xi^i = c_0$ in S^{xn} is designated by $S^{xn-1}_{c_0}$. It is clear that the surface $S^{xn-1}_{c_0}$ is the image of $S^{n-1}_{c_0}$ by the transformation $\boldsymbol{x}(\boldsymbol{s})$. We also assume that all expressions in this section are considered at the points $\boldsymbol{s} \in S^{n-1}_{c_0}$.

8.2 Application of Theorem to Popular Elliptic Models 235

Modification of Formula (8.128). Now we modify the formula (8.128) assuming that the monitor function $f(s)$ satisfies (8.129) and (8.130).

First of all, noting that in this case $\partial f[s(\boldsymbol{\xi})]/\partial \xi^j = 0$ when $j \neq i$ and $s(\boldsymbol{\xi}) \in S_{c_0}^{n-1}$, we conclude that

$$g_{\xi x}^{jm} f_{\xi^m} g_{\xi x}^{kp} f_{\xi^p}{}^x \Upsilon_{jk}^i = (f_{\xi^i})^2 g_{\xi x}^{im} g_{\xi x}^{ik}{}^x \Upsilon_{jk}^i$$

$$= (g_{\xi x}^{ii})^2 (f_{\xi^i})^2 \frac{\mathrm{d}}{\mathrm{d} \boldsymbol{n}_i} \left(\frac{1}{\sqrt{g_{\xi x}^{ii}}} \right), \quad (8.131)$$

$$i, j, k, m, p = 1, \ldots, n, \ i \text{ fixed},$$

since (8.9) with $s^i = \xi^i$, $i = 1, \ldots, n$. Further, from (8.130)

$$g_{sx}^{kp} f_{s^p} \frac{\partial \xi^i}{\partial s^k} = g_{\xi x}^{ii} f_{\xi^i} = g'(0) g_{\xi x}^{ii} \frac{\partial}{\partial \xi^i} \varphi[s(\boldsymbol{\xi})] = g'(0) \sqrt{g_{\xi x}^{ii}} \sqrt{\nabla(\varphi)}, \quad (8.132)$$

$$i = 1, \ldots, n, \ i \text{ fixed},$$

$$\nabla(f) = f_{s^k} f_{s^l} g_{sx}^{kl} = (f_{s^i})^2 g_{\xi x}^{ii} = [g'(0)]^2 \nabla(\varphi), \quad (8.133)$$

$$i, k, l = 1, \ldots, n, \ i \text{ fixed}.$$

So, using (8.131) in (8.128), we find

$$\frac{\mathrm{d}}{\mathrm{d} \boldsymbol{n}_i} \left(\frac{1}{\sqrt{g_{\xi x}^{ii}}} \right)$$

$$= -\frac{1 + (f_{\xi^i})^2 g_{\xi x}^{ii}}{g_{\xi x}^{ii}} \left\{ g_{sx}^{kl} f_{s^l} \frac{\partial \xi^i}{\partial s^k} \Delta_B^r[f] - (n-1) \sqrt{g_{\xi x}^{ii}} K_m(S_{c_0}^{xn-1}) \right\} \quad (8.134)$$

$$= -\frac{1 + [g'(0)]^2 \nabla(\varphi)}{\sqrt{g_{\xi x}^{ii}}} \{ g'(0) \sqrt{\nabla(\varphi)} \Delta_B^r[f] - (n-1) K_m(S_{c_0}^{xn-1}) \},$$

$$i, k, l = 1, \ldots, n, \ i \text{ fixed},$$

where $K_m(S_{c_0}^{xn-1})$ is the designation for the mean curvature of $S_{c_0}^{xn-1}$ in S^{xn}.

Now we compute the expression $\Delta_B^r[f]$ when f satisfies (8.129). Since the Beltramian operator is invariant of parametrizations we, in particular, get

$$\Delta_B^r[f] = \frac{1}{\sqrt{g^s}} \frac{\partial}{\partial \xi^j} (\sqrt{g^s} g_s^{jk} f_{s^k}), \quad j, k = 1, \ldots, n.$$

From (6.6), (6.7), (8.129) and (8.132) we sequentially find

$$f_{s^k} = g'(0)\varphi_{s^k}, \quad k = 1, \ldots, n,$$

$$f_{s^j s^k} = g''(0)\varphi_{s^j}\varphi_{s^k} + g'(0)\varphi_{s^j s^k},$$

$$g^s = g^{xs}[1 + \nabla(f)] = g^{xs}\{1 + [g'(0)]^2 \nabla(\varphi)\}, \quad j,k = 1, \ldots, n,$$

$$g_s^{jk} = g_{sx}^{jk} - \frac{(g')^2}{1 + [g'(0)]^2 \nabla(\varphi)} g_{sx}^{jl}\varphi_{s^j} g_{sx}^{kp}\varphi_{s^p} \qquad (8.135)$$

$$= \frac{1}{1 + [g'(0)]^2 \nabla(\varphi)}\{g_{sx}^{jk} + [g'(0)]^2[\nabla(\varphi)g_{sx}^{jk} - g_{sx}^{jl}\varphi_{s^j} g_{sx}^{kp}\varphi_{s^p}]\},$$

$$j,k = 1, \ldots, n,$$

where

$$g^{rs} = \det(g_{jk}^{xs}), \quad g^s = \det(g_{jk}^{rs}).$$

These relations result in

$$\nabla_{kj}^x(f) = f_{s^k s^j} - f_{s^l}{}^x\Upsilon_{kj}^l = g''(0)\varphi_{s^k}\varphi_{s^j} + g'(0)\nabla_{kj}^x(\varphi),$$

$$j,k = 1, \ldots, n,$$

and consequently

$$g_s^{jk}\nabla_{kj}^x(f) = \frac{1}{1 + [g'(0)]^2\nabla(\varphi)}\{\nabla(\varphi)g''(0) + g'(0)g_{sx}^{jk}\nabla_{kj}^x(\varphi)$$

$$- (n-1)[g'(0)]^2\nabla(\varphi)^{3/2} K_m(S_{c_0}^{xn-1})\}, \qquad (8.136)$$

$$j,k = 1, \ldots, n,$$

since (4.105). In accordance with (8.127) and (8.122) we also have

$$\Delta_B^r[f] = \frac{1}{1 + \nabla(f)} g_s^{lm}\nabla_{s^l s^m}^x(f), \quad l,m = 1, \ldots, n.$$

Therefore, availing us of (8.136), we find

$$g'(0)\sqrt{\nabla(\varphi)}\Delta_B^r[f] + (n-1)K_m(S_{c_0}^{xn-1})$$

$$= \frac{g'(0)\sqrt{\nabla(\varphi)}}{(1 + [g'(0)]^2\nabla(\varphi))^2}\{\nabla(\varphi)g''(0) + g'(0)g_{sx}^{jk}\nabla_{kj}^x(\varphi)\}$$

$$+ (n-1)\left(1 - \frac{[g'(0)]^4[\nabla(\varphi)]^2}{[1 + [g'(0)]^2\nabla(\varphi)]^2}\right)K_m(S_{c_0}^{n-1}), \quad j,k = 1, \ldots, n.$$

Substituting this equation in (8.134) gives, similarly to (8.96),

$$\frac{d}{dn_i}\left(\frac{1}{\sqrt{g_{\xi x}^{ii}}}\right) = -\frac{1}{\sqrt{g_{\xi x}^{ii}(1 + [g'(0)]^2\nabla(\varphi))}} d[s, g'(0), g''(0)], \qquad (8.137)$$

$$i = 1, \ldots, n, \quad i \text{ fixed},$$

where

$$d[s, g'(0), g''(0)] = g'(0)\sqrt{\nabla(\varphi)}[g'(0)g_{sx}^{jk}\nabla_{kj}^{x}(\varphi)$$
$$+g''(0)\nabla(\varphi)] + (n-1)\{1 + 2[g'(0)]^2 \nabla(\varphi)\} K_m(S_{c_0}^{n-1}).$$

Thus, similar to the case of the monitor surface over a domain, we conclude that the positive (negative) sign of the quantity $b[s, g'(0), g''(0)]$ point out the grid clustering (rarefaction) near the coordinate surface $\varphi(s) = 0$ in S^{xn}. If this coordinate surface is a boundary surface of S^{xn} the by a suitable choice of $g'(0)$ and $g''(0)$ we can realize the necessary behavior of the grid surfaces near this boundary segment $S_{c_0}^{xn-1}$. The monitor function can be specified, for example, by equation (8.80). Analogously to (8.81) and (8.82), if $g'(0) \neq 0$ is chosen then

$$d[s, g'(0), g''(0)] > 0$$

if

$$g''(0) > \max_{s \in S_{c_0}^{n-1}} p[s, g'(0)] \; ; \tag{8.138}$$

similarly

$$d[s, g'(0), g''(0)] < 0$$

if

$$g''(0) < \min_{s \in S_{c_0}^{n-1}} p[s, g'(0)] \; , \tag{8.139}$$

where

$$p[s, g'(0)] = -\frac{1}{g'(0)[\nabla(\varphi)]^{3/2}}\{[g'(0)]^2 \sqrt{\nabla(\varphi)} \Delta_B[\varphi]$$
$$+(n-1)[1 + 2[g'(0)]^2 \nabla(\varphi)] K_m(S_{c_0}^{xn-1})\} \; , \; j, k = 1, \ldots, n \; .$$

Two-Dimensional Case. When $n = 2$ then

$$\boldsymbol{x}_{s^1} = \frac{1}{\varphi_{s^2}}(\varphi_{s^2}\boldsymbol{r}_{s^1} - \varphi_{s^1}\boldsymbol{r}_{s^2})$$

and consequently

$$g_{11}^{xs} = \frac{1}{(\varphi_{s^2})^2}\{(\varphi_{s^2})^2 g_{11}^{rs} - 2\varphi_{s^2}\varphi_{s^1} g_{12}^{rs} + (\varphi_{s^1})^2 g_{22}^{rs}\} \; .$$

Also, using (4.68) and (4.69) we find

$$\boldsymbol{n}_2 = \boldsymbol{b}/|\boldsymbol{b}| \; ,$$

where

$$\boldsymbol{b} = (\boldsymbol{x}_{s^1} \cdot \boldsymbol{r}_{s^2})\boldsymbol{r}_{s^1} - (\boldsymbol{x}_{s^1} \cdot \boldsymbol{r}_{s^1})\boldsymbol{r}_{s^2}$$

$$= \frac{1}{\varphi_{s^2}}[(\varphi_{s^2}g_{12}^{rs} - \varphi_{s^1}g_{22}^{rs})\boldsymbol{r}_{s^1} - (\varphi_{s^2}g_{11}^{rs} - \varphi_{s^1}g_{12}^{rs})\boldsymbol{r}_{s^2}]$$

$$= \frac{g^{rs}}{\varphi_{s^2}}\varphi_{s^i}g_{sr}^{ik}\boldsymbol{r}_{s^k}, \quad i, k = 1, 2.$$

In the two-dimensional case equations (8.134) are transformed to

$$\frac{d}{dn_i}\left(\frac{1}{\sqrt{g_{\xi x}^{ii}}}\right) = -\frac{1 + [g'(0)]^2 \nabla(\varphi)}{\sqrt{g_{\xi x}^{ii}}}\{g'(0)\sqrt{\nabla(\varphi)}\Delta_B^r[f] - \sigma_i\}, \tag{8.140}$$

$$i, k, l = 1, 2, \quad i \text{ fixed},$$

since

$$K_m^{x\xi}(\xi^i) = \sigma_i, \quad i = 1, 2,$$

where σ_i is the geodesic curvature of the coordinate line $\xi^i = c_0$.

Control by Weight Functions.

Multidimensional Case. The grid system (8.97) is extended to the following system

$$\Delta_B^x[\xi^i] = P^i(\boldsymbol{s}), \quad i = 1, \ldots, n, \tag{8.141}$$

for the purpose of generating grids on the regular surface S^{xn}. Consequently equations (8.98) for the rate of change of the relative spacing of the grid surfaces $\xi^i = const$ on S^{xn} obtained through (8.141) are transformed to the following form

$$\frac{d}{dn_i}\left(\frac{1}{\sqrt{g_{\xi x}^{ii}}}\right) = -\frac{1}{\sqrt{g_{\xi x}^{ii}}}\left(\frac{P^i}{\sqrt{g_{\xi x}^{ii}}} + K_m^{x\xi}(\xi^i)\right), \tag{8.142}$$

$$i = 1, \ldots, n, \quad i \text{ fixed}.$$

Thus, analogously to the case of a domain, the sign of rate of clustering of the grid hypersurfaces $\xi^i = const$ on S^{xn} near a boundary segment $\xi^i = c_0$ is defined by the sign of the quantity

$$\frac{P^i}{\sqrt{g_{\xi x}^{ii}}} + (n-1)K_m^{x\xi}(\xi^i), \quad i = 1, \ldots, n, \quad i \text{ fixed},$$

at the points of this segment. The necessary sign is realized by the choice of the function $P^i(\boldsymbol{s})$.

Two-Dimensional Case. In the two-dimensional case the equations (8.142) for the analysis of the grid behavior near the boundary curve $\xi^i = c_0$ are as follows

$$\frac{d}{d\boldsymbol{n}_1}\left(\frac{1}{\sqrt{g^{11}_{\xi x}}}\right) = -\frac{1}{\sqrt{g^{11}_{\xi x}}}\left(\frac{P^1}{\sqrt{g^{11}_{\xi x}}} + \sigma_1\right),$$
$$\frac{d}{d\boldsymbol{n}_2}\left(\frac{1}{\sqrt{g^{22}_{\xi x}}}\right) = -\frac{1}{\sqrt{g^{22}_{\xi x}}}\left(\frac{P^2}{\sqrt{g^{22}_{\xi x}}} + \sigma_2\right),$$
(8.143)

since (8.141).

Control by a Diffusion Function.

Multidimensional Case. Equations (8.101) can be generalized in the following manner

$$\frac{\partial}{\partial s^j}\left(w\sqrt{g^{xs}}g^{jk}_{sx}\frac{\partial \xi^i}{\partial s^k}\right) = 0, \quad i,j,k = 1,\ldots,n, \quad (8.144)$$

where $w(s) > 0$ is a diffusion coefficient, for generating grids on a regular surface S^{xn}.

Note these equations, for $n \neq 2$, produce the same grid coordinates obtained by the grid equations (5.4) with respect to the monitor metric

$$g^s_{ij} = w^\alpha g^{xs}_{ij}, \quad \alpha = 2/(n-2), \quad i,j = 1,\ldots,n, \quad (8.145)$$

which forms a monitor Riemannian manifold over S^{xn}. Indeed

$$g^{ij}_s = w^{-\alpha} g^{ij}_s, \quad i,j = 1,\ldots,n,$$
$$g^s = w^{n\alpha} g^{xs},$$

therefore the equations (5.4) with respect to the metric (8.145) are equivalent to (8.144) since

$$\Delta_B[\xi^i] = \frac{1}{\sqrt{g^s}}\frac{\partial}{\partial s^j}\left(\sqrt{g^s}g^{jk}_s\frac{\partial \xi^i}{\partial s^k}\right) = \frac{1}{\sqrt{g^s}}\frac{\partial}{\partial s^j}\left(\sqrt{g^{xs}}w^{n\alpha/2-\alpha}g^{jk}_{sx}\frac{\partial \xi^i}{\partial s^k}\right)$$
$$= w^{n\alpha/2}\frac{1}{\sqrt{g^{xs}}}\frac{\partial}{\partial s^j}\left(w\sqrt{g^{xs}}g^{jk}_{sx}\frac{\partial \xi^i}{\partial s^k}\right), \quad i,j,k = 1,\ldots,n.$$

From (8.144) we find

$$\Delta^x_B[\xi^i] \equiv \frac{1}{\sqrt{g^{xs}}}\frac{\partial}{\partial s^j}\left(\sqrt{g^{xs}}g^{jk}_{sx}\frac{\partial \xi^i}{\partial s^k}\right) = -\frac{1}{w}\frac{\partial w}{\partial s^j}g^{jk}_{sx}\frac{\partial \xi^i}{\partial s^k},$$
$$i,j,k = 1,\ldots,n.$$

Substituting this equation for $\Delta_B[\xi^i]$ in (8.112) we obtain

$$\frac{d}{d\boldsymbol{n}_i}\left(\frac{1}{\sqrt{g^{ii}_{\xi x}}}\right) = \frac{1}{wg^{ii}_{\xi x}}\frac{\partial w}{\partial s^j}g^{jk}_{sx}\frac{\partial \xi^i}{\partial s^k} - \frac{n-1}{\sqrt{g^{ii}_{\xi x}}}K^{x\xi}_m(\xi^i),$$
$$i = 1,\ldots,n, \quad i \text{ fixed}.$$
(8.146)

So, analogously to the case of a domain, the rate of clustering near a boundary segment can be realized by the choice of the diffusion function $w(s)$.

8. Control of Grid Clustering

Two-Dimensional Case. In the two-dimensional case equations (8.140), since $K_m^{x\xi}(\xi^i) = \sigma_i$, $i = 1, 2$, have the following form

$$\frac{d}{d n_1}\left(\frac{1}{\sqrt{g_{\xi x}^{11}}}\right) = \frac{1}{w g_{\xi x}^{11}} \frac{\partial w}{\partial s^j} g_{sx}^{jk} \frac{\partial \xi^1}{\partial s^k} - \frac{1}{\sqrt{g_{\xi x}^{11}}} \sigma_1,$$

$$\frac{d}{d n_2}\left(\frac{1}{\sqrt{g_{\xi x}^{22}}}\right) = \frac{1}{w g_{\xi x}^{22}} \frac{\partial w}{\partial s^j} g_{sx}^{jk} \frac{\partial \xi^2}{\partial s^k} - \frac{1}{\sqrt{g_{\xi x}^{22}}} \sigma_2,$$

(8.147)

which allows one to specify the diffusion function $w(s)$ to provide grid clustering or grid rarefaction near boundary segments.

9. Numerical Implementation of Grid Generator

The comprehensive system of the generalized Laplace equations in the boundary value problem (5.4) allows one to generate grids on surfaces or in domains in a unified manner, regardless of their dimension. In particular, the transformed elliptic equations obtained from the generalized Laplace equations by changing mutually dependent and independent variables can be applied to produce grids in spatial blocks by means of the successive generation of grids on curvilinear edges, faces, and parallelepipeds, using the solution at a step $i < n$ as the Dirichlet boundary condition for the following step $i + 1 \leq n$. Thus both the interior and the boundary grid points of a domain or surface can be calculated by the similar elliptic solver.

This chapter reviewes a numerical algorithm for grid generation based on the comprehensive grid equations reviewed in Chaps. 5. and 7 and the formulas of monitor functions discussed in Chap. 8.

9.1 One-Dimensional Equation

This section describes a numerical algorithm for generating grids on a curve S^{x1} specified by a parametrization

$$\boldsymbol{x}(s) : [0, 1] \to R^k, \quad \boldsymbol{x} = (x^1, \ldots, x^k) \ .$$

The monitor function for controlling grid properties is prescribed as

$$\boldsymbol{f}(\boldsymbol{x}) : G^k \to R^l, \quad \boldsymbol{f} = (f^1, \ldots, f^l) \ ,$$

where G^k is a domain in R^k containing S^{x1}. As a result the monitor curve S^{r1} over S^{x1} is parametrized by

$$\boldsymbol{r}(s) : [0, 1] \to R^{l+k}, \quad \boldsymbol{r}(s) = (\boldsymbol{x}(s), \boldsymbol{f}[\boldsymbol{x}(s)]) \ .$$

9.1.1 Numerical Algorithm

The numerical grid on S^{x1} is computed after solving the Dirichlet boundary value problem for the equation (7.76) with respect to $s(\xi)$, namely, the grid nodes $\boldsymbol{x}_j$, $j = 0, 1, \ldots, N$, on S^{x1} are defined by the relation

$$\boldsymbol{x}_j = \boldsymbol{x}(s(jh)) \ , \quad j = 0, 1, \ldots, N \ , \quad h = 1/N \ ,$$

9. Numerical Implementation of Grid Generator

or by

$$x_j = x(s_j), \quad j = 0, 1, \ldots, N, \quad h = 1/N,$$

here s_j, $j = 0, 1, \ldots, N$, is a difference function obtained by the numerical solution on a uniform grid $\xi_j = jh$, $j = 0, 1, \ldots, N$, of the Dirichlet problem for the equation (7.76), i.e.

$$\frac{d}{d\xi}\left(\frac{ds}{d\xi}\sqrt{g^{rs}}\right) = 0, \quad 0 < \xi < 1, \quad (9.1)$$

$$s(0) = 0, \quad s(1) = 1,$$

where

$$g^{rs} = \boldsymbol{x}_s \cdot \boldsymbol{x}_s + \boldsymbol{f}_s \cdot \boldsymbol{f}_s = \frac{d\boldsymbol{x}}{ds} \cdot \frac{d\boldsymbol{x}}{ds} + \frac{d\boldsymbol{f}[\boldsymbol{x}(s)]}{ds} \cdot \frac{d\boldsymbol{f}[\boldsymbol{x}(s)]}{ds}.$$

Iterative Scheme. The nonlinear problem (9.1) is solved by an iterative process which is engendered by the numerical solution of the following parabolic problem with respect to a function $s(\xi, t)$

$$\frac{\partial s}{\partial t} - \frac{\partial}{\partial \xi}\left(\frac{\partial s}{\partial \xi}\sqrt{g^{rs}}\right) = 0, \quad 0 \leq \xi \leq 1, \quad 0 \leq t \leq T, \quad (9.2)$$

$$s(0, t) = 0, \quad s(1, t) = 1, \quad s(\xi, 0) = s_0(\xi).$$

The problem (9.2) is approximated on the uniform grid $(ih, n\tau)$ with respect to s_i^n, $i = 0, 1, \ldots, N$, $n = 0, 1, \ldots$, by the following natural stencil

$$\frac{s_i^{n+1} - s_i^n}{\tau} = \frac{1}{h^2}[(s_{i+1}^{n+1} - s_i^{n+1})v_{i+1/2}^n - (s_i^{n+1} - s_{i-1}^{n+1})v_{i-1/2}^n], \quad (9.3)$$

$$s_0^n = 0, \quad s_N^n = 1, \quad s_i^0 = s_0(ih), \quad h = 1/N,$$

where

$$v_{i+1/2}^n = \frac{1}{2}\left(\sqrt{g^{rs}(s_i^n)} + \sqrt{g^{rs}(s_{i+1}^n)}\right), \quad i = 0, 1, \ldots, N-1. \quad (9.4)$$

The scheme (9.3) is implicit. Its solution is obtained from the algorithm which is expounded by the application to the following difference reference problem

$$A_i^{n+1} s_{i-1}^{n+1} - C_i^{n+1} s_i^{n+1} + B_i^{n+1} s_{i+1}^{n+1} = -F_i^n, \quad i = 1, 2, \ldots, N-1,$$
$$s_0^{n+1} = a, \quad s_N^{n+1} = b. \quad (9.5)$$

The solution of (9.5) is found through the following recursive formulas

$$s_i^{n+1} = \alpha_{i+1}^{n+1} s_{i+1}^{n+1} + \beta_{i+1}^{n+1}, \quad i = 1, \ldots, N-1, \quad s_N^{n+1} = b, \quad (9.6)$$

where

$$\alpha_{i+1}^{n+1} = \frac{B_i^{n+1}}{C_i^{n+1} - \alpha_i^{n+1} A_i^{n+1}}, \quad i = 1, \ldots, N-1, \quad \alpha_1^{n+1} = 0,$$

$$\beta_{i+1}^{n+1} = \frac{A_i^{n+1} \beta_i^{n+1} + F_i^n}{C_i^{n+1} - \alpha_i^{n+1} A_i^{n+1}}, \quad i = 1, \ldots, N-1, \quad \beta_1^{n+1} = a. \quad (9.7)$$

Thus assuming in (9.5) $a = 0$, $b = 1$, and

$$A_i^{n+1} = v_{i-1/2}^n, \quad B_i^{n+1} = v_{i+1/2}^n, \quad C_i^{n+1} = v_{i-1/2}^n + v_{i+1/2}^n + \theta,$$
$$F_i^n = \theta s_i^n, \quad \theta = h^2/\tau, \quad i = 1, \ldots, N-1,$$
(9.8)

we obtain a solution of (9.3) at a step $n+1$ if it is known at the previous step n. Note the values of the initial function s_i^0, $i = 0, 1, \ldots, N$, are specified by the user. Typically it is assumed that

$$s_i^0 = ih, \quad i = 0, \ldots, N, \quad h = 1/N.$$

As an approximate numerical solution of (9.1) there is taken the solution s_i^n, $i = 0, 1, \ldots, N$, of (9.3) at a step number n if

$$\max_{0 \le i \le N} \frac{|s_i^{n+1} - s_i^n|}{\tau} \le \varepsilon,$$
(9.9)

for some sufficiently small $\varepsilon > 0$.

Step–by–Step Algorithm. The algorithm described above is presented here in a step-by-step manner.
Step 1.
Define an initial grid distribution of the parametric interval [0,1] by introducing a monotone difference function s_i^0, $i = 0, \ldots, N$, such that $s_0^0 = 0$, $s_N^0 = 1$.
Step 2.
Compute the difference function $v_{i+1/2}^0$, $i = 0, \ldots, N-1$, by formula (9.4).
Step 3.
Compute the difference functions A_i^1, B_i^1, C_i^1, F_i^0, $i = 1, \ldots, N-1$, by formulas in (9.8).
Step 4.
Compute the coefficients α_i^1 and β_i^1, $i = 1, \ldots, N$, by formulas in (9.7) with $a = 0$.
Step 5.
Compute the difference solution s_i^1, $i = 0, \ldots, N$, of the first step through the formula (9.6) taking into account $s_0^1 = 0$, $s_N^1 = b = 1$.
Step 6.
Return to step 2 assuming $s_i^0 = s_i^1$, $i = 0, \ldots, N$, where s_i^1 is the solution obtained at the step 5.

Continue until the tolerance requirement (9.9) is observed.

9.2 Two-Dimensional Equations

In this section a numerical algorithm for generating grids in two-dimensional domains and surfaces is described.

9.2.1 Algorithms for Generating Grids in Two-Dimensional Domains

Grid Equations. Let us first discuss the grid algorithm for a two-dimensional domain S^2. If $\boldsymbol{f}(\boldsymbol{s}) = [f^1(\boldsymbol{s}), \ldots, f^l(\boldsymbol{s})]$, $\boldsymbol{s} \in S^2$ is a monitor function controlling the grid behavior then the monitor surface S^{r2} over S^2 is parametrized by the following mapping

$$\boldsymbol{r}(\boldsymbol{s}): S^2 \to R^{2+l}, \quad \boldsymbol{r}(\boldsymbol{s}) = [s^1, s^2, f^1(\boldsymbol{s}), \ldots, f^l(\boldsymbol{s})].$$

We shell use for the logical domain Ξ^2 the unit square: $\Xi^2 = \{0 \leq \xi^1, \xi^2 \leq 1\}$. Let the transformation $\boldsymbol{s}(\boldsymbol{\xi})$ for generating a grid in S^2 be specified on the boundary of Ξ^2, i.e. there is a map

$$\boldsymbol{\varphi}(\boldsymbol{\xi}): \partial \Xi^2 \to \partial S^2, \quad \boldsymbol{\varphi} = (\varphi^1, \varphi^2) \tag{9.10}$$

which is continious on $\partial\Xi^2$ and smooth on each segment of $\partial\Xi^2$. Note the one-dimensional transformation on any segment of $\partial\Xi^2$ can be computed by the algorithm described in Sect. 9.1.1. The grid in S^2 is then found by the numerical solution of the Dirichlet problem for the system of equations (7.99) with the identification $s^i = x^i$, $i = 1, 2$. The boundary value problem for this system has the following form

$$D^{\xi}[s^m] = -D^{\xi}[f^k]\frac{\partial f^k}{\partial s^m}, \quad m = 1, 2, \quad k = 1, \ldots, l, \tag{9.11}$$

$$\boldsymbol{s}(\boldsymbol{\xi})\Big|_{\partial \Xi^2} = \boldsymbol{\varphi}(\boldsymbol{\xi}),$$

where

$$D^{\xi}[v] = g^{\xi}_{22}\frac{\partial^2 v}{\partial \xi^1 \partial \xi^1} - 2g^{\xi}_{12}\frac{\partial^2 v}{\partial \xi^1 \partial \xi^2} + g^{\xi}_{11}\frac{\partial^2 v}{\partial \xi^2 \partial \xi^2}, \tag{9.12}$$

$$g^{\xi}_{ij} = \frac{\partial \boldsymbol{s}}{\partial \xi^i} \cdot \frac{\partial \boldsymbol{s}}{\partial \xi^j} + \frac{\partial \boldsymbol{f}[\boldsymbol{s}(\boldsymbol{\xi})]}{\partial \xi^i} \cdot \frac{\partial \boldsymbol{f}[\boldsymbol{s}(\boldsymbol{\xi})]}{\partial \xi^j}, \quad i,j = 1, 2. \tag{9.13}$$

Another system followed from equations (7.9) and relations (2.21) is as follows:

$$D^{\xi}[s^m] = g^{\xi}\Delta_B[s^m], \quad m = 1, 2, \tag{9.14}$$

where

$$g^{\xi} = \det(g^{\xi}_{ij}), \quad i,j = 1, 2.$$

Since (7.2), (4.28), and (2.21)

$$\Delta_B[s^m] = -g^{ij}_s \Upsilon^m_{ij} = -g^{ij}_s g^{mp}_s f^k_{s^i s^j} f^k_{s^p}$$

$$= \frac{1}{g^s} g^{mp}_s f^k_{s^p} D^s[f^k], \quad i,j,m,p = 1, 2, \quad k = 1, \ldots, l, \tag{9.15}$$

where

$$D^s[v] = g_{22}^s \frac{\partial^2 v}{\partial s^1 \partial s^1} - 2g_{12}^s \frac{\partial^2 v}{\partial s^1 \partial s^2} + g_{11}^s \frac{\partial^2 v}{\partial s^2 \partial s^2} ,$$

$$g_{ij}^s = \delta_j^i + \frac{\partial f^k}{\partial s^i} \frac{\partial f^k}{\partial s^j} , \quad i,j = 1,2 , \quad k = 1,\ldots,l ,$$

$$g^s = \det(g_{ij}^s) , \quad i,j = 1,2 .$$

Availing us of (9.15) and the obvious equation

$$g^\xi = J^2 g^s ,$$

where

$$J = \det\left(\frac{\partial s^i}{\partial \xi^j}\right), \quad i,j = 1,2 ,$$

equations (9.14) are transformed to the following form

$$D^\xi[s^m] = -J^2 g_s^{mp} f_{s^p}^k D^s[f^k] , \quad m,p = 1,2 , \quad k = 1,\ldots,l . \tag{9.16}$$

In particular, when the monitor function $\boldsymbol{f(s)}$ is a scalar-valued function $f(\boldsymbol{s})$, i.e. $l = 1$, then using (7.62) in (9.15) we obtain

$$\Delta_B[s^m] = -\frac{1}{(g^s)^2} D^s[f] f_{s^m} , \quad m = 1,2 , \tag{9.17}$$

so in this case equations (9.16) are as follows:

$$D^\xi[s^m] = -\frac{J^2}{g^s} D^s[f] f_{s^m} , \quad m = 1,2 . \tag{9.18}$$

Parabolic Equations. The nonlinear boundary value problem (9.11) is solved by an iterative process. For this purpose (9.11) is replaced by the following parabolic problem with respect to the functions $s^m(\xi^1, \xi^2, t)$, $m = 1, 2$:

$$\begin{aligned}
\frac{\partial s^m}{\partial t} &= D[s^m] + D[f^k]\frac{\partial f^k}{\partial s^m} , \quad m = 1,2 , \quad k = 1,\ldots,l , \\
\boldsymbol{s}(\boldsymbol{\xi},t) &= \boldsymbol{\varphi}(\boldsymbol{\xi}) , \quad \boldsymbol{\xi} \in \partial \Xi^2 , \quad t \geq 0 , \\
\boldsymbol{s}(\boldsymbol{\xi},0) &= \boldsymbol{s}_0(\boldsymbol{\xi}) , \quad \boldsymbol{\xi} \in \Xi^2 .
\end{aligned} \tag{9.19}$$

The solution $\boldsymbol{s}(\boldsymbol{\xi},t)$ satisfying (9.19) aspires to the solution of (9.11) when $t \to \infty$. Therefore an approximate solution of (9.11) is obtained from the solution (9.19) computed for some sufficiently large value $t = T_0$.

Initial Transformation. The initial transformation

$$\boldsymbol{s}(\boldsymbol{\xi},0) = \boldsymbol{s}_0(\boldsymbol{\xi}) : \Xi^2 \to S^2 .$$

can be found by propagating the values of $\boldsymbol{\varphi}(\boldsymbol{\xi}) = [\varphi^1(\boldsymbol{\xi}), \varphi^2(\boldsymbol{\xi})]$ into the interior of the domain Ξ^2, for example, through the formula of Lagrange two-dimensional transfinite interpolation. This formula has the following recursive form for the components $s^i(\boldsymbol{\xi},0)$ of the mapping $\boldsymbol{s}_0(\boldsymbol{\xi})$:

$$F^i(\xi^1, \xi^2) = \alpha_{01}^i(\xi^1)\varphi^i(0, \xi^2) + \alpha_{11}^i(\xi^1)\varphi^i(1, \xi^2) ,$$

$$s^i(\xi^1, \xi^2, 0) = F^i(\xi^1, \xi^2) + \alpha_{02}^i(\xi^2)[\varphi^i(\xi^1, 0) - F^i(\xi^1, 0)] \qquad (9.20)$$

$$+ \alpha_{12}^i(\xi^2)[\varphi^i(\xi^1, 1) - F^i(\xi^1, 1)] , \quad i = 1, 2 ,$$

where the blending functions $\alpha_{kj}^i(s)$, $0 \leq s \leq 1$, fixed) are subject to the following restrictions

$$\alpha_{0j}^i(0) = \alpha_{1j}^i(1) = 1 , \quad \alpha_{0j}^i(1) = \alpha_{1j}^i(0) = 0 . \qquad (9.21)$$

In particular, for the simplest expressions of the blending functions

$$\alpha_{0j}^i(s) = 1 - s , \quad \alpha_{1j}^i(s) = s ,$$

satisfying (9.21) we find from (9.20)

$$F^i(\xi^1, \xi^2) = (1 - \xi^1)\varphi^i(0, \xi^2) + \xi^1 \varphi^i(1, \xi^2) ,$$

$$s^i(\xi^1, \xi^2, 0) = F^i(\xi^1, \xi^2) + (1 - \xi^2)[\varphi^i(\xi^1, 0) - F^i(\xi^1, 0)] \qquad (9.22)$$

$$+ \xi^2[\varphi^i(\xi^1, 1) - F^i(\xi^1, 1)] , \quad i = 1, 2 .$$

Iterative Scheme. Let us introduce, for convenience, new dependent variables $s(\boldsymbol{\xi}, t) = s^1(\boldsymbol{\xi}, t)$, $v(\boldsymbol{\xi}, t) = s^2(\boldsymbol{\xi}, t)$, $g_{ij} = g_{ij}^\xi$, $i, j = 1, 2$. With respect to the corresponding difference functions s_{ij}^n, v_{ij}^n, $0 \leq i, j \leq N$, $0 \leq n$, the problem (9.19) is approximated on the uniform grid $(ih, jh, n\tau)$ by the scheme

$$\frac{s_{ij}^{n+1/2} - s_{ij}^n}{\tau} = \frac{1}{h^2}[g_{22}(\boldsymbol{s}_{ij}^n)\boldsymbol{l}_{ij}(s^{n+1/2}) - 2g_{12}(\boldsymbol{s}_{ij}^n)\boldsymbol{m}_{ij}(s^n)$$

$$+ g_{11}(\boldsymbol{s}_{ij}^n)\boldsymbol{s}_{ij}(s^n)] + \boldsymbol{p}_{ij}(s^n) , \qquad (9.23)$$

$$1 \leq i, j \leq N - 1 , \quad n \geq 0 ,$$

$$\frac{v_{ij}^{n+1/2} - v_{ij}^n}{\tau} = \frac{1}{h^2}[g_{22}(\boldsymbol{s}_{ij}^n)\boldsymbol{l}_{ij}(v^{n+1/2}) - 2g_{12}(\boldsymbol{s}_{ij}^n)\boldsymbol{m}_{ij}(v^n)$$

$$+ g_{11}(\boldsymbol{s}_{ij}^n)\boldsymbol{s}_{ij}(v^n)] + \boldsymbol{q}_{ij}(v^n) , \qquad (9.24)$$

$$1 \leq i, j \leq N - 1 , \quad n \geq 0 ,$$

$$\frac{s_{ij}^{n+1} - s_{ij}^{n+1/2}}{\tau} = \frac{1}{h^2}[g_{11}(\boldsymbol{s}_{ij}^n)\boldsymbol{s}_{ij}(s^{n+1} - s^n)] , \qquad (9.25)$$

$$1 \leq i, j \leq N - 1 , \quad n \geq 0 ,$$

$$\frac{v_{ij}^{n+1} - v_{ij}^{n+1/2}}{\tau} = \frac{1}{h^2}[g_{11}(\boldsymbol{s}_{ij}^n)\boldsymbol{s}_{ij}(v^{n+1} - v^n)] , \qquad (9.26)$$

$$1 \leq i, j \leq N - 1 , \quad n \geq 0 ,$$

9.2 Two-Dimensional Equations

where

$$g_{11}(\boldsymbol{s}_{ij}^m) = \left(\frac{s_{i+1j}^m - s_{i-1j}^m}{2h}\right)^2 + \left(\frac{v_{i+1j}^m - v_{i-1j}^m}{2h}\right)^2$$

$$+ \frac{\boldsymbol{f}(\boldsymbol{s}_{i+1j}^m) - \boldsymbol{f}(\boldsymbol{s}_{i-1j}^m)}{2h} \cdot \frac{\boldsymbol{f}(\boldsymbol{s}_{i+1j}^m) - \boldsymbol{f}(\boldsymbol{s}_{i-1j}^m)}{2h},$$

$$1 \leq i, j \leq N-1,$$

$$g_{12}(\boldsymbol{s}_{ij}^m) = \frac{s_{i+1j}^m - s_{i-1j}^m}{2h} \frac{s_{ij+1}^m - s_{ij-1}^m}{2h} + \frac{v_{i+1j}^m - v_{i-1j}^m}{2h} \frac{v_{ij+1}^m - s_{ij-1}^m}{2h}$$

$$+ \frac{\boldsymbol{f}(\boldsymbol{s}_{i+1j}^m) - \boldsymbol{f}(\boldsymbol{s}_{i-1j}^m)}{2h} \cdot \frac{\boldsymbol{f}(\boldsymbol{s}_{ij+1}^m) - \boldsymbol{f}(\boldsymbol{s}_{ij-1}^m)}{2h},$$

$$1 \leq i, j \leq N-1,$$

$$g_{22}(\boldsymbol{s}_{ij}^m) = \left(\frac{s_{ij+1}^m - s_{ij-1}^m}{2h}\right)^2 + \left(\frac{v_{ij+1}^m - v_{ij-1}^m}{2h}\right)^2$$

$$+ \frac{\boldsymbol{f}(\boldsymbol{s}_{ij+1}^m) - \boldsymbol{f}(\boldsymbol{s}_{ij-1}^m)}{2h} \cdot \frac{\boldsymbol{f}(\boldsymbol{s}_{ij+1}^m) - \boldsymbol{f}(\boldsymbol{s}_{ij-1}^m)}{2h},$$

$$1 \leq i, j \leq N-1,$$

$$\boldsymbol{l}_{ij}(w) = w_{i+1j} - 2w_{ij} + w_{i-1j}, \quad 1 \leq i, j \leq N-1,$$

$$\boldsymbol{m}_{ij}(w) = \frac{1}{4}(w_{i+1j+1} - w_{i-1j+1} - w_{i+1j-1} + w_{i-1j-1}),$$

$$1 \leq i, j \leq N-1,$$

$$\boldsymbol{s}_{ij}(w) = w_{ij+1} - 2w_{ij} + w_{ij-1}, \quad 1 \leq i, j \leq N-1,$$

$$\boldsymbol{p}_{ij}(\boldsymbol{s}^m) = b_{ij}^k(\boldsymbol{s}^m)\frac{\partial f^k}{\partial s_1}, \quad k = 1,\ldots,l, \quad 1 \leq i, j \leq N-1,$$

$$\boldsymbol{q}_{ij}(\boldsymbol{s}^m) = b_{ij}^k(\boldsymbol{s}^m)\frac{\partial f^k}{\partial s_2}, \quad k = 1,\ldots,l, \quad 1 \leq i, j \leq N-1,$$

$$b_{ij}^k(\boldsymbol{s}^m) = g_{22}(\boldsymbol{s}_{ij}^m)\boldsymbol{l}_{ij}[f^k(\boldsymbol{s}^m)] - 2g_{12}(\boldsymbol{s}_{ij}^m)\boldsymbol{m}_{ij}[f^k(\boldsymbol{s}^m)]$$

$$+ g_{11}(\boldsymbol{s}_{ij}^m)\boldsymbol{s}_{ij}[f^k(\boldsymbol{s}^m)], \quad k = 1,\ldots,l, \quad 1 \leq i, j \leq N-1.$$

Algorithm for Computation. The solution of the scheme described above is obtained by applying successively formulas (9.6) and (9.7) for the solution of the reference difference problem (9.5). Namely, assuming

$$A_{ij}^{n+1/2} = g_{22}(s_{ij}^n) , \quad B_{ij}^{n+1/2} = g_{22}(s_{ij}^n) ,$$

$$C_{ij}^{n+1/2} = 2g_{22}(s_{ij}^n) + \theta , \tag{9.27}$$

$$F_{ij}^n = \theta s_{ij}^n - 2g_{12}(s_{ij}^n)\boldsymbol{m}_{ij}(s^n) + g_{11}(s_{ij}^n)\boldsymbol{s}_{ij}(s^n) + \boldsymbol{p}_{ij}(s^n) ,$$

we obtain, in accordance with (9.6) and (9.7), a solution of (9.23) for each fixed number j, $0 < j < N$,

$$s_{ij}^{n+1/2} = \alpha_{i+1j}^{n+1/2} s_{i+1j}^{n+1/2} + \beta_{i+1j}^{n+1/2} ,$$

$$i = 1, \ldots, N-1 , \quad s_{Nj}^{n+1/2} = \varphi_1(1, jh) , \tag{9.28}$$

where

$$\alpha_{i+1j}^{n+1/2} = \frac{B_{ij}^{n+1/2}}{C_{ij}^{n+1/2} - \alpha_{ij}^{n+1/2} A_{ij}^{n+1/2}} ,$$

$$i = 1, \ldots, N-1 , \quad \alpha_{1j}^{n+1/2} = 0 , \tag{9.29}$$

$$\beta_{i+1j}^{n+1/2} = \frac{A_{ij}^{n+1/2} \beta_{ij}^{n+1/2} + F_{ij}^n}{C_{ij}^{n+1/2} - \alpha_{ij}^{n+1/2} A_{ij}^{n+1/2}} ,$$

$$i = 1, \ldots, N-1 , \quad \beta_{1j}^{n+1/2} = \varphi_1(0, jh) . \tag{9.30}$$

Analogously the solution of (9.24), for each index j, $1 \leq j \leq N-1$, is expressed by the following recursive formula

$$v_{ij}^{n+1/2} = \alpha_{i+1j}^{n+1/2} v_{i+1j}^{n+1/2} + \beta_{i+1j}^{n+1/2} ,$$

$$i = 1, \ldots, N-1 , \quad v_{Nj}^{n+1/2} = \varphi_2(1, jh) , \tag{9.31}$$

where $\alpha_{i+1j}^{n+1/2}$ and $\beta_{i+1j}^{n+1/2}$ are computed by (9.30) and (9.31), respectively, with

$$\alpha_{1j}^{n+1/2} = 0 , \quad \beta_{1j}^{n+1/2} = \varphi_2(0, jh) ,$$

$$A_{ij}^{n+1/2} = g_{22}(s_{ij}^n) , \quad B_{ij}^{n+1/2} = g_{22}(s_{ij}^n) ,$$

$$C_{ij}^{n+1/2} = 2g_{22}(s_{ij}^n) + \theta , \tag{9.32}$$

$$F_{ij}^n = \theta v_{ij}^n - 2g_{12}(s_{ij}^n)\boldsymbol{m}_{ij}(v^n) + g_{11}(s_{ij}^n)\boldsymbol{s}_{ij}(v^n) + \boldsymbol{q}_{ij}(v^n) .$$

In order to compute (9.25) by (9.6) we assume, for each fixed i, $1 \leq i \leq N-1$,

$$A_{ij}^{n+1} = g_{11}(s_{ij}^n) , \quad B_{ij}^{n+1} = g_{11}(s_{ij}^n) , \quad C_{ij}^{n+1} = 2g_{11}(s_{ij}^n) + \theta ,$$

$$F_{ij}^{n+1/2} = \theta s_{ij}^{n+1/2} - g_{11}(s_{ij}^n)\boldsymbol{s}_{ij}(s^n) . \tag{9.33}$$

In accordance with (9.6), (9.7) we find the solution of (9.25)

9.2 Two-Dimensional Equations

$$s_{ij}^{n+1} = \alpha_{ij+1}^{n+1} s_{ij+1}^{n+1} + \beta_{ij+1}^{n+1},$$

$$j = 1, \ldots, N-1, \quad s_{iN}^{n+1} = s_1(ih, 1), \tag{9.34}$$

where

$$\alpha_{ij+1}^{n+1} = \frac{B_{ij}^{n+1}}{C_{ij}^{n+1} - \alpha_{ij}^{n+1} A_{ij}^{n+1}}, \quad j = 1, \ldots, N-1, \quad \alpha_{i1}^{n+1} = 0, \tag{9.35}$$

$$\beta_{ij+1}^{n+1} = \frac{A_{ij}^{n+1} \beta_{ij}^{n+1} + F_{ij}^{n+1/2}}{C_{ij}^{n+1} - \alpha_{ij}^{n+1} A_{ij}^{n+1}},$$

$$j = 1, \ldots, N-1, \quad \beta_{i1}^{n+1} = s_1(ih, 0). \tag{9.36}$$

Analogously the solution of (9.26) for each fixed index i, $1 \leq i \leq N-1$, is expressed by the recursive formula

$$v_{ij}^{n+1} = \alpha_{ij+1}^{n+1} v_{ij+1}^{n+1} + \beta_{ij+1}^{n+1}, \quad v_{iN}^{n+1} = s_2(ih, 1), \tag{9.37}$$

where α_{ij}^{n+1} and β_{ij}^{n+1} are computed by (9.36), and (9.37), respectively, in which

$$\alpha_{i1}^{n+1} = 0, \quad \beta_{i1}^{n+1} = s_2(ih, 0),$$

$$A_{ij}^{n+1} = g_{11}(s_{ij}^n), \quad B_{ij}^{n+1} = g_{11}(s_{ij}^n), \quad C_{ij}^{n+1} = 2g_{11}(s_{ij}^n) + \theta, \tag{9.38}$$

$$F_{ij}^{n+1/2} = \theta v_{ij}^{n+1/2} - g_{11}(s_{ij}^n) s(v^n).$$

An approximate solution of (9.19) is the solution s_{ij}^n at a step n such that

$$\max_{0 \leq i,j \leq N} \frac{1}{\tau} \sqrt{(s_{ij}^{n+1} - s_{ij}^n)^2 + (v_{ij}^{n+1} - v_{ij}^n)^2} \leq \varepsilon, \tag{9.39}$$

for some sufficiently small ε.

Analogously there are computed equations (9.14), (9.16), and (9.18).

Step–by–Step Algorithm. The numerical algorithm described above is a sequential process. The details of the algorithm are presented here in a step-by-step manner.

Step 1.
Define an initial grid distribution, using the formulas (9.20) or (9.22), by introducing two difference functions s_{ij}^0 and v_{ij}^0, $0 \leq i,j \leq N$ such that

$$(s_{0j}^0, v_{0j}^0) = s(0, ih), \quad (s_{Nj}^0, v_{Nj}^0) = s(1, jh),$$

$$(s_{i0}^0, v_{i0}^0) = s(ih, 0), \quad (s_{iN}^0, v_{iN}^0) = s(ih, 1).$$

Step 2.
Compute the functions $g_{11}(s_{ij}^0)$, $g_{12}(s_{ij}^0)$, $g_{22}(s_{ij}^0)$, $g(s_{ij}^0)$, $\boldsymbol{m}_{ij}(s^0)$, $\boldsymbol{m}_{ij}(v^0)$, $\boldsymbol{p}_{ij}(s^0)$, $\boldsymbol{q}_{ij}(s^0)$, $b_{ij}^l(s^0)$.
Step 3.
Compute $s_{ij}^{0+1/2}$ using (9.28 – 9.30). Step 4.

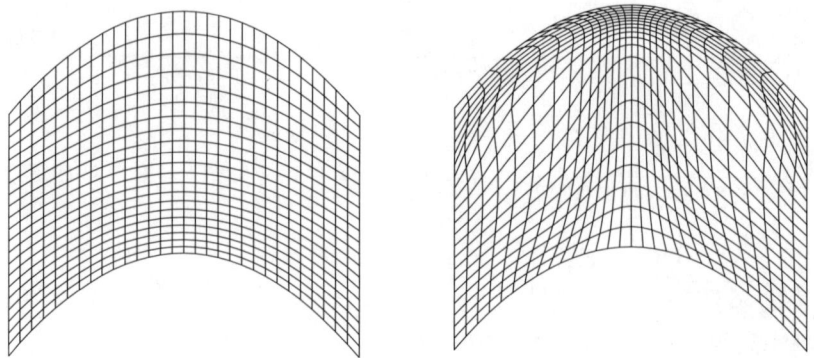

Fig. 9.1. Node rarefaction of the grid generated without a monitor function near a convex part of the boundary curve (*left*) and node concentration of the grid genearted with the use of a monitor function near the same boundary segment (*right*)

Compute $v_{ij}^{0+1/2}$ using (9.31 – 9.32).
Step 5.
Compute s_{ij}^{0+1} using (9.34 – 9.36).
Step 6.
Compute v_{ij}^{0+1} using (9.37 – 9.38).
Step 9.
Return to step 2 assuming $(s_{ij}^0, v_{ij}^0) = (s_{ij}^1, v_{ij}^1)$, where s_{ij}^1 and v_{ij}^1 are taken from steps 5 and 6.

Continue until the tolerance (9.39) is observed.

Figure 9.1 demonstrates grids in a two-dimensional domain generated by the algorithm described.

9.2.2 Algorithm for Generating Grids on Two-Dimensional Surfaces

Let us now discuss the numerical grid algorithm for a surface S^{x2} represented as

$$\boldsymbol{x}(\boldsymbol{s}) : S^2 \to R^3 \, , \quad \boldsymbol{x} = (x^1, x^2, x^3) \, , \quad \boldsymbol{s} = (s^1, s^2) \, . \tag{9.40}$$

If $\boldsymbol{f}(\boldsymbol{x}) = [f^1(\boldsymbol{x}), \ldots, f^l(\boldsymbol{x})]$, $\boldsymbol{x} \in G \supset S^{x2}$, is a monitor function controlling the grid property then the monitor surface S^{r2} over S^{x2} is parametrized as follows:

$$\boldsymbol{r}(\boldsymbol{s}) : S^2 \to R^{3+l} \, , \quad \boldsymbol{r}(\boldsymbol{s}) = \{\boldsymbol{x}(\boldsymbol{s}), f^1[\boldsymbol{x}(\boldsymbol{s})], \ldots, f^l[\boldsymbol{x}(\boldsymbol{s})]\} \, . \tag{9.41}$$

As well as in the case of a two-dimensional domain we choose the unit square for the logical domain Ξ^2. We also assume that the boundary transformation

$$\boldsymbol{\varphi}(\boldsymbol{\xi}) : \partial \Xi^2 \to \partial S^2 \, , \quad \boldsymbol{\varphi} = (\varphi^1, \varphi^2) \, , \tag{9.42}$$

which is continious on $\partial\Xi^2$ and smooth on each segment of $\partial\Xi^2$, has been specified on the boundary grid points of $\partial\Xi^2$, for example, by computing it through the algorithm described in Sect. 9.1.1.

The grid on S^{x2} is obtained by mapping with $\boldsymbol{x}(\boldsymbol{s})$ the grid nodes computed in S^2 through the numerical solution of the Dirichlet problem with respect to $\boldsymbol{s}(\boldsymbol{\xi})$ for the grid equations (7.44). In analogy with (9.11) the boundary value problem is formulated as

$$D[s^m] = -g^{r\xi}\left[\Delta_B[f^k]\frac{\partial f^k}{\partial s^i}g_{sx}^{im} + g_s^{pj}\Upsilon_{pj}^m\right],$$

$$i,j,m,p = 1,2, \quad k = 1,\ldots,l, \tag{9.43}$$

$$\boldsymbol{s}(\boldsymbol{\xi})\Big|_{\partial\Xi^2} = \boldsymbol{\varphi}(\boldsymbol{\xi}),$$

where

$$D[v] = g_{22}^{\xi}\frac{\partial^2 v}{\partial\xi^1\partial\xi^1} - 2g_{12}^{\xi}\frac{\partial^2 v}{\partial\xi^1\partial\xi^2} + g_{11}^{\xi}\frac{\partial^2 v}{\partial\xi^2\partial\xi^2}, \tag{9.44}$$

$$g_{ij}^{\xi} = \frac{\partial \boldsymbol{x}}{\partial\xi^i}\cdot\frac{\partial \boldsymbol{x}}{\partial\xi^j} + \frac{\partial \boldsymbol{f}[\boldsymbol{x}(\boldsymbol{s}(\boldsymbol{\xi}))]}{\partial\xi^i}\cdot\frac{\partial \boldsymbol{f}[\boldsymbol{x}(\boldsymbol{s}(\boldsymbol{\xi}))]}{\partial\xi^j}, \quad i,j = 1,2, \tag{9.45}$$

$$\Upsilon_{pj}^m = g_{sx}^{mi}\left(\frac{\partial^2 \boldsymbol{x}}{\partial s^p \partial s^j}\cdot\frac{\partial \boldsymbol{x}}{\partial s^i}\right), \quad i,j,m,p = 1,2, \tag{9.46}$$

g_{sx}^{im}, $i,m = 1,2$, is the (im)th element of the matrix inverse to the matrix (g_{ij}^{xs}),

$$g_{ij}^{xs} = \frac{\partial \boldsymbol{x}}{\partial s^i}\cdot\frac{\partial \boldsymbol{x}}{\partial s^j}, \quad i,j = 1,2, \tag{9.47}$$

g_s^{ij}, $i,j = 1,2$, is the (ij)th element of the matrix inverse to the matrix (g_{ij}^s),

$$g_{ij}^s = g_{ij}^{xs} + \frac{\partial \boldsymbol{f}[\boldsymbol{x}(\boldsymbol{s})]}{\partial s^i}\cdot\frac{\partial \boldsymbol{f}[\boldsymbol{x}(\boldsymbol{s})]}{\partial s^j}, \quad i,j = 1,2. \tag{9.48}$$

The problem (9.43) is solved by the iterative process which is similar to that applied to the case of a two-dimensional domain. Thus, analogously to (9.19), we find a solution of (9.43) as the numerical solution of the parabolic problem with respect to $s^m(\xi^1,\xi^2,t)$, $m = 1,2$, for a sufficiently large t,

$$\frac{\partial s^m}{\partial t} - D[s^m] = g^{r\xi}\left[\Delta_B[f^k]\frac{\partial f^k}{\partial s^i}g_{sx}^{im} + g_s^{pj}\Upsilon_{pj}^m\right],$$

$$i,j,m,p = 1,2, \quad k = 1,\ldots,l,$$

$$\boldsymbol{s}(\boldsymbol{\xi},t) = \boldsymbol{\varphi}(\boldsymbol{\xi}), \quad \boldsymbol{\xi}\in\partial\Xi^2,$$

$$\boldsymbol{s}(\boldsymbol{\xi},0) = \boldsymbol{s}_0(\boldsymbol{\xi}), \quad \boldsymbol{\xi}\in\Xi^2.$$

$$\tag{9.49}$$

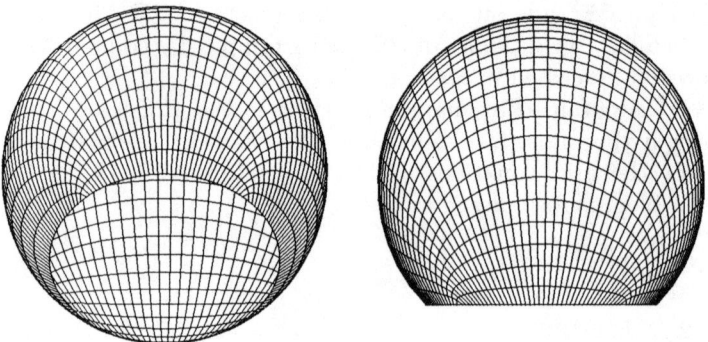

Fig. 9.2. Grid rarefaction near a concave part of the boundary curve of a cut sphere with the use of a monitor function

In general, the boundary value problem (9.49) is solved numerically through the scheme (9.23 – 9.26) in which $\boldsymbol{p}_{ij}$ and $\boldsymbol{q}_{ij}$ is the product of h^2 and the right-hand part of equations in (9.49) for $m = 1$ and $m = 2$, respectively, computed at the corresponding step level.

It is obvious that in the case of the parametrization $\boldsymbol{x}(\boldsymbol{s})$ specified in the form, similar to

$$\boldsymbol{x}(\boldsymbol{s}) = (s^1, s^2, x^3(s^1, s^2)) ,$$

the transformation $\boldsymbol{s}(\boldsymbol{\xi})$ can be obtained from the algorithm prescribed above for the two-dimensional domain S^2 in which the monitor function $\boldsymbol{f} = (f^1, \ldots, f^l)$ is replaced by the monitor function $\boldsymbol{g} = (x^3, f^1, \ldots, f^l)$.

Figures 9.2 and 9.3 illustrate surface grids generated by the algorithm.

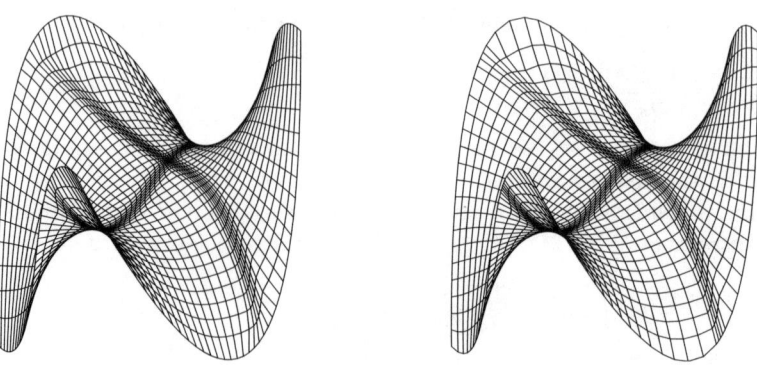

Fig. 9.3. An adaptive surface grid generated with the use of two monitor functions. The boundary grid in the left figure is generated without the use of the monitor functions

9.3 Three-Dimensional Equations

For generating grids in a three-dimensional domain $X^3 \subset R^3$ with the use of a monitor function

$$\boldsymbol{f}(\boldsymbol{x}), \quad \boldsymbol{x} \in X^3, \quad \boldsymbol{f}(\boldsymbol{x}) = [f^1(\boldsymbol{x}), \ldots, f^l(\boldsymbol{x})], \quad \boldsymbol{x} = (x^1, x^2, x^3),$$

we formulate the following boundary value problem with the equations (7.99)

$$D[x^i] + D[f^k]\frac{\partial f^k}{\partial x^i} = 0, \quad i = 1, 2, 3, \quad k = 1, \ldots, l,$$

$$\boldsymbol{x}(\boldsymbol{\xi})\Big|_{\partial \Xi^3} = \boldsymbol{\varphi}(\boldsymbol{\xi}),$$
(9.50)

where

$$D[v] = g^\xi g_\xi^{ij} \frac{\partial^2 v}{\partial \xi^i \partial \xi^j}$$

$$= [g_{22}^\xi g_{33}^\xi - (g_{23}^\xi)^2]\frac{\partial^2 v}{\partial \xi^1 \partial \xi^1} + 2[g_{23}^\xi g_{13}^\xi - g_{12}^\xi g_{33}^\xi]\frac{\partial^2 v}{\partial \xi^1 \partial \xi^2}$$

$$+2[g_{12}^\xi g_{23}^\xi - g_{22}^\xi g_{13}^\xi]\frac{\partial^2 v}{\partial \xi^1 \partial \xi^3} + [g_{11}^\xi g_{33}^\xi - (g_{13}^\xi)^2]\frac{\partial^2 v}{\partial \xi^2 \partial \xi^2}$$

$$+2[g_{12}^\xi g_{13}^\xi - g_{11}^\xi g_{23}^\xi]\frac{\partial^2 v}{\partial \xi^2 \partial \xi^3} + [g_{11}^\xi g_{22}^\xi - (g_{12}^\xi)^2]\frac{\partial^2 v}{\partial \xi^3 \partial \xi^3},$$

$$g_{ij}^\xi = \frac{\partial \boldsymbol{x}}{\partial \xi^i} \cdot \frac{\partial \boldsymbol{x}}{\partial \xi^j} + \frac{\partial \boldsymbol{f}[\boldsymbol{x}(\boldsymbol{\xi})]}{\partial \xi^i} \cdot \frac{\partial \boldsymbol{f}[\boldsymbol{x}(\boldsymbol{\xi})]}{\partial \xi^j}, \quad i, j = 1, 2, 3.$$

Analogously to the solution (9.19) we find a solution of (9.50) as a limit with $t \to \infty$ of the solution of the corresponding parabolic problem

$$\frac{\partial x^m}{\partial t} = D[x^m] + D[f^k]\frac{\partial f^k}{\partial x^m}, \quad m = 1, 2, 3, \quad k = 1, \ldots, l,$$

$$\boldsymbol{x}(\boldsymbol{\xi}, t) = \boldsymbol{\varphi}(\boldsymbol{\xi}), \quad \boldsymbol{\xi} \in \partial \Xi^3, \quad t \geq 0,$$
(9.51)

$$\boldsymbol{x}(\boldsymbol{\xi}, 0) = \boldsymbol{x}_0(\boldsymbol{\xi}), \quad \boldsymbol{\xi} \in \Xi^3.$$

Initial Transformation. The initial transformation

$$\boldsymbol{x}(\boldsymbol{\xi}, 0) = \boldsymbol{x}_0(\boldsymbol{\xi}) : \Xi^3 \to X^3.$$

can be found by propagating the values of $\boldsymbol{\varphi}(\boldsymbol{\xi})$ into the interior of the domain Ξ^3, for example, through the formula of Lagrange transfinite interpolation. In particular for the simplest expressions of the blending functions

$$\alpha_{0j}^i(s) = 1 - s, \quad \alpha_{1j}^i(s) = s,$$

we find from (1.10)

$$F_1^i(\xi^1,\xi^2,\xi^3) = (1-\xi^1)\varphi^i(0,\xi^2,\xi^3) + \xi^1\varphi^i(1,\xi^2,\xi^3) ,$$

$$F_2^i(\xi^1,\xi^2,\xi^3) = F_1^i(\xi^1,\xi^2,\xi^3) + (1-\xi^2)[\varphi^i(\xi^1,0,\xi^3)$$

$$-F_1^i(\xi^1,0,\xi^3)] + \xi^2[\varphi^i(\xi^1,1,\xi^3) - F_1^i(\xi^1,1,\xi^3)] ,$$

$$x^i(\xi^1,\xi^2,\xi^3) = F_2^i(\xi^1,\xi^2,\xi^3) + (1-\xi^3)[\varphi^i(\xi^1,\xi^2,0)$$

$$-F_2^i(\xi^1,\xi^2,0)] + \xi^3[\varphi^i(\xi^1,\xi^2,1) - F_2^i(\xi^1,\xi^2,1)] ,$$

$$i = 1,2,3 ,$$

(9.52)

A numerical algorithm for solving the problem (9.51) is formulated analogously to the two-dimensional algorithms reviewed by formulas (9.23–9.26), namely, by splitting the process of the numerical solution into a series of one-dimensional algorithms. An example of a three-dimensional grid is demonstrated in Fig. 9.4.

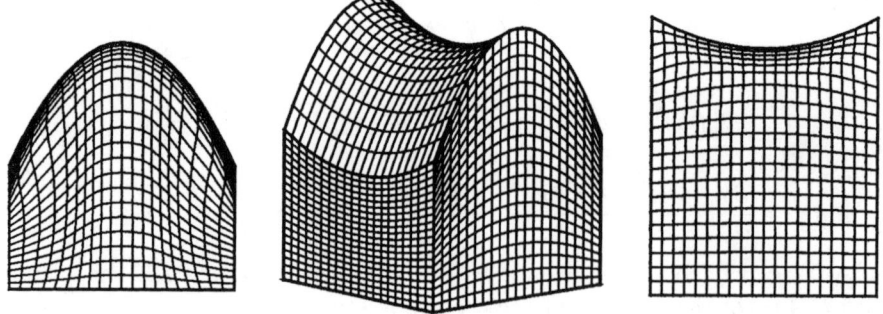

Fig. 9.4. Grid clustering near the upper boundary segment provided by a monitor function. The left-handed and right-handed pictures are central sections of the geometry

References

Alalykin, G.B., Godunov, S.K., Kireyeva, L.L., Pliner, L.A. (1970): *On Solution of One-Dimensional Problems of Gas Dynamics in Moving Grids.* Nauka, Moscow (Russian)

Albone, C.M. (1992): Embedded meshes of controllable quality synthesised from elementary geometric features. AIAA Paper 92-0633

Albone, C.M., Joyce, M.G. (1990): Feature-associated mesh embedding for complex configurations. AGARD Conference Proceedings 464.13

Allwright, S. (1989): Multiblock topology specification and grid generation for complete aircraft configurations. In Schmidt, W. (ed.): *AGARD Conference Proceedings 464, Applications of Mesh Generation to Complex 3-D Configurations.* Loen, Norway. Advisory Group for Aerospace Research and Development, NATO

Amsden, A.A., Hirt, C.W. (1973): A simple scheme for generating general curvilinear grids. J. Comput. Phys. **11**, 348–359

Anderson, D.A. (1983): Adaptive grid methods for partial differential equations. In Ghia, K.N., Ghia U. (eds.): *Advances in Grid Generation.* ASME, Houston, pp. 1–15

Andrews, A.E. (1988): Progress and challenges in the application of artificial intelligence to computational fluid dynamics. AIAA Journal **26**, 40–46

Arina, R., Casella, M. (1991): A harmonic grid generation technique for surfaces and three-dimensional regions. In Arcilla, A.S., Hauser, J., Eiseman, P.R., Thompson, J.F. (eds.): *Numerical Grid Generation in Computational Fluid Dynamics and Related Fields.* North-Holland, New York, pp. 935–946

Atta, E.H., Vadyak, J. (1982): A grid interfacing zonal algorithm for three-dimensional transonic flows about aircraft configurations. AIAA Paper 82-1017

Atta, E.H., Birchelbaw, L., Hall, K.A. (1987): A zonal grid generation method for complex configurations. AIAA Paper 87-0276

Baker, T.J. (1995): Prospects and expectations for unstructured methods. In: *Proceedings of the Surface Modeling, Grid Generation and Related Issues in Computational Fluid Dynamics Workshop.* NASA Conference Publication 3291, NASA Lewis Research Center, Cleveland, OH, pp. 273–287

Baker, T.J. (1997): Mesh adaptation strategies for problems in fluid dynamics. Finite Elements Anal. Design. **25**, 243–273

Barfield, W.D. (1970): An optimal mesh generator for Lagrangian hydrodynamic calculations in two space dimensions. J. Comput. Phys. **6**, 417–429

Belinsky, P.P., Godunov, S.K., Ivanov, Yu.V., Yanenko I.K. (1975): The use of one class of quasiconformal mappings to generate numerical grids in regions with curvilinear boundaries. Zh. Vychisl. Maths. Math. Fiz. **15**, 1499–1511 (Russian)

Belk, D.M., Whitefield, D.L. (1987): Three-dimensional Euler solutions on blocked grids using an implicit two-pass algorithms. AIAA Paper 87-0450

Benek, J.A., Buning, P.G., Steger, J.L. (1985): A 3-d chimera grid embedding technique. AIAA Paper 85-1523

References

Benek, J.A., Steger, J.L., Dougherty, F.C. (1983): A flexible grid embedding technique with application to the Euler equations. AIAA Paper 83-1944

Berger, M.J., Oliger, J. (1983): Adaptive mesh refinement for hyperbolic partial differential equations. Manuscript NA-83-02, Stanford University, March

Brackbill, J.U. (1993): An adaptive grid with directional control. J. Comput. Phys. **108**, 38–50

Brackbill, J.U., Saltzman, J. (1982): Adaptive zoning for singular problems in two directions. J. Comput. Phys. **46**, 342–368

Carey, G.F. (1997): *Computational Grids. Generation, Adaptation, and Solution Strategies.* Taylor and Francis, London

Chan, W.M., Buning, P.G. (1995): Surface grid generation methods for overset grids. Computer and Fluids. **24**(5), 509–522

Chan, W.M., Steger, L.G. (1992): Enhancement of a three-dimensional hyperbolic grid generation scheme. Appl. Maths. Comput. **51**(1), 181–205

Chiba, N., Nishigaki, I., Yamashita, Y., Takizawa, C., Fujishiro, K. (1998): A flexible automatic hexahedral mesh generation by boundary-fit method. Comput. Methods Appl. Mech Engng. **161**, 145–154

Chu, W.H. (1971): Development of a general finite difference approximation for a general domain. J. Comput. Phys. **8**, 392–408

Chumakov, G.A., Chumakov, S.G. (1998): A method for the 2-D quasi-isometric regular grid generation, J. Comput. Phys. **133**, 1–28

Cordova, J.Q., Barth, T.J. (1988): Grid generation for general 2-D regions using hyperbolic equations, AIAA Paper 88-0520

Crowley, W.P. (1962): An equipotential zoner on a quadrilateral mesh. Memo, Lawrence Livermore National Lab., 5 July 1962

Danaev, N.T., Liseikin, V.D., Yanenko, N.N. (1980): Numerical solution on a moving curvilinear grid of viscous heat-conducting flow about a body of revolution. Chisl. Metody Mekhan. Sploshnoi Sredy **11**(1), 51–61 (Russian)

Dannenhoffer, J.F. (1995): Automatic blocking for complex three-dimensional configurations. In: *Proceedings of the Surface Modeling, Grid Generation, and Related Issues in Computational Fluid Dynamics Workshop.* NASA Lewis Research Center, Cleveland OH, May, p. 123

Dvinsky, A.S. (1991): Adaptive Grid Generation from Harmonic Maps on Riemannian Manifolds. J. Comput. Phys. **95**, 450–476

Edwards, T.A. (1985): Noniterative three-dimensional grid generation using parabolic partial differential equations. AIAA Paper 85-0485

Eells, J., Lenaire, L. (1988): Another report on harmonic maps. Bull. London Math. Soc. **20**(5), 385–524

Eiseman, P.R. (1980): Geometric methods in computational fluid dynamics. ICASE Report 80-11 and Von Karman Institute for Fluid Dynamics Lecture Series Notes

Eiseman, P.R. (1985): Grid generation for fluid mechanics computations. Ann. Rev. Fluid Mech. **17**, 487–522

Eiseman, P.R. (1987): Adaptive grid generation. Comput. Methods. Appl. Mech. Engng. **64**, 321–376

Eriksson, L.E. (1982): Generation of boundary-conforming grids around wing-body configurations using transfinite interpolation. AIAA Journal **20**, 1313–1320

Eriksson, L.E. (1983): Practical three-dimensional mesh generation using transfinite interpolation. Lecture Series Notes 1983-04, von Karman Institute for Fluid Dynamics, Brussels

Field, D.A. (1995): The legacy of automatic mesh generation from solid modeling. Comp. Aided Geom. Design **12**, 651–673

Georgala, J.M., Shaw, J.A. (1989): A discussion on issues relating to multiblock grid generation. In Schmidt, W. (ed.): *AGARD Conference Proceedings 464, Applications of Mesh Generation to Complex 3-D Configurations*. Loen, Norway. Advisory Group for Aerospace Research and Development, NATO

George, P.L., Borouchaki, H. (1998): *Delaunay Triangulation and Meshing*. Editions Hermes, Paris

Giannakopoulos, A.E., Engel, A.J. (1988): Directional control in grid generation. J. Comput. Phys. **74**, 422–439

Godunov, S.K., Prokopov, G.P. (1967): Calculation of conformal mappings in the construction of numerical grids. J. Comput. Maths. Math. Phys. **7**, 1031–1059 (Russian)

Godunov, S.K., Prokopov, G.P. (1972): On utilization of moving grids in gasdynamics computations. J. Vychisl. Matem. Matem. Phys. **12**, 429–440 (Russian) [English transl.: USSR Comput. Math. and Math. Phys. **12** (1972), 182–195]

Godunov, S.K., Romenskii, E.I., Chumakov, G.A. (1990): Grid generation in complex domains by means of quasi-conformal mappings. Proc. Institute of Mathematics, Novosibirsk, Nauka, **18**, 75–84 (Russian)

Gordon, W.J., Hall, C.A. (1973): Construction of curvilinear coordinate systems and applications to mesh generation. Int. J. Numer. Meth. Engng. **7**, 461–477

Gordon, W.J., Thiel, L.C. (1982): Transfinite mappings and their application to grid generation. In Thompson, J.F. (ed.): *Numerical Grid Generation*. North-Holland, New York, pp. 171–192

Hawken, D.F., Gottlieb, J.J., Hansen, J.S. (1991): Review of some adaptive node-movement techniques in finite-element and finite-difference solutions of partial differential equations. J. Comput. Phys. **95**, 254–302

Hedstrom, G.W., Rodrigue C.M. (1982): Adaptive-grid methods for time-dependent partial differential equations. Lect. Notes Math. **960**, 474–484

Ho-Le (1988): Finite element mesh generation methods: a review and classification. Computer-Aided Design **20**, 27–38

Holcomb, J.F. (1987): Development of a grid generator to support 3-dimensional multizone Navier–Stokes analysis. AIAA Paper 87-0203

Huang, W. (2001): Variational mesh adaptation: isotropy and equidistribution. J. Comput. Phys. **174**, 903–924

Huang, W., Ren, Y., Russel, R.D. (1994): Moving mesh PDEs based on the equidistribution principle. SIAM J. Numer. Anal. **31**, 709–730

Jacquotte, O.-P. (1987): A mechanical model for a new grid generation method in computational fluid dynamics. Comput. Meth. Appl. Mech. Engng. **66**, 323–338

Jeng, Y.N., Shu, Y.-L. (1995): Grid combination method for hyperbolic grid solver in regions with enclosed boundaries. AIAA Journal **33**(6), 1152–1154

Khakimzyanov, G.S., Shokin, Yu.I., Barakhnin, V.B., Shokin, N.Yu. (2001): *Numerical Modeling of Flows with Surface Waves*. Novosibirsk, Sibirian Division of the Russian Academy of Sciences

Khamayseh, A., Mastin, C.W. (1996): Computational conformal mapping for surface grid generation. J. Comput. Phys. **123**, 394–401

Kim, B., Eberhardt, S.D. (1995): Automatic multiblok grid generation for high-lift configuration wings. *Proceedings of the Surface Modeling, Grid Generation, and Related Issues in Computational Fluid Dynamics Workshop*. NASA, Lewis Research Center, Cleveland OH, May, p. 671

Knupp, P., Steinberg, S. (1993): *Fundamentals of Grid Generation*. CRC Press, Boca Raton

Krugljakova, L.V., Neledova, A.V., Tishkin, V.F., Filatov, A.Yu. (1998): Unstructured adaptive grids for problems of mathematical physics (survey). Math. Modeling **10**(3), 93–116 (Russian)

Lee, K.D., Loellbach, J.M. (1989): Geometry-adaptive surface grid generation using a parametric projection. J. Aircraft **2**, 162–167

Lee, K.D., Huang, M., Yu, N.J., Rubbert, P.E. (1980): Grid generation for general three-dimensional confugurations. In Smith, R.E. (ed.): *Proc. NASA Langley Workshop on Numerical Grid Generation Techniques*. Oct., p. 355

Lin, K.L., Shaw, H.J. (1991): Two-dimensional orthogoal grid generation techniques. Comput. Struct. **41**(4), 569–585

Liseikin, V.D. (1991a): On generation of regular grids on n-dimensional surfaces. J. Comput. Math. Math. Phys. **31**, 1670–1689 (Russian). [English transl.: USSR Comput. Math. Math. Phys. **31**(11) (1991), 47–57]

Liseikin, V.D. (1991b): Techniques for generating three-dimensional grids in aerodynamics (review). Problems Atomic Sci. Technology. Ser. Math. Model. Phys. Process **3**, 31–45 (Russian)

Liseikin, V.D. (1992): On a variational method of generating adaptive grids on n-dimensional surfaces. Soviet Math. Docl. **44**(1), 149–152

Liseikin, V.D. (1996): Construction of structured adaptive grids – a review. Comput. Math. Math. Phys., **36**(1), 1–32

Liseikin, V.D. (1998a): Algebraic adaptation based on stretching functions. Russ. J. Numer. Anal. Math. Modeling **13**(4), 307–324

Liseikin, V.D. (1998b): A method of algebraic adaptation. Comput. Math. Math. Phys., **38**(10), 1624–1640

Liseikin, V.D. (1999): *Grid Generation Methods*. Springer, Berlin

Liseikin, V.D. (2001a): *Layer Resolving Grids and Transformations for Singular Perturbation Problems*. VSP, Utrecht

Liseikin, V.D. (2001b): Application of notions and relations of differential geometry to grid generation. Russ. J. Numer. Anal. Math. Modeling **16**(1), 57–75

Liseikin, V.D. (2002a): Analysis of grids derived by a comprehensive grid generator. Russ. J. Numer. Anal. Math. Modeling **17**(2), 183–202

Liseikin, V.D. (2002b): On geometric analysis of grid properties. Russ. Docl. Academ. Nauk **383**(2), 167–170

Liseikin, V.D. (2003): On analysis of clustering of numerical grids produced by elliptic models. Russ. J. Numer. Anal. Math. Modeling **18**(2)

Mastin, C.W. (1992): Linear variational methods and adaptive grids. Computers Math. Applic. **24**(5/6), 51–56

McNally, D. (1972): FORTRAN program for generating a two-dimensional orthogonal mesh between two arbitrary boundaries. NASA, TN D-6766, May

Miki, K., Takagi, T. (1984): A domain decomposition and overlapping method for the generation of three-dimensional boundary-fitted coordinate systems. J. Comput. Phys. **53**, 319–330

Nakamura, S. (1982): Marching grid generation using parabolic partial differential equations. Appl. Math. Comput. **10**(11), 775–786

Nakamura, S., Suzuki M. (1987): Noniterative three-dimensional grid generation using a parabolic-hyperbolic hybrid scheme. AIAA Paper 87-0277

Noack, R.W. (1985): Inviscid flow field analysis of maneuvering hypersonic vehicles using the SCM formulation and parabolic grid generation. AIAA Paper 85-1682

Noack, R.W., Anderson D.A. (1990): Solution adaptive grid generation using parabolic partial differential equations: AIAA Journal **28**(6), 1016–1023

Reed, C.W., Hsu, C.C., Shiau, N.H. (1988): An adaptive grid generation technique for viscous transonic flow problems. AIAA Paper 88-0313

Rizk, Y.M., Ben-Shmuel, S. (1985): Computation of the viscous flow around the shuttle orbiter at low supersonic speeds. AIAA Paper 85-0168

Rubbert, P.E., Lee. K.D. (1982): Patched coordinate systems. In Thompson, J.F. (ed.): *Numerical Grid Generation*, North-Holland, New York, p. 235

Schonfeld, T., Weinerfelt, P., Jenssen, C.B. (1995): Algorithms for the automatic generation of 2d structured multiblock grids. *Proceedings of the Surface Modeling, Grid Generation, and Related Issues in Computational Fluid Dynamics Workshop.* NASA, Lewis Research Center, Cleveland OH, May, p. 561

Shaw, J.A., Weatherill, N.P. (1992): Automatic topology generation for multiblock grids. Appl. Math. Comput. **52**, 355–388

Shephard, M.S., Grice, K.R., Lot, J.A., Schroeder, W.J. (1988a): Trends in automatic three-dimensional mesh generation. Comput. Strict. **30**(1/2), 421–429

Smith, R.E. (1981): Two-boundary grid generation for the solution of the three-dimensional Navier–Stokes equations. NASA TM-83123

Smith, R.E. (1982): Algebraic grid generation. In Thompson, J.F. (ed.): *Numerical Grid Generation.* North-Holland, New York, pp. 137–170

Smith, R.E., Eriksson, L.E. (1987): Algebraic grid generation. Comp. Meth. Appl. Mech. Eng. **64**, 285–300

Soni, B.K., Huddleston, D.K., Arabshahi, A., Yu, B. (1993): A study of CFD algorithms applied to complete aircraft configurations. AIAA Paper 93-0784

Sorenson, R.L. (1986): Three-dimensional elliptic grid generation about fighter aircraft for zonal finite-difference computations. AIAA Paper 86-0429

Sparis, P.D. (1985): A method for generating boundary-orthogonal curvilinear coordinate systems using the biharmonic equation. J. Comput. Phys. **61**(3), 445–462

Starius, G. (1977): Constructing orthogonal curvilinear meshes by solving initial value problems. Numer. Math. **28**, 25–48

Steger, J.L. (1991): Grid generation with hyperbolic partial differential equations for application to complex configurations. In Arcilla, A.S., Hauser, J., Eiseman, P.R., Thompson, J.F. (eds.): *Numerical Grid Generation in Computational Fluid Dynamics and Related Fields.* North-Holland, New York, pp. 871–886

Steger, J.L., Chaussee, D.S. (1980): Generation of body fitted coordinates using hyperbolic differential equations. SIAM. J. Sci. Stat. Comput. **1**(4), 431–437

Steger, J.L., Rizk, Y.M. (1985): Generation of three-dimensional body-fitted coordinates using hyperbolic partial differential equations. NASA, TM 86753, June

Steger, J.L., Sorenson, R.L. (1979): Automatic mesh-point clustering near a boundary in grid generation with elliptic partial differential equations. J. Comput. Phys. **33**, 405–410

Steinberg, S., Roache, P.J. (1986): Variational grid generation. Numer. Meth. Partial Differential Equations **2**, 71–96

Steinbrenner, J.P., Chawner, J.R., Fouts, C.L. (1990): Multiple block grid generation in the interactive environment. AIAA Paper 90-1602

Stewart, M.E.M. (1992): Domain decomposition algorithm applied to multielement airfoil grids. AIAA Journal **30**(6), 1457

Tai, C.H., Chiang, D.C., Su, Y.P. (1996): Three-dimensional hyperbolic grid generation with inherent dissipation and Laplacian smoothing. AIAA Journal **34**(9), 1801–1806

Takagi, T., Miki, K., Chen, B.C.J., Sha, W.T. (1985): Numerical generation of boundary-fitted curvilinear coordinate systems for arbitrarily curved surfaces. J. Comput. Phys. **58**, 69–79

Takahashi, H., Shimizu, H. (1991): A general purpose automatic mesh generation using shape recognition technique. Comput. Engng. ASME **1**, 519–526

Tamamidis, P., Assanis, D.N. (1991): Generation of orthogonal grids with control of spacing. J. Comput. Phys. **94**, 437–453

Thacker, W.C. (1980): A brief review of techniques for generating irregular computational grids. Int. J. Numer. Meth. Engng. **15**(9), 1335–1341

Thoman, D.C., Szewczyk, A.A. (1969): Time-dependent viscous flow over a circular cylinder. Phys. Fluids Suppl. **II**, 76

Thomas, P.D. (1982): Composite three-dimensional grids generated by elliptic systems. AIAA Journal **20**(9), 1195–1202
Thomas, P.D., Middlecoff, J.F. (1980): Direct control of the grid point distribution in meshes generated by elliptic equations. AIAA Journal **18**(6), 652–656
Thomas, M.E., Bache, G.E., Blumenthal, R.F. (1990): Structured grid generation with PATRAN. AIAA Paper 90-2244
Thompson, J.F. (1984a): Grid generation techniques in computational fluid dynamics. AIAA Journal **22**(11), 1505–1523
Thompson, J.F. (1984b): A survey of dynamically-adaptive grids in the numerical solution of partial differential equations. AIAA Paper 84-1606
Thompson, J.F. (1985): A survey of dynamically-adaptive grids in the numerical solution of partial differential equations. Appl. Numer. Math. **1**, 3–27
Thompson, J.F. (1987a): A composite grid generation code for general 3-d regions. AIAA Paper 87-0275
Thompson, J.F. (1987b): A general three-dimensional elliptic grid generation system on a composite block structure. Comput. Meth. Appl. Mech. Engng. **64**, 377–411
Thompson, J.F. (1996): A reflection on grid generation in the 90s: trends, needs influences. In Soni, B.K., Thompson, J.F., Hauser, J., Eiseman, P.R. (eds.): *Numerical Grid Generation in CFD.* Mississippi State University, **1**, pp. 1029–1110
Thompson, J.F., Weatherill, N.P. (1993): Aspects of numerical grid generation: current science and art. AIAA Paper 93-3539
Thompson, J.F., Thames, F.C., Mastin, C.W. (1974): Automatic numerical generation of body-fitted curvilinear coordinate system for field containing any number of arbitrary two-dimensional bodies. J. Comput. Phys. **15**, 299–319
Thompson, J.F., Warsi, Z.U.A., Mastin C.W. (1982): Boundary-fitted coordinate systems for numerical solution of partial differential equations – a review. J. Comput. Phys. **47**, 1–108
Thompson, J.F., Warsi, Z.U.A., Mastin C.W. (1985): *Numerical Grid Generation. Foundations and Applications.* North-Holland, New York
Visbal, M., Knight, D. (1982): Generation of orthogonal and nearly orthogonal coordinates with grid control near boundaries. AIAA Journal **20**(3), 305–306
Vogel, A.A. (1990): Automated domain decomposition for computational fluid dynamics. Computers and Fluids **18**(4), 329–346
Warsi, Z.U.A. (1981): *Tensors and Differential Geometry Applied to Analytic and Numerical Coordinate Generation.* MSSU-EIRS-81-1, Aerospace Engineering, Mississippi State University
Warsi, Z.U.A. (1982): Basic differential models for coordinate generation. In Thompson, J.F. (ed.): *Numerical Grid Generation.* North-Holland, New York, pp. 41–78
Warsi, Z.U.A. (1986): Numerical grid generation in arbitrary surfaces through a second-order differential-geometric model. J. Comput. Phys. **64**, 82–96
Warsi, Z.U.A. (1990): Theoretical foundation of the equations for the generation of surface coordinates. AIAA Journal **28**(6), 1140–1142
Warsi, Z.U.A., Thompson, J.F. (1990): Application of variational methods in the fixed and adaptive grid generation. Comput. Math. Appl. **19**(8/9), 31–41
Weatherill, N.P., Forsey, C.R. (1984): Grid generation and flow calculations for complex aircraft geometries using a multi-block scheme. AIAA Paper 84-1665
White, A.B. (1990): Elliptic grid generation with orthogonality and spacing control on an arbitrary number of boundaries. AIAA Paper 90-1568

Widhopf, G.D., Boyd, C.N., Shiba, J.K., Than, P.T., Oliphant, P.H., Huang, S-C., Swedberg, G.D., Visich, M. (1990): RAMBO-4G: An interactive general multi-block grid generation and graphics package for complex multibody CFD applications. AIAA Paper 90-0328

Winslow, A.M. (1967): Equipotential zoning of two-dimensional meshes. J. Comput. Phys. **1**, 149–172

Winslow, A.M. (1981): Adaptive mesh zoning by the equipotential method. UCID-19062, Lawrence Livermore National Laboratories

Wulf, A., Akrag, V. (1995): Tuned grid generation with ICEM CFD. *Proceedings of the Surface Modeling, Grid Generation, and Related Issues in Computational Fluid Dynamics Workshop*. NASA, Lewis Research Center, Cleveland OH, May, p. 477

Yanenko, N.N., Danaev, N.T., Liseikin, V.D. (1977): A variational method for grid generation. Chisl. Metody Mekhan. Sploshnoi Sredy **8**(4), 157–163 (Russian)

Zegeling, P.A. (1993): *Moving-Grid Methods for Time-Dependent Partial Differential Equations*. CWI Tract 94, Centrum voor Wiskund en Informatica, Amsterdam

Index

Arc length parameter 54

Basic
- normal vector 63
- parallelepiped 62
- vector 62

Beltrami's parameter
- first 78
- mixed 77
- second 78
- second differential 113, 144

Beltramian operator 112

Calculus of variation 40

Cell
- deformation 10, 29
- edge 8
- reference 8
- standard 8

Christoffel symbols 48
- of the first kind 48, 68
- of the second kind 48, 69, 130

Code 30

Compatibility 12

Consistent discretization 11

Coordinate
- Cartesian 18, 131
- curvilinear 59
- grid 111, 115
- hypersurface 60
- line 60
- local system 67
- logical 111
- orthogonal 49
- parametric 111

Covariant derivative 74

Critical point 130

Curvature
- Gaussian 86
- geodesic 81
- mean 80, 86, 102
- principal 102, 106

Curve
- length 41
- parametrization 41, 53
- quality 54

Domain
- decomposition 31
- parametric 33, 111, 137
- physical 9

Energy density 130

Equation
- algebraic 10
- generalized Laplace 114
- hyperbolic 25
- parabolic 23
- parametric 59
- Serret–Frenet 55

Equidistribution principle 116

Euclidean
- metric 131
- space 131

Euler theorem 11

Function
- admissible 27
- blending 22
- control 24
- harmonic 130
- monitor 68, 137
- weight 68, 116, 137

Functional
- diffusion-adaptation 125
- energy 130
- of grid smoothness 127

Fundamental form
- first 61
- second 85, 101

Gauss relation 48

Grid
- boundary-conforming 12, 18
- boundary-fitting 12
- Cartesian 18
- coordinate 18
- deformation 10
- moving 19
- nodes 8
- organization 11
- quality 21
- size 10
- structured 17, 18

Intermediate transformation 114
Intersection 101
Intrinsic geometry 61
Inverse 34, 38, 54

Jacobi matrix 33
Jacobian 36

Left-handed orientation 36
Length 54

Manifold
- monitor 68, 111, 122, 137
- Riemannian 65
- standard monitor 138
Maximum principle 23
Measure
- of line bending 56
- of relative clustering 80
- of relative spacing 79
Method
- algebraic 21
- differential 21
- finite-difference 11
- finite-volume 12
- hybrid grid 26
- hyperbolic 25
- variational 21
Metric
- monitor 123, 137
- spherical 117

Orthonormal basis 56

Problem
- boundary value 116
- well-posed 27

Product
- cross 44
- dot 36
- tensor 47

Radius of curvature 55
Rate of twisting 57
Right-handed orientation 36

Source term 24
Specification
- explicit 5
- implicit 5
Surface
- minimal 102
- monitor 68, 102, 123, 137
- multidimensional 59
- regular 59
- warping 101

Tangent n-dimensional plane 60
Tensor
- component 72
- contravariant 62, 73
- contravariant metric 42
- covariant 73
- covariant metric 40, 61
- invariant 77
- metric 36
- metric contravariant 137
- metric covariant 137
- mixed 73
- of mixed derivatives 74
- of order zero 73
- product 82
- rank 72
- surface metric 61
Torsion 56, 57
Triad 45
Turbulence 10

Vector
- basic 83
- basic normal 61
- binormal 55
- curvature 54
- normal 37, 46, 61
- tangent 60
- tangential 35, 36, 54
- unit normal 81

Printing: Saladruck Berlin
Binding: Stürtz AG, Würzburg